Food Safety Management Programs

Debby Newslow

Food Safety Management Programs

Applications, Best Practices, and Compliance

CRC Press
Taylor & Francis Group
Boca Raton London New York

CRC Press is an imprint of the
Taylor & Francis Group, an **informa** business

CRC Press
Taylor & Francis Group
6000 Broken Sound Parkway NW, Suite 300
Boca Raton, FL 33487-2742

© 2014 by Taylor & Francis Group, LLC
CRC Press is an imprint of Taylor & Francis Group, an Informa business

No claim to original U.S. Government works

Version Date: 20131118

International Standard Book Number-13: 978-1-4398-2679-9 (Hardback)

This book contains information obtained from authentic and highly regarded sources. Reasonable efforts have been made to publish reliable data and information, but the author and publisher cannot assume responsibility for the validity of all materials or the consequences of their use. The authors and publishers have attempted to trace the copyright holders of all material reproduced in this publication and apologize to copyright holders if permission to publish in this form has not been obtained. If any copyright material has not been acknowledged please write and let us know so we may rectify in any future reprint.

Library of Congress Cataloging-in-Publication Data

Newslow, Debby L., author.
    Food safety management programs : applications, best practices, and compliance / Debby Newslow.
        pages cm
    Summary: "The value of an effective well-defined, implemented, and maintained management system is priceless. Based on the requirements of ISO 900 I, ISO 22000, and PAS 220, this text provides a collection of participation-based tools that can be applied to the development, implementation, and maintenance of a management-based system in the food industry. The book clearly explains how to choose the right system for an organization, what the requirements are, how to identify gaps in the system, and ways to integrate these resources into individual operations. The author also compares ISO and HACCP programs as well as includes case studies from professionals"-- Provided by publisher.
    Includes bibliographical references and index.
    ISBN 978-1-4398-2679-9 (hardback)
    1. Food industry and trade--Quality control. 2. Food industry and trade--Management. 3. Food--Safety measures. I. Title.

TP372.5.N478 2004
363.19'26--dc23
                                                                          2013045574

Visit the Taylor & Francis Web site at
http://www.taylorandfrancis.com

and the CRC Press Web site at
http://www.crcpress.com

*Dedicated to all those that shared their knowledge
and experiences through the years*

And a very special thank you goes to

*Jon G. Porter for being a great mentor,
friend, and a pioneer in food safety*

*John Weihe for sharing his wealth of knowledge,
positive focus, and for always believing in me*

Frank Carbone for his unrelenting encouragement and friendship

My Dad for his guidance and sharing his strong will

And

For my Mom who continues to be an inspiration at age 92

Contents

Preface

For more than 20 years, we have heard talk about International Standardization Organization (ISO)-based standards and whether or not there is a place for these standards in the food industry. The majority of the food industry felt we had all the regulations and requirements necessary to produce safe products of the highest quality. Is this really true? ISO compliance was only required if the product was shipped overseas. ISO required too much documentation. It cost too much. It just wasn't needed. It was difficult to understand the internal advantages of having a structured system. Over the years, some companies who decided to implement an ISO-based standard system have seen many advantages; others still wonder if it was worthwhile. It is hard to put a dollar savings on something like a recall that didn't happen!

So now, in today's world, we ask again, "Is there a place in the food industry for structured management systems?" There is no doubt that a structured, effective management system can enhance an organization. It is also evident that an inadequate, poorly designed system can be a true detriment to an organization. This statement is based on many years of experience and first-hand exposure to many different food sectors, ranging from growing and producing the raw products and materials to manufacturing sites, restaurants, and grocery stores that have direct contact with the consumer. The value of an effective, well-defined, implemented, and maintained food management system is priceless. This type of system focuses on food safety, continuous improvement, and meeting customer requirements. It is supported by commitment from top management and structured processes managed through measurable objectives. It is important to consider a structured management system as a management tool; when it is integrated effectively into a process, it provides the necessary foundation and structure. By utilizing documentation and objective evidence, consistency can be maintained throughout the entire operation, while management focuses on continuous improvement and meeting or exceeding customers' needs and expectations. A well-defined and formalized system clearly drives management responsibility, a policy statement, measureable objectives, and expected performance as the basis for a sound management system. Continuous improvement is inherent in the system, especially through effective processes that include top management support, management review meetings, internal audits, corrective and preventive actions. This is demonstrated in Figure P.1.

If one of these key elements (leg of the stool) weakens, then the system will begin to falter. All three legs of the stool must remain strong in order to support the compliance and continuous improvement of an effective management system. This stool will surface a few more times in this text because it provides a clear picture of the strength and stability of an effective management system.

The key question is to identify which system is best for the specific operation(s). Other questions follow, such as "Where do we begin? How well do we really

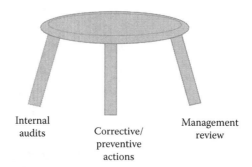

Internal audits   Corrective/preventive actions   Management review

FIGURE P.1   An effective management system.

understand what is required? What must we do, or what have we done already?" The answers to these questions can't be found in one gigantic subject matter book or even a series of books. Answers are unique to each individual operation. Since I grew up minutes from Fenway Park, I developed a love for baseball at a young age. It is all about teamwork and applying learned skills. I enjoy using baseball as an analogy. One could read every book ever written on baseball and baseball techniques, but all the reading material in the world is not going to make a major league ball player out of every one of us. The information will definitely help in making decisions and fine-tuning skills and goals. Having the knowledge and applying the skills is critical. Just as critical is having a team of individuals with the knowledge and skills in which each team member has a distinct role.

It is critical to identify and apply effective "tools." The key words are "effective" and "tools." The intent of this text is not to answer every question but to provide the tools that can be applied to each individual operation so that questions can be answered in a manner that is best for that particular operation, adding value to its processes and preparing a foundation for establishing and maintaining an effective management-based food safety and hopefully quality system. It is recommended that the organization include the development of a quality management system in its journey. Some standards require this; with others, there is a choice.

Which system is best for your particular operation? The choices are many, including FSSC 22000, ISO 9001, ISO 22000, or possibly a non-ISO system with similar requirements such as SQF (Safe Quality Food) Code, the BRC (British Royal Consortium) Global Standard for Food Safety or another GFSI approved scheme. The answer may be in developing a food safety program that is structured to meet the 12 steps of Hazard Analysis and Critical Control Point (HACCP) (the five presteps and seven basic principles) as defined by *Codex Alimentarius*. Should the environmentally focused standard ISO 14001 and Health and Safety (OHSAS) 18000 be included? The answer may likely be a combination of more than one standard such as ISO 9001 and FSSC 22000 or a food safety management standard such as SQF or BRC that includes quality parameters. Chapter 1 provides a background into the GFSI (Global Food Safety Initiative) approved food safety schemes. These will vary over time just as specific requirements do; thus, wherever possible, a website is identified. The reader is encouraged to visit these websites frequently and take advantage of the excellent guidance and background information provided. References to several of the standards are included throughout the text with further references summarized and included in specific chapters. It is important to note that these are not all inclusive but are meant as guidelines. Again, it is critical that the reader go directly to the website of the chosen standard to get the complete and up-to-date information.

As an industry professional, an auditor, and a consultant, I have experienced many different systems, not only a wide range of food-related systems, but also applications outside of the food industry. I have encountered the best of the best and other systems that were poorly defined with failure imminent. In most instances, the latter were those that had management teams who did not really see the value of a management system; they did as little as possible to get a certificate that made their customers happy.

In today's world of food safety, customers are not only demanding certification, but they also demand evidence of effectiveness and sustainability. Food safety is very serious. "The Center for Disease Control and Prevention (CDC) identifies [approximately] 1000 disease outbreaks annually." It is estimated that about 48 million people (one in six Americans) get sick from something they ate. "Contaminated food sends [approximately] 128,000 victims to the hospital, and it kills 3,000 children and adults" (*Madeline Drexler, "Why your food isn't safe," Good Housekeeping*, October 2011). Keep in mind that these numbers do not include the cases that go unreported. Jon Porter, a pioneer in food safety, always said, "every 200th meal will get you." We may have a stomachache or a bad night of what we might call a 24-hour bug, but in reality, it is most likely something we ate. The illness passes, life goes on. The real cause of the illness never gets investigated or reported. Food safety is serious. Every organization or establishment that in any way produces or serves food for consumption has a direct responsibility to ensure that it is safe. The food safety standards and the information in this text are meant to be proactive. We can no longer learn from our mistakes, we cannot allow mistakes to happen. In today's world of food safety, we must be proactive and prevent mistakes from occurring. We must make every effort to prevent hazards or make certain that the hazards are controlled and do not result in a food safety event.

I have seen excellent systems evaporate due to lack of resources and management commitment. Others prevail and improve over the years with focused and effective management commitment and support. Some systems faltered and then are revived. How does this happen? Effective and sustainable management systems are directly related to the application of resources, management commitment, and the application of the right tools in the right manner.

The focus of this text is to provide the insight and shared experiences that can be applied to the development, implementation, and maintenance of an effective food safety management system. There are many references to "quality" standards because, although they are not always required, "quality" just makes good business sense. An effective system may be based on a combination of the requirements of ISO 9001, ISO 22000, ISO 14000, OHSAS 18000, and GFSI-approved schemes (i.e., FSSC 22000, BRC, SQF, IFS, etc.). The safety of food products is fundamental. Developing an effective food safety management system that supports structure and discipline and focuses on providing the consumer with a safe product of the highest quality is paramount.

Also, another critical point is that the information in this text focuses on current food safety standards. These standards are very similar in content, but also different in some significant areas. The choice of the standard is up to the organization. This text is not meant to promote specific standards and ignore others. The reference material is meant to focus on emphasis and clarity. The information communicated in this text can be applied to any standard. The reader must be familiar with the specific requirements of the standard of choice.

Not every element from each standard has been addressed in this text; however, the information is presented in a manner that will provide a strong foundation for whichever standard is chosen by the organization. It is imperative that management ensure a strong understanding within the organization of the requirements of the standard of choice. The information in this text is based on many years of experience from subject matter experts throughout many different food industry sectors. The information and guidance is meant to supplement and aid in enhancing an understanding of the standards. It is up to the reader to determine how to apply (integrate) this information into his or her system.

Borrowing a successful format from my first book, *The ISO 9000 Quality System: Applications in Food and Technology* (Wiley, 2001), the content of this text will not only be based on my experience, but will also contain contributions from many individuals who have a variety of experience in science-based industries such as food, chemical, and tobacco. These subject matter experts have had significant roles within management systems, enjoying successes and dealing with struggles inherent in maintaining and improving mature and effective management systems.

In conclusion, this text provides experience-based information that can be translated and integrated into any operation. Processes are unique to each operation. One of the first steps in system development is to learn to "integrate" the requirements of the chosen standard into the operation, not reengineer your operation to become compliant. It is essential that the foundation be based on this concept. This text clearly explains the requirements, the method for integrating these requirements into your operation, and the process for identifying gaps and addressing these gaps in a manner that is not only compliant but also best for the organization. This is essential for the development of an effective, efficient, value-added, and sustainable management system. No matter which food sector is represented by the organization, its specific program requirements, or its processes, can all derive benefits from integrating this management system into its operation. Food safety standards promote "from farm to fork."

The main focus must be to integrate this system applying common sense based on experience. A good friend of mine is now plant manager for the third facility (three different companies) that he has supported and led through the management system certification process. I remember celebrating the first certification more than 20 years ago and asking him if it was worth it. Although his current company does not allow him to be recognized by name, his response was "absolutely and I will not operate without it." Today his current facility is not only approved to ISO 9001 and FSSC 22000, but the team has won numerous other awards, demonstrating team accomplishments that were built on the structure and discipline of the management system concept.

The road to compliance is not sugar coated, but it is worth the effort. Russ Marchiando, packaging plant manager at Wixon, Inc. in St. Francis, Wisconsin, describes the development and implementation process as "a journey not a destination." Thank you in advance for choosing this textbook as one of your sources for information on food safety management standards. I welcome any comments, suggestions, or questions (www.newslow.com).

INDUSTRY PROFESSIONALS WHO SHARED THEIR EXPERIENCES

Guido Abreu, quality system analyst, Dosal Tobacco Corporation

Patrick Bele, former food program manager, United States, for Bureau Veritas.

Dan Bernkopf, vice president, FSQA Applications SafetyChain Software

Rick Biros, founder and former president, Carpe Diem, publisher of *Food Quality Magazine*, 1994–2006, and now the founder and president of Innovative Publishing Company, LLC.

Mike Burness, vice president, Global Quality and Food Safety, Chiquita Brands International, Inc.

Andy Fowler, former vice president administration and environmental compliance, Bacardi & Company Limited

Dorothy Gittings, an independent consultant and third-party auditor, D.L. Gittings Consulting Services, Inc.

Joe Hembd, management systems manager, Chesterman Company

Bob Kaegi, director, Protein Applications & Seasonings, Wixon, Inc.

Alan Lane, former packaging manager, now an independent consultant

Dr. Tatiana Lorea, manager, Food Safety Education and Training, Ecolab, Food & Beverage Division.

Russ Marchiando, packaging plant manager, Wixon, Inc.

Thomas Marchisello, industry development director, Grocery Retail, GS1 US.

Victor Muliyil, technical manager for North America Food Safety Services, SGS

Yolanda Nader, CFO and CEO, Dosal Tobacco Company

Laura Dunn Nelson, industry relations director, Alchemy Systems

Russ Patty, an independent consultant (former quality manager of a beverage operation)

Eric Putnam, food safety, quality and training systems manager, Wixon, Inc.

Erasmo Salazar, food safety lead assessor and food technical adviser, LRQA

Dennis Sasseville, an independent consultant and third-party auditor, president of Cobalt Advisors, Bedford, NH

Dr. Ron Schmidt, retired professor, University of Florida

Tim Sonntag, vice president, Quality & Technical Services, Wixon, Inc.

Charlie Stecher, an independent consultant and auditor, Anchar, Inc.

Bob Wihl, independent consultant, D.L. Newslow & Associates, Inc.

John Yarborough, CEO, solutions manager, LLP

Author note: At the time that this text was finalized and sent to press, the GFSI approved schemes may have experienced revisions different than those that are quoted in our text. Basic requirements have not been changed, but some may have been clarified and/or enhanced. Applications and best practices apply; however, as stated throughout this text, it is important for everyone, no matter what stage of FSMS compliance (defining, implementing, maintaining, etc.) to always monitor the website of the standard of choice to ensure that revisions are identified, understood and applied as deemed necessary by the top management team, food safety leader, and third party registrar.

Acknowledgments

The content of this book is a culmination of many years of experience, learning and working with many professionals in food and food-related industries. I would like to acknowledge and thank everyone who I have had the pleasure to meet and work with over the years. The knowledge and wisdom from so many individuals helped to provide the foundation for this text.

I would like to thank all of my colleagues and confidantes working every day in the trenches of food safety who took their time to share experiences publicly and those who, due to company policy, were unable to be recognized (you know who you are!). Their names are included in a separate section.

I would also like to send a special thank you to Dennis Sasseville and Erasmo Salazar who took their time to contribute an entire chapter for this text. Their expertise, combined with the subject matter of their chapter(s), is invaluable for the theme of this text.

I would also like to thank a wonderful team of individuals who spent countless hours proofreading, typing, revising content, and putting up with the stress of so much material and tight deadlines, not to mention working with a sleep-deprived author. I would like to thank the following individuals whose dedication included long days, weekends, and in some instances, lonely spouses; without their support, this text would never have been possible: Melissa Hays, Martha Newslow, Jade Piearson, Angela Hester, Heather Garvey, Justin Duhaime, and Savannah Skipper.

I would also like to thank Dorothy Gittings, Nancy McDonald, Dennis Sasseville, and Mary Hinkle for the time they spent proofreading and providing expert feedback on chapter content and structure.

A special "woof" to all my critters who sat up with me keeping me company on those really long days and nights of writing, especially Sophia Rose, Maggie Leigh, and Scout who stayed right by my side no matter what.

I would like to take this opportunity to thank those who made this text possible, beginning with the patient and knowledgeable team at CRC Press/Taylor & Francis Group. Without the continued support, guidance, and patience of Steve Zollo, Kari Budyk, Glenon Butler, and Vijay Bose, I would have been lost in a black hole months ago. I must be honest, when I agreed to do this text, I had no idea how involved and how much information would come together and how quickly the requirements would change. It was great fun and very challenging, but the toughest part was knowing when to end the adventure.

Also, thank you to my friends and relatives who recognized and understood the critical time frame of this project and accepted whatever constraints were necessary. A very special thank you to my 92-year-old mom who sacrificed much of our time together to make this text possible. I learned from your determination and tenacity and your "can do" spirit. You always said there is no such word as "can't."

Author

Debby L. Newslow is the president of D. L. Newslow & Associates, Inc., located in Orlando, Florida. She is the author of *The ISO 9000 Quality System: Applications in Food and Technology*, released by John Wiley & Sons Publishing in February 2001 and also the author of an HACCP-based chapter included in the 2003 John Wiley & Sons release of the *Food Safety Compendium*. She has just completed her second book titled *Food Safety Management Programs—Applications, Best Practices, and Compliance* published by CRC Press/Taylor & Francis Group, which will be available for purchase in the fourth quarter of 2013.

A food science and technology degree from the University of Florida prepared Debby to begin a career as a quality control manager with T.G. Lee Foods, a division of Dean Foods. In 1985, she joined The Coca-Cola Company in its Minute Maid Division (CCF) as a research and development food scientist. In 1987, she transferred to quality assurance as a corporate auditor and project specialist, where she was instrumental in developing a GMP audit program, creation of a company standard quality assurance manual, and assisted with ISO 9002:1994 certifications of three different process operations.

In 1995, Debby joined Lloyd's Register Quality Assurance (LRQA) as a certified quality assurance lead auditor. She has been involved with ISO certification and consulting activities in a wide range of industries such as pet food, dairy, beverage, soup, spice, candy, sugar, citrus, meats, poultry, seafood, tobacco, chemicals, packaging, and cosmetics. She has assisted and consulted with more than 400 different companies, ranging from 2–500 plus employees.

Debby currently subcontracts for leading registrars as an ISO 9001 lead auditor qualified to perform audits in food (including high-risk foods), chemical, and packaging operations for ISO 9001, 22000, and HACCP. Her main interests are assisting food, chemical, and various other industries with business system development and maintenance activities. This includes ISO compliance (9000, 14000, OHSAS 18001, 22000, 17025, and 22716), HACCP, SQF, BRC, GMP, food hygiene, HACCP prerequisite programs, crisis management, and other related systems/processes. When not assisting companies with development and maintenance activities, she teaches and conducts workshops based on these topics.

Debby has published numerous articles in such trade journals as *Cultured Dairy Products Journal*; *Dairy, Food, and Environmental Sanitation*; *Dairy Foods*; and *Food Quality*. A presenter at meetings and short courses sponsored by such organizations as the University of Florida, IFT, SME, Citrus Engineers, Agricultural Consultants, Food Safety Summit, Food Quality, and the American Spice Trade Association, she has also served on the executive board of Florida IFT and is currently on the editorial board of Food Protection. She has served on executive boards of the Citrus Division and Quality Assurance Division of IFT and Cultured Dairy Products Association and has been a member of the development team for the ASQ-HACCP certification exam and the NRFSP HACCP manager exam.

Debby is an IRCA certified quality assurance lead assessor, an American Society of Quality (ASQ) certified quality auditor, certified HACCP auditor, certified quality manager, and certified food safety manager. She is active in IFT, FLA IFT, International Association for Food Protection, Florida Association for Food Protection (FAFP), and ASQ. In 2003, Debby was the recipient of the Roy Wilson Sparkle Award presented by the Florida Division of IFT for her contributions to sanitation in the citrus industry. In 2007, she received the Bob Olsen Award for her distinguished contributions and services to Florida Section IFT, and in 2008, she received the Sanitarian of the Year Award from the FAFP. She was also conferred the Food Safety Leadership Award—Training, which was presented by the NSF at the May 2013 Food Safety Summit.

1 Introduction to Food Safety

An introduction to food safety: Where have we come and where are we going is a view point that I am not always sure is understood. Historically, it all began at the beginning of the twentieth century with the passage of the Federal Meat Inspection Act (FMIA) in 1906. Upton Sinclair's book *The Jungle*, which detailed significant food safety issues in meat packing facilities, had a direct impact on the passage of the FMIA. The Pure Food and Drug Act, which later became the Federal Food, Drug, and Cosmetic Act, was also enacted in the same era in US history. At the turn of the twenty-first century, we saw an emphasis on food safety in which regulatory agencies were requiring (or proposing) the Hazard Analysis Critical Control Point (HACCP) process for what was considered high-risk segments of the food industry (e.g., meats and poultry, seafood, 100% juice products). HACCP dates back to the early 1960s when the concept was developed by Dr. Howard Bauman at the Pillsbury Company in association with the National Aeronautics and Space Administration and the US Army Laboratories in Natick, Massachusetts, in response to the needs of the space program. Out of necessity, food products developed for astronauts had to be free of pathogens. In 1973, the federal government used the concept for HACCP as the basis for low-acid and acidified canned food regulations (21 CFR 113). Through the 1970s and 1980s, some manufacturers did try to apply the HACCP concept, but for the most part, efforts were not successful. Most tried to apply the concept to all *control points*, which burdened the system causing its application to be impossible to manage effectively. It was not uncommon for a company to have 100 plus critical control points.

As we have evolved into this century with the Internet and technology, every year brings a new sense of urgency to the world of food safety. Microorganisms continue to emerge becoming more tolerant to their environment with survival mechanisms for harsh environments consistently improving. Microbiological detection methods and the reporting structure for food-borne illnesses have improved greatly along with the ability to track pathogens to their source. Margaret A. Hamburg, MD, Commissioner of Food and Drugs, stated in an article published at foodsafety.gov that "each year, food borne illness strikes 48 million Americans, hospitalizing a hundred thousand and killing thousands." These figures are huge but are very likely a low estimate. The food supply must be as safe as it can be. No exceptions!

HACCP

As the industry moved forward and began focusing on food safety, organizations began to apply the HACCP concept internally to their operation. Basic HACCP focuses on the Codex Alimentarius Guidelines, which incorporate the seven principles of HACCP plus five presteps into a document meant to be applied worldwide as a guideline for establishing and maintaining HACCP programs for all aspects of the food industry. The basics of HACCP are defined by Codex Alimentarius (2003): "Hazard Analysis and Critical Control Point (HACCP) system and guidelines for its application" (www.codexalimentarius.org). The National Advisory Committee on Microbiological Criteria for Foods was established in 1985 in the United States and held its first meeting in 1988. The committee went on to issue a similar HACCP guideline. Although these guidelines provided a good foundation, the structure and discipline of a management system approach was difficult to achieve.

The HACCP process is designed to be a failure mode effect analysis for food that applies science, experience, and common sense to identify, prevent, evaluate, and control hazards that are significant for food safety. This concept identifies existing and potential hazards: biological, pathogenic, chemical, and physical in the ingredients and process of the manufacturing of consumable goods from the farm to the table, establishing controls for public health. HACCP provides the framework to produce foods safely and prove they are safe. An effective HACCP program must be built on the foundation of effective prerequisite programs.

FDA FOOD SAFETY MODERNIZATION ACT

On January 4, 2011, President Obama signed into law the FDA Food Safety Modernization Act (FSMA), which is the most encompassing reform of food safety laws in more than 70 years. This act focuses its aim on ensuring that the US food supply is safe, applying a focus on prevention rather than responding to a contamination outbreak. The time has passed for the food industry to learn from its mistakes; now we must be proactive, not reactive. Effectiveness must be measured by evaluating our processes, products, ingredients, packaging materials, processing aids, and product contact surfaces using the eyes of prevention to identify existing and potential hazards, then putting actions in place that either eliminate or correct the situation. We must have confidence that we are producing a product that is safe for consumption. Every associate has a responsibility to ensure the production of a safe product, taking every measure necessary to ensure this safety. It is time for top management to seriously evaluate their processing systems to define a proactive food safety program that identifies and ensures the control of existing and potential food safety hazards, based on current scientific principles, standards, and concerns.

As this text goes to press, because of the complexity of FDA FSMA, the food industry is waiting for some rules to be clarified. In the meantime, consumers above and beyond government laws are demanding safe products. Food manufacturers are receiving extreme pressure from their customers to develop a proactive program that forces management to evaluate their products and processes, and suppliers to ensure that products are free of hazards and safe for consumption. In today's world of food safety, we define the word *safe* as a food that is free of any food hazards. *Food safety* is defined as a finished product that does not contain a food hazard when used according to its intended use.

MANAGEMENT SYSTEMS

This text will refer to the GFSI-approved schemes as food safety management systems (FSMS). Some may feel that this terminology is not exactly correct and that only FSSC 22000 (Food Safety System Certification) is a true management system. This will be explained in more detail later in this chapter; however, for the purpose of this text, these definitions are not challenged. The important point is that these food safety schemes, when implemented effectively and compliance is sustained, provide a proactive tool to ensure the production of a safe product.

According to the official ISO website, the term management system "describes the set of procedures an organization needs to follow in order to meet its objectives."

In a small organization there may not be an official system, just "our way of doing things." Often "our way of doing things" is not written down, instead it is in the head of the staff. However, the larger the organization the more likely it is that there are written instructions about how things are done. This makes sure that nothing is left out and that everyone is clear about who needs to do what, when and how. When an organization systemizes how it does things, this is known as a management system (ISO n.d.).

ISO 9001:2008: A QUALITY MANAGEMENT SYSTEM STANDARD

Prior to providing an overview of the individual food safety standards, it is important to first discuss ISO 9001:2008. ISO 9001 truly provides the foundation for most, if not all, management system-based standards. It is important to initially understand the basis and foundation of this standard. *Quality management system* is a term that has evolved over the years with the inception of the ISO 9000 standards in 1987. ISO 9001 is an international standard directed at the quality management process of an organization. However, this statement can be misleading because this standard focuses on an organization's complete system. Although the title is *quality* management system, these requirements focus on the complete system and its ability to meet customer requirements while continuously improving.

ISO 9001 evolved with the need for consistency and harmonization in international trade dating back as early as NATO documents AQAP-1 in 1968. In 1983, the International Organization for Standardization established Technical Committee 176 to develop an international standard that would focus on quality and management. Quality was again the focus on the organization's consistent ability to meet defined requirements that ensured that customer requirements were clearly understood, defined, and met. It was the work of Technical Committee 176 that resulted in the publication of the original ISO 9000 series in 1987.

ISO 9000 is a term that references a group of international standards that are directed at the management process of an organization. *ISO* means *equal* in Greek. The International Organization for Standardization was founded in 1946 in Geneva, Switzerland. This organization is made up of approximately 90 countries, including the United States.

To date, the ISO 9000 series of standards were revised in 1994, 2000, and 2008. As of this writing, the current version is ISO 9001:2008 and will be referred to as

ISO 9001 from this point forward. Note: It is recommended that the reader access http://www.iso.org for the current version.

The ISO 9001 quality management standard focuses on the existence, implementation, and effectiveness of a quality management system, not the individual product. The standard is generic in text and can be implemented by manufacturing and service organizations regardless of the type of product or service.

The first registration to an ISO 9000 standard in the United States occurred in 1991. It was 1994 though, before a food manufacturing company actually achieved registration. In the years that followed, much of the food industry remained skeptical of the true advantage for developing systems compliant with this standard.

The management responsibility section of ISO 9001 provides a clear directive to top management of the organization to become involved and make decisions based on the effectiveness and suitability of the system, required resources, continuous improvement, and the system's ability to meet its defined *measurable objectives*. Keep in mind that improvement is very difficult without structured *measurements* to track it.

Many food companies have translated the requirements of ISO 9001 into their own internal management system. Systems were defined, implemented, and monitored internally rather than seeking formal registration from a third-party registrar. It has been stated that meeting this structure and discipline through internal compliance resulted in considerable internal benefits to the company.

ISO 9001 is a management tool, which, when integrated into a process, provides the foundation and structure through documentation, along with objective evidence to promote consistency throughout the entire operation while focusing on continuous improvement and meeting customer's needs and expectations. A well-defined and formalized system clearly defines management responsibility, policy statement, measurable objectives, and expected performance as the basis for a sound management system. Continuous improvement is inherent in the system, especially through effective processes from top management review meetings, internal audits, and corrective action/preventive action (CAPA).

Effective processes for internal audits, corrective/preventive actions, and management review are essential for the development and maintenance of every management system.

FOOD SAFETY MANAGEMENT SYSTEM STANDARDS (FSMS)

In addition to changes in regulations and the involvement of federal and state governments, FSMS standards have evolved. Depending on where a company does business, developing a compliant system ranges from voluntary to mandatory. Although individual food safety management standards do have some differences (the preference for specific standards is, in many instances, food sector specific), all focus on applying auditable food safety requirements to a defined system. The approved GFSI schemes also may vary on their defined scope, certification processes, and specific criteria.

In the following text, this chapter reviews some basics for each standard, possible thoughts to apply when choosing a specific standard and also a brief look at some of the benefits of each. Current websites as of the time that this text is going to print are included. It is recommended that the reader reviews the *current* websites making a decision that is best for the individual organization. In addition, each subsequent chapter provides compliance and best practice guidelines. In some instances, a specific standard may be quoted to provide an additional foundation and clarification information. At the end of each chapter is a chart that identifies related elements of International Featured Standards (IFS), FSSC 22000, Safe Quality Food (SQF), and British Retail Consortium (BRC) that may apply to the topic of the chapter, but not have been referenced in its text. IFS, FSSC 22000, SQF, and BRC are currently the most commonly adopted schemes by manufacturers and also those that can be applied throughout most of the food sectors. Websites for the current (as of the release of this text) GFSI-approved schemes are included with this information. The remaining schemes, in most instances, focus on specific food sectors; however, content of this text may also indirectly be applied to these schemes. As an organization makes a decision to move forward toward compliance to a specific standard, it is critical that the management team ensures that its implementation, management, and food safety teams are familiar with that standard. Information contained in this text is meant to provide the reader with a sound foundation toward the implementation, maintenance, and continually improve the food safety (and quality) management system no matter which standard is chosen.

GLOBAL FOOD SAFETY INITIATIVE

The Global Food Safety Initiative (GFSI), a nonprofit foundation managed by the Consumer Goods Forum, was established in May 2000. Membership to the GFSI Board of Directors is by invitation only. The GFSI Foundation Board of Directors includes representatives from major global retailers, manufacturers, and food service operators who oversee basic management and direction. The GFSI Foundation Board of Directors have developed a document that is not a food safety management standard, but a guide that defines the requirements that must be met by the food safety management standards that are recognized by GFSI. The GFSI website (http://mygfsi.com/) defines its vision to drive continuous improvement in food safety while strengthening consumer confidence worldwide. The primary focus of its objectives is to reduce food safety risks by delivering "equivalence and convergence between effective FSMS."

Victor Muliyil (technical manager for North America Food Safety Services, SGS) and Supreeya Sansawat (global food business manager SGS) in their white paper "Comparing Global Food Safety Initiative (GFSI) Recognized Standards" October 2012, provide some very good insight on the planned path for organizations that develop, implement, maintain, and continually improve an FSMS compliant with a GFSI-approved scheme.

> Holding a certification to a GFSI-approved scheme is fast becoming an industry standard as more and more organizations within the industry are expecting entire supply chains for a given product to prove this capability. Organizations audited and certified to a GFSI-approved scheme increase their chances of being a chosen supplier to retailers

and/or manufacturers who are demanding this of their suppliers. The vision of the GFSI benchmark—"once certified, accepted everywhere"—is moving in the direction of being realized, both across the industry and across the world. With a number of GFSI-approved food safety schemes now available, competitive suppliers are already certified, seeking certification or developing their processes and identifying the best certification scheme for their organization. The hope is that as certification becomes even more widespread, consumer confidence will be fully restored and food scares will become a thing of the past.

The process for GFSI to review an FSMS standard and confirm that it meets the requirements of the most current version of the GFSI guidelines is known as *benchmarking*. *Benchmarking* is defined as a "procedure by which a food safety–related scheme is compared to the GFSI Guidance Document." The following link contains the most current version of the GFSI guidance document: (http://www.mygfsi.com/gfsifiles/Guidance_Document_Sixth_Edition_Version_6.1.pdf).

The GFSI Board does not get involved in the actual certification process. Certification is achieved through a third-party audit by an approved registration body against one of the approved GFSI schemes.

Originally, four food safety standards (SQF, BRC, Dutch HACCP, and IFS) were *benchmarked* as approved GFSI schemes. Note that IFS did originally stand for International Food Standards, but with the release of Version 6, it is now referred to as the International Featured Standards. In 2009, ISO 22000 in combination with PAS 220 (Publicly Available Specification) was benchmarked and approved, becoming the fifth approved scheme at that time. The standard representing the combination of ISO 22000 and PAS 220 became known as FSSC 22000. The PAS 220 standard has been withdrawn with content now provided in ISO 22002-1. Additional PAS documents such as PAS 223 for managing food safety for packaging and PAS 222 for the manufacture of food and feed for animals have now been created and released. Other industry-specific PAS documents are planned for creation and release in coming months and years.

At the time this text went to print, there were eight benchmarked accepted GFSI schemes; however, since this may change, it is recommended that the astute reader monitor the GFSI website (http://mygfsi.com/) for the most current information. A comparison between a sampling of the FSMS standards is included as part of each chapter that addresses requirements and best practices. This comparison may include a reference to a specific section of the standard or a website link depending on the most appropriate approach based on the chapter discussion. As stated previously, for those standards that apply to a specific food sector(s), the web link, in some instances, is considered the most effective means for referencing specific requirements.

It is important to again stress that the focus of this text is to provide a foundation of information and best practices based on experience so that an organization can define, implement, and maintain an effective FSMS that is value added for its operation. Currently, there are eight GFSI-approved schemes, food safety management standards that an organization may choose. There are other standards available that address food safety management but at the time of this printing are not recognized by GFSI. There is also ISO 9001, which some food companies worldwide have applied to their systems prior to the development of a specific food safety management standard, which ensures that their quality system was compliant, including the manufacture of safe foods. Years ago, a good friend of mine told me that a food company cannot have quality without food safety. This is a statement that gets stronger with age. ISO 9001 is mentioned throughout this text for, in fact, this author is a true believer that this standard has provided a sound foundation for today's world of management system standards. The information presented in this text is not skewed toward one particular standard, but it is meant to enhance the reader's overall knowledge and to provide a *best practice* approach based on experience and common sense. It is recommended that each organization review the information provided and use this to make a choice that is best for the organization adding value and structure to its operation. A review of the current (at the time this text is released) GFSI-recognized schemes follows, but again it must be emphasized that the choice is up to the organization based on its processes, products, food industry sector, and many times, the preference of its customers. The GFSI guidelines and the standards themselves undergo what sometimes seems as frequent revisions. It has been said that once a revision is officially released, then the group in charge of the standard starts work on its next revision. In order to ensure an accurate understanding of the requirements, both in choosing a standard and then staying current with the requirements, it is critical to monitor the websites for changes, comments, and basic information. The websites of the current standards, additional information, and many references are identified frequently throughout this text. The reader is encouraged to monitor this to ensure that the most current requirements are known, understood, and, where applicable, applied to a compliant system.

FSMS standards do have a lot in common with each other. Major differences usually focus on a particular industry sector. As mentioned previously, the purpose of this text is to provide insight on defining, implementing, and maintaining an effective FSMS. The various schemes may or may not be specifically quoted in various sections of this text, but the insight provided does relate to FSMS compliance overall. It is up to the individual organizations to review the available standards to determine which one is most appropriate for its operation and then apply the compliance requirements accordingly.

GFSI RECOGNIZED SCHEMES

The following is a brief description, including a web link, for those standards that have been benchmarked as schemes that meet the current (release of this text) defined GFSI guidelines. Note that at the time this text went to print the GFSI website did list some of these standards as in process for review:

BRC (British Retail Consortium) Global Standard for Food Safety (Sixth Edition): The BRC Global Standard is a global safety and quality certification program widely applied throughout over 100 countries by suppliers and global retailers. The BRC Global Standard for Food Safety was one of the first standards to be recognized and approved in meeting the GFSI guidelines for food safety.

The BRC Global Standard for Food Safety was developed with the objective of specifying the safety, quality, and operational criteria required for food manufacturers to comply with regulations and to protect consumers. All companies involved in the end-to-end (farm to fork) supply chain are required to have a clear understanding of

- Descriptive requirements for process and hygienic control that provide clear guidelines as to how food safety should be addressed
- A simple certification process that requires only an on-site audit (there is no requirement for a desk study step)
- Includes an option of an enrollment audit for sites working toward the full standard
- Includes an option for voluntary unannounced audits to show your high level of commitment to food safety and quality
- Requires recertification audits to be done within a fixed timescale to ensure continuous certification
- Focus is on quality, food safety, and legality

FIGURE 1.1   The benefits experienced with BRC global standard for food safety certification. (From Muliyil, V. and Sansawat, S., Comparing Global Food Safety Initiative (GFSI) recognized standards, p.6, October, 2012.)

the products they produce and distribute and have the systems in place to identify and control food safety hazards.

As previously stated, this is a very popular standard worldwide and has been for many years. Through experience, there have been many benefits experienced by an organization when choosing this standard. However, many of these benefits also relate to other approved schemes. *Muliyil and Sansawat* (p. 6) have identified benefits that would be considered specific for this standard summarized in Figure 1.1.

The BRC Global Standard for Food Safety plus additional current information related to this standard can be found at http://www.brcglobalstandards.com/GlobalStandards/Home.aspx.

Canada GAP: Canada GAP (Good Agricultural Practices) is the Canadian Horticultural Council's On-Farm Food Safety Program. This program consists of national food safety standards and a certification system for the safe production, storage, and packing of fresh fruits and vegetables. This program is focused on assisting producers, packers, and storage locations in applying the basic seven principles of HACCP to develop, implement, and maintain food safety programs. This standard was developed with its focus on six crop-specific programs. Compliance provides evidence to the organization's customers that it has proactive, preventive-type programs in place to minimize the possibility of contamination with a food hazard. The program must be implemented, compliant, and maintained effectively with evidence (records) available to demonstrate continual compliance. More information on this standard and its requirements can be found at http://www.canadagap.ca/.

Food Safety System Certification (FSSC) 22000 Food Products: FSSC 22000 contains a complete certification scheme for food safety systems based on existing standards for certification (ISO 22000:2005, ISO 22002-1:2009, and FSSC 22000 additional requirements). ISO 22000:2005 is not an approved GFSI scheme because of its weakness in defining specific requirements for prerequisite programs; however, manufacturers already certified against ISO 22000:2005 can add ISO 22002-1 to its scope (and pass the audit) to become certified to the approved GFSI scheme. Organizations that want to integrate quality into their management system scope can accomplish this through compliance with the requirements of ISO 9001. FSSC 22000 was developed for the certification of food safety systems, of organizations in the food chain that process or manufacture human food products, animal products, perishable vegetable products, products with a long shelf life, or food ingredients (additives, vitamins, bio-cultures). FSSC 22000 has planned inclusions of additional scopes as the proposed technical specifications for sector prerequisite programs (PRPs) (i.e., PAS 222 and PAS 223) are developed and implemented.

There are many benefits to choosing the FSSC 22000 standard, but again it depends on the organization's structure and also the food industry sector. FSSC 22000 is a comprehensive standard built on the ISO 9001 format, which encourages integrating the requirements into the organization's process rather than reengineering and changing everything to become compliant. Understanding the requirements and identifying the gaps between a current system and the standard are essential to creating the road-map to compliance. ISO 22002-1 is a great guideline for the development of effective prerequisite programs. We use that standard to assist operations that just want strong and effective prerequisite programs. Muliyil and Sansawat (p. 8) have summarized FSSC benefits based on their experience in Figure 1.2.

As mentioned several times, it is critical for the reader to review current websites either for the standard of choice or in order to make a decision on which standard is best for the organization. FSSC 22000 made a key revision (see preface for more details) just prior to this text going to print, which included a requirement for the organization to ensure compliance of laboratory activities to the ISO 17025 standard. ISO 17025 is not discussed in this text; however, it is a standard used for the certification of laboratories. At this time, FSSC 22000 does not require certification to ISO 17025 but wants to ensure that the laboratory does meet these requirements. As a reader, it is a critical step toward ensuring a complete understanding of the requirements to review current websites for the most current version of the standards.

An important aspect of FSSC 22000 is that it recognizes itself as a *management system certification scheme* stating clearly in its document (Note-11-4390-FSSC) that "IFS, BRC, and SQF are process/product certification schemes" not "management system certification schemes." It goes on to state that the FSSC audit is longer in

---

- Provides a good framework against which an organization can develop its FSMS, as it is not too descriptive and has the flexibility to allow the organization to choose the best way to control its own system
- Includes comprehensive requirements detailing how the organization can conduct an effective HACCP analysis or HACCP studies
- Promotes continuous improvement in food safety
- Targets its focus on food safety and legal compliance
- Easily integrates with an organization's existing management system or other systems in place, that is, quality management systems, environmental management systems, etc.
- Allows small, less structured organizations to implement an externally developed system
- Adopted by many major brands, thus beneficial for ingredient suppliers to be aligned with their customers
- Acceptance by the European Cooperation for Accreditation (EA), which means that most accreditation bodies now accept FSSC 22000

FIGURE 1.2 The benefits experienced with FSSC 22000 certification. (From Muliyil, V. and Sansawat, S., Comparing Global Food Safety Initiative (GFSI) recognized standards, p.8, October, 2012.)

duration and audits with more depth. It also states that "another significant difference between a management system audit and a process/product audit is that management system audits focus ... much stronger on management commitment, effectiveness, and continuous improvement." More information on this topic and also on additional requirements identified for the FSSC standard can be found on its website (http://www.fssc22000.com/en/).

It is not the purpose of this text to show preferences to one standard over another, but the focus is to provide a foundation of information based on experience, which includes best practices. It is stressed in this chapter and is continually stressed throughout this text that it is up to the organization to review the standards and to make the best choice for its operation and its food sector. Websites for each of the standards are referenced frequently throughout this text. These websites are kept current making each standard's website the best source of current information and requirements of the individual standards.

Global Aquaculture Alliance (GAA) Seafood Processing Standard focuses on aquaculture seafood. The GAA is an international nonprofit trade association dedicated to advancing environmentally and socially responsible aquaculture. GAA recognizes that aquaculture is the only sustainable means of increasing seafood supply to meet the food needs of the world's growing population, which led to the development of the BAP standards. The BAP standard addresses those elements important to aquaculture and also provides a means to measure compliance. The individual standards do vary depending on the process providing an effective variation to address the complete seafood chain. More information about this standard can be obtained from the GAA website at http://www.gaalliance.org/.

GLOBALG.A.P. is a private sector body that sets voluntary standards for the certification of production processes of agricultural (including aquaculture, livestock, fresh fruits, vegetables) products around the globe. The focus of the GLOBALG.A.P standard is on providing a foundation that ensures that products are handled safely in a manner that promotes food safety. More information related to the GLOBALG.A.P standard can be found at http://www.globalgap.org/.

Global Red Meat Standard (GRMS) is a scheme specifically developed by the Danish Agriculture and Food Council in cooperation with the Danish Cooperative of slaughterhouses and the Danish Meat Institute for the meat industry. Its cornerstone is product safety, focusing on critical areas affecting the maintenance of high meat safety requirements. This standard was designed to address all aspects of meat processing with a strong emphasis on basic good manufacturing practice (GMP) requirements. More information related to this standard can be found at http://www.grms.org/.

International Featured Standards (IFS) is a standard for auditing food safety and quality of processes, primary production, and products of food manufacturers. This standard was originally developed by German retailers and has grown to be the standard of choice for retailers in many of the EU countries including Germany, France, and Italy. IFS is also increasing in awareness and application in North America with several large companies moving forward to compliance. This standard has recently released its Version 6, which provides additional focus on food processing companies. Muliyil and Sansawat (pp. 9–10) have identified several benefits of IFS certification that are reflected in Figure 1.3.

- A simple certification process that only requires an on-site audit (there is no requirement for a desk study step).
- Focus is on quality, food safety, and legality. Once certified, there is no need for a reaudit and certification for 1 year (this applies to all levels awarded).
- Consists of a global network of strategically placed offices covering Europe, the Americas, and Asia supporting retailers, suppliers, and certification bodies with operational, training, and business development.
- Suppliers are given a 12-month time period to make corrective actions (when not directly related to food safety or regulatory compliance) allowing for budget planning and continuous improvements.
- Audits can be conducted electronically with software support, which also provides reporting with year-on-year results, certification audit importing/analysis, and global category specific benchmarking.
- Food safety and quality certification are both covered in one audit saving money by reducing the potential for further audits.
- Criteria is risk based, and there are no prescriptive elements.
- IFS audit portal is both a database and a reporting and notification tool.
- IFS offers an integrity program providing quality assurance and a formal compliant management system for retailer confidence.

FIGURE 1.3 The benefits experienced with IFS certification. (From Muliyil, V. and Sansawat, S., Comparing Global Food Safety Initiative (GFSI) recognized standards, pp. 9–10, October, 2012.)

More information related to the IFS standards can be found at http://www.ifs-certification.com/.

PrimusGFS is a voluntary private scheme that establishes requirements for the certification of products in the agricultural sector on a worldwide level. The scope of PrimusGFS is focused on the food safety of products in the agricultural sector designated for human consumption. PrimusGFS applies an integrated supplier approach and does address the full supply chain from prefarm to postfarm gate production. More information related to PrimusGFS can be obtained at its website http://www.primusgfs.com/.

Safe Quality Food (SQF) code has been recently revised combining the previous two standard approaches (SQF 1000 and SQF 2000) into one standard that includes modules for all industry sectors. For the purpose of this text and as related to most types of food manufacturing, Module 2 and Module 11 apply. Module 2 defines the system elements. Module 11 defines the requirements for the prerequisite programs. The SQF standard does provide a table that identifies which module applies to which food industry sectors. Before beginning this journey, be sure to confirm which modules apply. This should be done either directly with the SQF Institute or through a reputable and qualified registrar.

SQF offers three certification levels:

1. Level 1 is basically compliant with basic GMP requirements. Level 1 is not recognized as an approved GFSI scheme.
2. Level 2 is a recognized GFSI scheme and includes the requirements for Level 1 plus compliant HACCP food safety requirements.
3. Level 3 includes the requirements of Level 2 plus compliance with quality management system requirements.

- Enhancement of the organization's FSMS
- Certification demonstrates commitment to producing and trading safe food
- Increases consumer confidence of the products produced
- Brand equity enhanced
- Certification prepares organizations for inspection by regulatory authorities and other stakeholders
- Improvement in new market and customer prospects
- Attainment of SQF level 3 provides the organization the opportunity to use the SQF quality mark on their products

FIGURE 1.4 The benefits experienced with SQF certification. (From Muliyil, V. and Sansawat, S., Comparing Global Food Safety Initiative (GFSI) recognized standards, p.13, October, 2012.)

Through their experience, Muliyil and Sansawat (p. 13) have identified several benefits that have resulted from SQF certification. These benefits are reflected in Figure 1.4.

More information related to the SQF Code can be found at http://www.sqfi.com/.

Again, it must be emphasized that the descriptions of the current (at the time this text goes to press) GFSI-approved schemes are meant to provide the reader with a very basic description of the various food safety standards available. These standards, including references to specific elements, are discussed throughout this text; however, it is essential that the astute reader visit either the individual websites for each standard or the GFSI website for a complete overview of the GFSI-recognized schemes. The recognized schemes may change, some may lose their approval rating for one reason or another, while new ones may be added meeting the requirements of the GFSI guideline and become recognized through the GFSI benchmarking process.

ISO 22000:2005

ISO 22000:2005 is an international standard that was developed with a specific focus on an FSMS and closely integrates with ISO 9001, a quality management system standard. ISO 22000 cannot stand alone as a GFSI-approved scheme; however, when combined with the requirements of ISO 22002-1, it becomes FSSC 22000, which is a GFSI-recognized scheme. ISO 22002-1 is titled "Prerequisite programmes on food safety—Part 1: Food manufacturing." This document replaces the original PAS 220. Additional PAS documents have been developed: PAS 223 (managing food safety for packaging) and PAS 222 (prerequisite programs for food safety in the manufacture of food and feed for animals). It is expected that, FSSC 22000 will extend its scope to include these along with future-focused industry-specific PAS documents. However, in the meantime, it is encouraged that if these or any other PAS documents apply specifically to the food segment of interest, they be used for guidance in defining and establishing prerequisite programs. Prerequisite programs, as the foundation for food safety programs, both specific requirements and those left up to the organization, are discussed in PRP-specific chapters and subchapters.

ISO 9001 AND FSSC 22000

These individual standards have been reviewed previously in this text; however, it is important to review two important questions that surface quite frequently:

We are certified to ISO 9001. As we move forward with FSSC 22000, do we need to keep our ISO 9001 certification?

If we go forward with FSSC 22000 certification, do we have to also go for ISO 9001 certification?

These questions can be answered together. Through experience, it has been found that it is not only more advantageous, but actually also more efficient to implement ISO 9001 or at least several elements of ISO 9001 simultaneously with FSSC 22000. It is much more difficult to separate quality from food safety than it is to include it. If an organization is implementing a document control program, it makes much better sense to include both quality and food safety documents rather than try to separate these two categories. It also sends a stronger message to associates when the elements are applied throughout the organization. This logic applies to most of the elements. It is tough to justify the training required directly for food safety but ignore the same logic when training is required for quality-only parameters. Most of these aspects overlap, and when the organization tries to separate them, many times critical omissions occur that result in potentially major situations.

Those organizations that are already certified to ISO 9001 have a direct advantage to integrate those specific food safety requirements (many of which may already be included) into the existing management system. Integrating the requirements of more than one management standard has numerous positive opportunities for the organization (Chapter 19). The following is a list of those elements that ISO 22000 and ISO 9001 have in common:

Policy statement	Measurable objectives
Management review	Document control
Record control	Human resources
Infrastructure	Maintenance
Internal audits	Traceability
Control of nonconforming product	Measurement and monitoring of product
Calibration	CAPA
Outsourced processes	Purchasing

Patrick Bele, former Food Program Manager, United States, for Bureau Veritas, encourages those already approved to ISO 9001 not to drop certification but to maintain compliance and integrate the food safety requirements into the organization's current management system. It just makes good business sense to build on the structure and discipline inherent with the ISO 9001 standard. Mr. Bele identified the following areas that would not be audited if the organization dropped its ISO 9001 approval including those areas listed that relate directly to customer and legal requirements:

- Quantity and net fill per product
- Labeling (except allergens)
- Packaging/Palletizing
- End-product shelf life
- On-time delivery
- Order fill
- Service-level performance
- Procurement process
- Design and development process
- Demand planning
- Schedule of operations

Some of these areas may be audited but *only* as they relate to food safety. It just doesn't make sense to separate food safety and quality. As stated previously, through experience, it is harder to separate food safety and quality, not to mention the confusion that can result when this occurs. Again, think of the message that goes to the associates. It is okay to ignore quality and quality doesn't matter, so if you make a mistake, just be very sure that it isn't related to food safety. Having a safe product is critical, but not paying attention to quality or in other words, not giving quality the structure and discipline required can lose customers, possibly costing the organization its business.

When asked about their choice on which management standard to apply to their system, *Yolanda Nader, the CFO and CEO of Dosal Tobacco Company*, stated that her organization was striving for compliance to GMPs as stringent as those applied to the food and the medical device industries:

> We were interested in establishing a QMS system to improve the operations of the business and that could also prepare the organization for GMP's and other regulatory requirements it would be facing in the future due to new legislation. Primarily focus was on QMS, the food safety standard was also selected as [a guideline possibly for the future because our team feels that] it is deemed to be what is anticipated to be promulgated by the regulatory agency for GMPs.

DEFINING, IMPLEMENTATION, AND THE DECISION TO BE CERTIFIED

Once the standard of choice is identified, then the next decision is whether to develop and implement the requirements, but not go for certification. Of course, this may not be a choice if customers are demanding certification. There are many companies going forward even if the customer isn't asking for certification. These organizations are doing this because it just makes good business sense. Keep in mind that not going all the way through certification is doing so strictly for the internal benefits; however, through experience, the discipline and the structure required to get through the certification process far outweigh any reason not to go forward. Developing a system and not taking it through the certification process is like going to law school and not taking the bar exam. Just like a law school graduate who doesn't pass the bar exam and can't be recognized as a lawyer, developing a system but not becoming certified does not provide the organization with the recognition and opportunities it has earned. A group of professionals were asked why their organizations made the *certification* decision and what benefits they have experienced because of this:

> System compliance provided my company with a structured proactive approach to food safety plus due to fierce global competition, management wanted to differentiate themselves from the clutter. (*Andy Fowler, former VP Administration and Environmental Compliance, Bacardi & Company Limited*)

> A major reason we pursued certification was to obtain an objective, independent review of not only the tactical elements of our food safety programs, but also the strategic application of the overall food safety management system. This was of particular interest to ensure that we were consistently integrating our food safety system directly into the business. Our decision was not linked to customer pressure. We have a process of continuous improvement built into our food safety management program, so independent certification was a logical next step. (*Mike Burness, Vice President, Global Quality and Food Safety, Chiquita Brands International, Inc*)

> We wanted to make sure that the consumer could feel confident in the safety of the product. With that confidence there is an increased confidence in the customer's brand and ultimately confidence in our brand. It was not, necessarily, customer pressure that led us toward certification, but rather more of a change in the philosophies and actions of the industry as a whole. This, coupled with a growing consumer activist eye made the writing on the wall very clear. We have seen a great advantage to doing the process. One of the great advantages is that we have learned to assess risk in areas that we formerly knew of no risk. There was also a great deal of overlap with the improvement of food safety. For example, as soon as we received our certification there were a number of new and existing customers looking to us for more work. (*Eric Putnam, Food Safety, Quality & Training Systems Manager, Wixon, Inc.*)

> We were pressured into it through our parent company by some of our largest customers. As the system has developed we have continued to see the value this system provides for our operations. Although we have not seen any direct payback from the customers who pressured us, establishing this system has provided us with a structured and disciplined management system which is evident through our improved audit processes (internal and external). We find that we are more organized. Our documentation is more thorough. Our compliance level has been greatly improved. Another advantage we have seen is in our communications, both internal and external. We find deficiencies in our processes in a timelier manner. Our follow up has improved providing us with an excellent foundation for continuous improvement. (*Joe Hembd, Management Systems Manager, Chesterman Company*)

> We have experienced many advantages because of us attaining certification. Because of the changing world and food safety requirements we may not be ahead of the curve but we at least are abreast of it. It has taught us that without a food safety program installed to catch and document your procedures you are one recall away from going out of business or having severe damage done to your company's reputation. It has given us credibility with our peers and customers. It has made the company safer for everyone including the employees. The challenges were 2 stages. Stage 1 was just understanding what was expected and getting all the documents and procedures in place. (*Anonymous*)

Our major customer required it. Yes we saw great advantages even though it was hard at first but as we continued our journey we found that we were better prepared to make decisions understanding the road map to compliance that we were developing. There were tremendous advantages to everyone moving forward on the same page. (*Russ Patty, independent consultant and former beverage company's quality manager*)

Alan Lane, former packaging manager, now an independent consultant, shared the following insight:

After being AIB certified, with a "Superior" rating, for many years, it was determined that recent bad publicity, (i.e. PCA debacle), and the desire to be on the leading edge, that the paperboard packaging company I was Operations Manager for, should pursue a GFSI approved scheme to not only differentiate ourselves from our competition, but also to better our management system and support our continuous improvement philosophy. While there was no particular customer pressure, there was the acknowledgement that if you were in the packaging food packaging industry, you were in the food business, regardless. Advantages of pursuing the GFSI, (in this case, SQF), were clearly a market advantage, but again, it was undertaken to ensure that the maximum benefit to the QMS was achieved. As continuous improvement is integral to the requirements, this was one initiative that supported that requirement. Because the decision was an executive decision to pursue SQF, there really was no down side.

The following are comments provided by some *industry subject matter experts* that are not allowed to be recognized because of company legal policies; however, they have requested that we do share their experiences and identify them as *Anonymous*:

Our organization went to a benchmarked approved GFSI schemes because it was requested by our largest customer and complying with our customer requests helps to ensure continuous long term business. However, that said we had incorporated a robust internal management system along with a strong top management commitment which made the transition to the third party standard much smoother than it did for some other local organizations. This initiative has been beneficial to improving our quality systems, food safety and has resulted in no findings during our unannounced audits from state and federal regulatory agencies. (*Anonymous*)

Our organization choose the FSSC 22000 food safety management system because top management felt that it would provide a cohesive framework to merge with our current ISO 9001 quality management system and our HACCP programs. This positioned our company to take advantage of global markets receptive to the approved benchmarked GFSI schemes. (*Anonymous*)

Once the process of implementing the FSSC 22000 System was in place and gaining momentum, the site team began to realize the magnitude of the impact that this had on the production process as well as the associates on the floor. The overall knowledge and awareness to food safety from each and every associate was something that not many of us had anticipated. Every associate understood exactly what their role was in ensuring the products being produced were food safe. Some of the overall advantages have been seen across different areas of the site. The safety and environmental team at the site has achieved world class scores on audits thanks in part to the process of becoming FSSC 22000 certified. (*Anonymous*)

We had to obtain BRC in order to keep our products in Sam's Club. The number one advantage that we have experienced is having more structured and accurate documentation. (*Anonymous*)

Our organization believed because of its structure with multiple facilities that embracing a global standard would assist us with implementation of a standardized and globally recognized food safety and quality system throughout our entire food industry sector. (*Anonymous*)

CHOOSING THE STANDARD

As discussed earlier in this chapter, the organization has a choice as to which standard it chooses. These choices can be based on many parameters, but it has been observed that the decision primarily depends on the food industry segment, familiarity with a specific standard within the organization, and customer preferences.

The question as to which standard was chosen and why was posed to this same group of professionals. The following is a sampling of responses received:

A couple of reasons for my organization to choose SQF were that it was recommended by multiple retail customers; at the time it had the most trained and available auditors; and the food safety and quality programs were similar to the management systems that we already had in place. (*Anonymous*)

Organizations choose to implement a management system for many reasons; *Guido Abreu, Quality System Analyst for Dosal Tobacco Corporation*, shares the following insight considered by top management in making their decision choosing ISO 9001:2008:

Dosal Tobacco Corporation [top management team] decided to implement a QMS in order to achieve business improvements and to prepare the company for new FDA regulations, among them includes establishing GMPs for the tobacco industry. While the initiative was not related to customer pressure, we have certainly found it to be an effective avenue to meet our customer's expectations. Key to this approach has been constant monitoring of consumer complaints and customer surveys. Dosal Tobacco Corporation chose to adopt ISO 9001:2008 because [we felt] it would help us establish a formal QMS, initiate the process of changing our established culture and preparing us to meet regulatory requirements. Our top management team chose to comply with the quality standard, and then mirror the food safety GMPs in preparation for the anticipated FDA requirements. We believe that the food safety standards would be more related to our industry than any others; however, initially it was felt that the full system structure and discipline of a compliant ISO 9001:2008 system would provide us the foundation required to journey through our culture development and our focus for GMP compliance.

We chose FSSC 22000 (ISO 22000:2005 and ISO 22002-1:2009) specifically for business reasons. First, as an international business, it was certainly important for us to adopt a standard that was accepted on a global level. While there are numerous standards that are accepted globally, some are *more* geographically prominent than others. We found FSSC to be the most widely accepted on a global scale. Secondly, we also evaluated many of the standards and felt as though FSSC was the best overall fit for our business. FSSC provided both the rigor from a food safety perspective and the flexibility from a business perspective, making it the right fit for us. Thirdly, since it is ISO based, FSSC is also a public standard. As such, it is essentially *open* for input and recommendations for improvement and updating. We thought this was important as there was no direct ownership of the standard. Additionally, because FSSC includes continuous improvement as part of the management system, it was most consistent with our existing philosophy regarding food safety excellence. (*Mike Burness, Vice President, Global Quality and Food Safety, Chiquita Brands International, Inc*)

Our company chose BRC then switched to IFS and then finally to FSSC 22000 to align with our parent company's internal food safety and quality program. Our organization overall experienced very positive improvements with the FSSC structure. Also, our plant manager had previous ISO 9001 experience so applying the ISO concept enhanced his already high level of commitment and system knowledge. (*Anonymous*)

Tom Marchisello, Industry Development Director, Grocery Retail, GS1 US, provides the following guidance related to choosing the management system that best fits an operation:

In all the organizations I led - relative to the implementation of food safety and quality management systems certified to a GFSI benchmarked standard - We historically implemented systems consistent with the regional generally accepted standard – for example - BRC; SQF; IFS; Dutch HACCP Code. I believe that over time, FSSC 22000 will become the predominant global standard and the others will become obsolete. In hindsight, FSSC 22000 would better drive global consistency because it is based on the ISO 9000 foundation.

In some instances, organizations don't have a choice. They must go with their parent company's choice or many times the same standard that the customer has used. These types of situations are not as challenging as they sound. Understand the standard of choice and then move forward toward certification:

We didn't really choose. We went with FSSC 22000 because that was the standard of choice of our parent company. Staying with that focus provided us some additional resources and opportunities that we would not have had if we had not made that decision. (*Russ Patty, an independent consultant and former quality manager of a beverage operation*)

Our parent company recommended BRC so management felt it was best to remain consistent. (*Anonymous*)

We chose FSSC 22000 because it fit with our current ISO QMS certification which provided us the opportunity to have the same registrar as the one we had had for years for ISO 9001 certification. (*Anonymous*)

We did not have a choice since our parent company was specifically requiring us to become certified to FSSC 22000. However, by choosing this standard it was compatible with our existing quality management system which provided us an excellent foundation to integrate requirements together. (*Joe Hembd, Management Systems Manager Chesterman Company*)

We chose FSSC 22000 because it is based in ISO and we were already certified in ISO 9001:2008; management felt that it was a more globally recognized standard and because it is a less prescriptive standard thus allowing us the room to use our systems to more effectively evaluate our processes, control our risks using the risk assessment program when necessary to provide justification for our decisions that are best for our organization and that best suits all stakeholders. (*Eric Putnam, Food Safety, Quality & Training Systems Manager, Wixon, Inc.*)

We chose SQF, primarily due to the knowledge that Walmart was endorsing this standard, but also because it seemed to be the favored standard in the US at that time. (*Alan Lane, former packaging manager now an independent consultant*)

CREATING AN EFFECTIVE AND VALUE-ADDED SYSTEM IS A MUST!

Risks arising from an improperly designed, implemented, and maintained food safety system can have financially crippling effects, from brand damage, decreased consumer confidence, product recalls, illness, death, and legal ramifications. With globalization and an increase in supplies from emerging countries, the likelihood of these risks grows and the processes to manage them effectively become more complex. An additional element in this struggle is the varying acceptance of food safety standards and certification, which vary from market to market. In an ideal world, everyone would accept one food safety certificate. Unfortunately, the multitude of methods available around the world presents a bewildering array of options that could potentially distract from their core objective: to improve food safety. (Groenveld and Pillay 2009)

CREATING AND SUSTAINING A FOOD SAFETY CULTURE

Frank Yiannas, Vice President of Food Safety for Walmart, states that "to effectively create or sustain a food safety culture, it is critical to have a systems thinking mindset. You must realize the interdependency of each of the various efforts your organization chooses to put into practice and how the totality of those efforts might influence people's thoughts and behaviors. In order to create a food safety culture, [the organization] needs to have a systems-based approach to food safety. The goal of the food safety professional should be to create a food safety culture—not a food safety program" (Yiannas 2010, p. 16).

NEXT STEP...

The next step is to start reviewing the requirements and best practices. Although this text includes an abundance of information on compliance, applying knowledge with actual experience, there is no substitute for actually starting the journey and learning as the system develops. The goal of this text is to provide a roadmap toward compliance, but remember the organization must make the decision to provide the resources; top management must be committed, supportive, and engage the entire associate team. The road to success and a value-added management system is paved, but the logistics of the journey must be identified, defined, implemented, and maintained in a manner that is best for the organization adding valued structure to the production of a safe product that meets customer and regulatory requirements through a compliant and continually improving management system.

The *three-legged stool* is referred to quite often in this text. Figure 1.5 identifies management review, CAPA, and internal audits as the three legs. These elements, when equally strong and effective, support the management system not only for compliance but also for sustainability and continuous improvement. An excellent system will (not might but will) self-destruct if any one of these three *legs* weakens. The reverse is also true in that a struggling system will survive and improve when each of these three legs

CAPA   Internal audits
                    Management review

FIGURE 1.5   Management system's three-legged stool.

are strong and effective. A variation of the system-focused three-legged stool related to the food safety/HACCP program is included in Chapter 16.

The safety of our food products must always be given top priority in all that we do. As the title of this chapter reflects, choosing the best standard for an operation and then defining, implementing, maintaining, and continuously improving the FSMS is a continual journey. It is very important along all paths of this journey that one monitors the specific websites related to the standard of choice along with keeping up-to-date on current food safety–related news and incidents. The standards do get revised (sometimes more frequently than we expect) along with the world of food safety continuing to evolve.

REFERENCES

Groenveld, C. and Pillay, V. 2009. Food safety goes global, *Food Quality*, February–March, 2009. www.foodquality.com/mag/02012009.03012009/fq_03012009_FE3.htm

International Organization for Standardization (ISO), n.d., *Management System Standards*, http://www.iso.org/iso/home/standards/management-standards.htm

Muliyil, V. and Sansawat, S. 2012. Comparing Global Food Safety Initiative (GFSI) recognized standards, October, 2012.

Yiannas, F. 2010. *Food Safety Culture Creating a Behavior-Based Food Safety Management System*, Springer, New York, 2010, pp. 16, 18.

2 General Requirements

At this time in the food industry, there are many reasons why a company may choose to develop, implement, and maintain a system compliant with one of the Global Food Safety Initiative (GFSI) approved, benchmarked schemes. Of course, one of the most recent drivers is the pressure being applied by major customers. However, the decision to move forward must be a decision not only because customers or regulators are forcing or encouraging the move, but because it makes good business sense for the organization. A well-defined, managed, and compliant system adds value and reduces overall costs for the organization. Actually, cost savings numbers are tough to identify. How does one measure the savings of something that did not happen? However, there are enough companies with recalls, incidents, and ruined reputations for management to fill in the blanks, understanding what could happen and might very well happen if the organization does not take control and focus on preventing issues rather than addressing issues after a crisis. In the past, we learned from our mistakes. In modern times, we cannot afford to learn from our mistakes; we must avoid these mistakes by being proactive. The structure and discipline provided by an effective food safety management system does not guarantee food safety, but it absolutely provides the foundation for a proactive process to ensure that many mistakes can be avoided and prevented. This type of system not only requires a team effort that is supported by top management, but also empowers and trains all associates to understand and play a significant part in food safety every day. Food safety affects everyone; each person has a role no matter what his/her area of responsibility includes.

OPTIMUM FOOD SAFETY MANAGEMENT SYSTEM

The optimum food safety management system is one that reduces risk through systematic reduction of variation. The benefits of a healthy food safety management system include minimized food risks, stronger relationships with other links in the food chain, and a clear message about your proactive approach to food safety. (Link 2008, p. 6)

The effective food management system must be implemented in a manner that ensures that all activities that impact food safety are defined. The structure of the food safety management system must fit the needs of the company. Although the food safety standards do have most of the requirements in common, there are definitely some specific differences, depending on the focus of the chosen standard. It is important for management to understand that second only to the production of a safe product is the assurance that the requirements are integrated into the system, limiting reengineering of the operation for the standard. There are some very specific requirements that must be met; however, even among these, a requirement may not relate to the organization's operation or the operation may have mitigated the risks, which have eliminated the hazard or reduced it to an acceptable level. Mitigation and risk assessment are discussed in more detail in "Risk Management" (Chapter 15).

MANAGEMENT SYSTEM SCOPE

When thinking about defining, implementing, and maintaining a system compliant with any standard, it is very important to understand not only the focus of the standard, but also the scope of the standard reflected by the organization's management system. Many times, when working with a company, defining the scope of the system often becomes a struggle for top management. Basically the requirement is very straightforward; how should the requirements of the standard be applied to the system? The main result must be an effective food safety management system that incorporates change management and continuous improvement processes that not only are effective, but also meet the requirements of the standard of choice.

The scope connecting the company with the requirements of the food safety standard of choice must be clearly defined and must meet the requirements of the standard. ISO 22000:2005 Section 4.1 has very specific requirements related to defining the scope of the system. These include, but are not limited to, specifying "the products or product categories, processes, and production sites that are addressed by the food safety management system." SQF edition 7 describes the "scope as the food sector categories and products processed and handled by the site."

GENERAL REQUIREMENTS FOR COMPLIANCE

ISO 22000:2005 Section 4.1 is basically a summary of the entire standard. Titled *General Requirements*, it states that the basic requirements for compliance must ensure that all food safety hazards that are reasonably expected to occur are identified, evaluated, and controlled in such a manner as to not directly or indirectly harm the consumer. A process is defined and implemented so that food safety–related product information could be (and is, when needed) communicated throughout the *food chain*. If the organization has an issue with food safety, then it has a process in place to communicate this in an effective manner that prevents any harm to the consumer. Communication is stressed both internally (within the organization) and externally. External communication means throughout the food chain that includes suppliers, customers, and consumers (Chapter 3). A customer may be the same as a consumer; however, a customer may also be a broker or a large chain store that purchases directly from the organization and then sells to the consumer.

Each of the food safety management standards requires either a formalized food safety manual (top-level manual) or a formalized process for documentation that identifies and links to system documentation and records required for the food safety standard. The GFSI guidelines require a top-level manual. As discussed in Chapter 1, in order for a scheme to be approved by GFSI, it must meet the requirements of its guidelines.

The advantages of a food safety manual (i.e., top-level manual) that answers the requirements of the specific standard are many. When the standard says *shall*, then the top-level food safety manual answers as to how this is being done specifically for the organization's system or it references the documentation that does define this. It is highly recommended that the organization create a food safety manual that clearly links to the requirements of the chosen standard and wherever possible defines the organization's response to the requirements of the chosen standard. This provides an excellent foundation and many times reduces the necessity to have extra documents (Chapter 6).

Example 1: The standard states "the organization shall provide the resources for the establishment, management, and maintenance of the work environment." The food safety manual answers this *shall* statement specifically for this operation by stating that "Top management assesses and ensures that the food safety management system has the necessary resources through several means, such as management review meetings, daily shift meetings, project planning meetings, staff meetings, and internal audits."

Example 2: ISO 22000:2005 5.8.1 states that "top management shall review the organization's food safety management system at planned intervals to ensure its continuing suitability, adequacy, and effectiveness." The food safety manual's response would be that top management reviews this food safety management system at a minimum of quarterly through the management review process to ensure its continuing suitability, adequacy, and effectiveness.

Example 3: Related to the same reference as Example 2, the food safety manual may state that top management reviews this food safety management system at planned intervals as defined in the procedure for management responsibility. Keep in mind that a theme that is emphasized throughout this text is to never define the same thing in more than one document or section of a document. If the topic is referenced, then state *according to* and send the reader to the document or section where it is defined.

SQF Code edition 7 clause 2.1.3 requires that a Food Safety Manual be documented and maintained by the organization. The manual must include: "(1) a summary of the organization's food safety policies and the methods it will apply to meet the requirements of the standard; (2) the policy statement and organizational chart; (3) the scope of the certification; and (4) a list of products covered under the certification. Additionally, the manual shall include references to the written procedures, prerequisite programs, food safety plan, and other relevant documentation." (Note: Above information was taken from the standard that is under copyright by the Food Marketing Institute.)

MANAGEMENT SYSTEM MANUAL

If an organization has established multiple management systems (i.e., quality, food safety, environmental, etc.), then it is strongly recommended that a *management system manual* be created that incorporates or, translates the requirements of all standards (integrates the complete requirements) into the organization's processes.

TOP MANAGEMENT AND MANAGEMENT COMMITMENT

As we proceed through understanding, defining, implementing, and maintaining an effective management system, a central theme is top management commitment. Top management is defined as management ultimately responsible for the system. This may be one person such as the president or the general manager, or it could be the top position and those who answer directly to that position (i.e., staff). Defining top management and management responsibility is discussed in more detail in Chapter 3. No matter which standard is chosen, it is critical to continually focus on the fact that top management of an organization must demonstrate a strong commitment not only to the implementation and management of this system, but also to the resource needs of the management system (Chapter 3).

Hindsight is 20-20, and it is always a great learning opportunity to discuss the experiences, challenges, victories, and the *if we had only done this* comments shared by system implementation associates once the system has been established and implemented. *Guido Abreu, Quality System Analyst for Dosal Tobacco Corporation* shared that it would have been beneficial to have had a "simple ISO orientation for the non-QMS professional, focusing on the different levels throughout the organization." This is a frequent statement and sometimes is based on the organization not having competent management system compliance training through the initial defining phase. It is very important to plan the journey or experience the *challenges* caused from lack of or ineffective planning. Journeys of any scope are much more enjoyable and less stressful when effective planning prior to the journey is performed. An additional example is to leave on a "spur of the moment" cross-country road trip, only to find all hotels full, restaurants closed, and gas stations too far apart. Mr. Abreu, in looking back, commented that "taking more time to plan and perform process flow analysis would have minimized the amount of documents produced. As the system began to materialize and mature, we quickly learned that a major continuous improvement project would focus on reviewing the current documents, revising and reducing them to enhance their focus and overall effectiveness."

CONTROLLED PROCESSES

The key enabler for food safety is control. For this reason, it is appropriate to briefly elaborate on controlled processes. A controlled process is work conducted in predetermined order and applied at specified levels, resulting in the consistent achievement of the desired outcome. Maintaining the order of the process steps and controlling the process parameters, minimize the variation of outcomes. If the order of work to be done and the level at which it is applied are left undefined, each individual will do their best, but not necessarily use the same process on the same job each time. The outcomes of such efforts will vary widely. In other words, following an established process will provide better overall results than the overall results of individuals putting forth their best efforts. This is why Dr. Deming reminds us that "Doing our best is not always good enough." (Link 2008, p. 6)

SYSTEMS APPROACH FOR MANAGEMENT

The systems approach to management is a central theme when discussing effective management systems. Dr. John Surak describes the following:

> The standard recognizes that processes do not operate in isolation. The output of one process usually becomes an input into another process. These processes link together to form a system. Therefore, if a company is to be effective and efficient in meeting its goals, the company must manage all of the processes as a system rather than trying to manage each process individually. (Surak 2001, p. 707)

KEY POINTS

a. Each organization must evaluate the standards and choose the one that makes the best sense for its operation and products.
b. Although initial interest may focus on customer pressure to develop and become certified to a third party-type food safety management standard, the value to the company overall becomes evident as the system matures.
c. The structure and discipline provided by an effective food safety management system does not guarantee food safety, but it absolutely provides the foundation for a proactive process to ensure that many mistakes can be avoided and prevented.
d. Positive, ongoing management commitment is critical to the ultimate success and effectiveness of the food safety management system.

COMMON FINDINGS/NONCONFORMANCES

System findings that are identified or linked to the overall system are rare, but may be identified if the total overall system has not been established. In some instances, a system finding can be a result of several findings related to several elements. Individually, the findings may not be significant, but together indicate a breakdown of the total system. Findings related to the overall system are graded as a major nonconformance almost 100% of the time. The following are examples of findings that could be reported against the system as a whole:

a. It could not be confirmed through the review of objective evidence that this organization had implemented a management system that was compliant to the current version of the chosen standard.
b. Evidence at this time was not available to confirm that requirements as stated in *the chosen* standard had been implemented and in compliance.
c. Findings during this evaluation indicated that the organization did not have a system compliant to the requirements of the chosen standard.

REFERENCES

Link, E. 2008. *An Audit of the System, Not of the People: An ISO 22000:2005 Pocket Guide for Every Employee*, Quality Pursuit, Inc., Fairport, NY.

Surak, J.G. 2001. International Organization for Standardization ISO 9000 and related standards. In: *Food Safety Handbook*, Schmidt, R. and Rodrick, G. (eds.), Wiley-Interscience, Hoboken, NJ.

More information related to systems and processes is discussed in Chapter 5 (Internal Auditing).

---

*Additional Food Safety Management System References (not all inclusive)

General requirements spread throughout the standards:

FSSC 22000:2010
    ISO 22000:2005 Section 4.0

SQF 7th Edition: Spread throughout the standard

BRC Issue 6: Spread throughout the standard

IFS Version 6: Section 2

*Note: Always review the standard or visit the website for your chosen standard to ensure that all related requirements are identified and addressed.

---

3 Management Responsibility

UNDERSTANDING MANAGEMENT COMMITMENT

Each standard addresses the issue of management responsibility in its own manner; however, they all focus on the commitment demonstrated by top management. Commitment by top management has always been the foundation to success, but actually identifying this specifically as a critical success factor has, in the past, been more implied than actually practiced. Times are changing; standards actually not only state that management commitment is critical, but also outline specific criteria that must be demonstrated. Compliance activities must become a way of life beginning with the highest level and cascading down to all associates if created, managed, and supported effectively.

Most organizations identify a top management team led by the highest ranking individual within the scope of the standard and whose areas of responsibility to report directly to that position. In the world of the food safety management standard, this management team may also be the food safety team. Depending on the standard of choice, size, and number of food safety/Hazard Analysis and Critical Control Point (HACCP) programs of the organization, the food safety team, which will be discussed in more detail in a later chapter, may also be what we formerly called the HACCP team, or the HACCP team may be a subgroup of the food safety team. The food safety standards define specific responsibilities that must be completed by the food safety team; thus, care must be applied when identifying the teams to ensure that these requirements are met. Before moving too quickly into management commitment and the ensuing discussion on management responsibility, it is necessary for the organization to define *top management* or the *top management team*. This can be done either in the food safety management system (FSMS) manual, in a procedure for management responsibility, or even in the management review meeting minutes; whatever source is used, it must be defined.

Food Safety System Certification (FSSC) 22000:2005 5.1 states that "top management must provide evidence of its commitment to the development and implementation of the food safety management system and to continually improving its effectiveness by doing the following:

a. Showing that food safety is supported by the business objectives of the organization;

b. Communicating to the organization the importance of meeting the requirements of this International Standard, any statutory and regulatory requirements, as well as customer requirements for food safety;

c. Establishing the food safety policy;

d. Conducting management reviews; and

e. Ensuring the availability of resources."

Similar requirements, like those listed earlier, are required in the Safe Quality Food (SQF) Code, edition 7 elements 2.1 Management Commitment, 2.1.1 Management Policy, and 2.1.2 Management Responsibility.

Basically this means that there must be visible support, positive evidence of commitment, resource allocation, and continuous improvement as a result of top management's involvement. These requirements ensure that the system has support from the top and does not become the responsibility of one individual or one department. In the past, many such initiatives became just another *quality* program.

Charlie Stecher, an independent consultant and third-party auditor, Anchar, Inc. shares his concern that as an auditor, he is surprised how frequently he finds how "little the organization's management knows about the standard. The different interpretations of the same element can even vary from company to company and from individual to individual within a company."

Attitude limitations and lack of ongoing support from middle and senior management have been frequently stated as a disappointment by associates struggling to get support from the top. Many times, associates hesitate to provide their total commitment because they perceive lack of commitment from management and front-line supervisors. Middle management, in some instances, may present an attitude of skepticism, which doesn't lend much support. Many times, this skepticism is prompted by lack of understanding as to what the system's structure and discipline really mean and the value that it can add to the organization overall. This paragraph is not meant as a negative comment against management support. Previously, this concern had been stated as a possible handicap to early progress. However, in most cases, as system implementation progressed, many of these same folks experienced a complete reversal as management and the entire associate team began to experience the advantages of having an effective, well-disciplined management system.

> Each employee or person within an organization has a personal responsibility for preparing or serving safe food. It also illustrates that food safety is interdependent. All employees within the whole of the organization or company have a shared responsibility to ensure food safety. And the sum of food safety efforts within an organization is critically dependent on and greater than its parts. But my all-time favorite definition, because of its simplicity, is Culture is the way we do things around here. Simply put, a food safety culture is how an organization or group does food safety. (Yiannas 2010, p.12)

IFS V6 1.1.1 states that "senior management [must] draw up and implement a corporate policy. This [must] consider as a minimum:

- Customer focus;
- Environmental responsibility;
- Sustainability;
- Ethics and personnel responsibility;

- Product requirements (includes: product safety, quality, legality, process and specification);
- Corporate policy [must] be communicated to all employees."

Successful implementation, long-term effectiveness, and continuous improvement to a management system standard are directly related to top management support. Every management system standard emphasizes and requires evidence that this support not only exists, but also that top management is actively involved in system activities. Lack of management support must never be an issue. This is essential to achieving ultimate success. If this importance is not truly believed and communicated by top management in a manner that is clearly evident throughout the system, then there will be many disappointments with only limited and temporary system usefulness and success. Top management support and involvement are absolutely essential for the system to reach its true potential and to continuously improve through the system maturing process.

Yolanda Nader, CFO and CEO of Dosal Tobacco Company, when asked how she communicated top management commitment through all levels of the organization stated that

> It is about demonstrating to the top management team and all associates, that having a structured and effective management system adds value to the entire organization demonstrating a team approach to compliance and continuous improvement. The system when implemented effectively does work, adding the structure and discipline that the best organizations use to keeping improving and getting better.

Ms. Nader also commented that "the buy in from the employees once they started to see progress was fantastic but was because associates saw first hand that their management had become believers in the process."

SUPPORT THROUGH ALL LEVELS OF THE ORGANIZATION

In some operations, members of middle management are often only partially supportive because that is what they think their supervisors want. This type of situation is a top management challenge that must be addressed. Associates at lower levels may perceive this as a type of lip service, also having a direct effect on the system's overall effectiveness. Changing the culture, the way one thinks about what is done, following documentation, and being accountable can be one of the most difficult hurdles to overcome. However, with management's support and the application of the management system's structure and discipline, accomplishing this can and does result in an effective management system.

Frank Yiannas, VP of Food Safety for Walmart shares his thoughts on *leadership* and *food safety culture*:

> A food safety culture starts at the top and flows downward. It does not flow from the bottom up. It is a leadership function to create a food safety vision, set expectations, and inspire others to follow. It's interesting to note that in the field of food safety, we often talk about food safety management. We rarely talk about food safety leadership. But management and leadership are different. According to Maxwell (1998), "the main

difference between the two is that leadership is about influencing people to follow, while management focuses on maintaining systems and processes." Leading companies with strong safety cultures not only have strong food safety management systems in place, they also have strong leaders committed to food safety who are able to influence others and lead the way to safer performance. (Yiannas 2010, p. 16)

As previously stated, many times the biggest adversaries of system compliance become the strongest supporters, after they realize and experience the system's benefits firsthand. Food industry subject matter experts share the following thoughts based on their individual experience:

> One of the most difficult challenges in the beginning was the difficulty in getting some middle managers and some operations' department supervisors to realize what management system can do for them, and how it provides a way of solving their department's problems. This negative attitude is certainly contagious, especially to the supervisors' direct reports. There is no doubt that our current system stresses documentation, and that it successfully holds people accountable to the management system much more than previously. This was somewhat threatening to these people and culture changes slowly; but once it was realized what such a system could do for them and their departments, they realized that their jobs actually became easier. (*Tim Sonntag, Vice President, Quality & Technical, Wixon, Inc.*)

> One of the hardest parts of our implementation effort was in getting managers and their respective staffs to take responsibility for the process and to be committed to ensuring that the management system was effective. (*Anonymous*)

> The biggest surprise experienced as a result of certification was the way our company's management came together to become intimately involved in the management system. There [was] support from the most unlikely of places. Some of the positions that I expected to have the most resistance... provided some of the best assistance. The level of management involvement as a result of certification [was] certainly... the biggest surprise. Once that level of involvement [was] secured, the drive for improvement became significantly easier... certification in reality simply provided the framework for [managers] to solve their own problems. (*Russ Marchiando, Packaging Plant Manager, Wixon, Inc.*)

In addition to providing proactive support to the production of a safe product, using the system to also track quality indicators just makes good business sense.

> Management involvement is crucial to the success of the system. While certification itself is a rewarding goal, it should not be the reason for undertaking the endeavor. Management involvement can be garnered by developing a strategic plan for system implementation as it relates to the business. An example of this could be the inclusion of economic quality indicators (e.g., cost of product manufactured/sold, consumer satisfaction indices, etc.) in management review. Relating the executional effectiveness of the system to these indices easily demonstrates the impact of a rigorous management system to upper management. (*Mike Burness, vice-president, Global Quality and Food Safety, Chiquita Brands International, Inc*).

> Executive management on-hands involvement was seriously missing during the initial implementation activities. This, overall, was detrimental to a timely and effective implementation; however, once management "saw the light", the process took off. (*Anonymous*)

According to previous comments from those who have experienced management system implementation, not only effective management commitment, but also visible evidence that the top management team truly believes in the system, is critical to the success and sustainability of an effective and continuously improving management system.

FOOD SAFETY POLICY STATEMENT

A top management team responsibility is to create a food safety policy statement that reflects the organization's commitment for not only the manufacture of a safe product reflective of the organization's role in the food chain but also its assurance of meeting all regulatory and customer requirements:

> The food safety policy is a collection of position statements or guiding principles intended to influence and determine decisions and actions with regard to the safety of food affected by your company's processes. (Link 2008, p. 19)

Most food safety management standards require that the food safety policy statement is monitored (reviewed) on a regular basis (usually through management review meetings) to ensure that it continues to reflect the organization's focus on food safety and remains suitable for its FSMS. If the standard of choice includes quality parameters or if top management chooses to integrate the management system with other standards (i.e., quality, environmental, etc.), then it makes sense to expand the policy statement to include those requirements. Once the policy statement is defined by top management, then evidence must be available to confirm that it has been communicated to associates representing all levels of the organization.

The IFS V6 FSMS standard requires that environmental, sustainability, and social responsibility be addressed in the food safety policy statement. The British Retail Consortium (BRC) FSMS requires that the policy includes the requirement to ensure that the certificate does not expire and that the organization has a purchased current version of the BRC Global Standard for Food Safety. The latter is stated as a requirement; however, in reality, no matter which standard is chosen, the organization must have a copy of that standard to ensure compliance to that standard.

MEASURABLE FOOD SAFETY GOALS AND OBJECTIVES

Top management must next identify food safety–related measurable goals and objectives that support the food safety policy statement. This can be a tough task because some team members try to pick only those measurable goals that are food safety related when in actuality, the majority of food safety goals (how do you know your product is safe?) do overlap with other goals. It is critical that measurable food safety goals are identified and communicated throughout the organization.

An observant general manager of a large citrus company once said to his team, "without measurable goals, we cannot improve." This is absolutely true. Good intentions are great, but without being able to track and measure progress, it is very difficult to truly know the status of any activity. Measuring the system's performance provides data related to achieving the goals along with providing a foundation for continuous improvement. Measurable objectives must support the food safety policy statement and also be realistic (not just easy to achieve) to the scope of the food safety management system. For example, if it was stated "Company X produces the safest product possible," how would the success in achieving this be measured? It is a great statement, but how does Company X measure its success in meeting this goal? A similar statement would be that "I am going to lose weight." Without a measurement and a goal, success is difficult to track. It is a great feeling, though, to have measurements and be able to say "I lost 20 pounds" or "I lost 25 inches" or "I went down two sizes." Without measurable goals, we cannot improve.

Some examples of food safety–related measurable objectives may include

- X number of food safety–related customer and/or consumer complaints
- X percentage of on-time performance of food safety–related preventive maintenance and work orders
- X percentage of on-time performance of internal audits and GMP audits
- X percentage of on-time completion of corrective action/preventive action (CAPA) related to food safety
- X percentage of food safety–related continuous improvement projects
- X percentage of on-time completed food safety–related training
- X achievement of on-time deliveries with 100% product in specification

The following provides insight on measuring performance:

Ideally, food safety performance expectations should be objective, observable, and related to specific tasks and behaviors.

Measure Knowledge vs. Measure Behavior. It has been said, what we know is of little consequence. It's what we do that is important. It's one thing to measure knowledge, but do employees actually do what they are expected and trained to do? The only way to know for sure is to measure specific behaviors on activities. (Yiannas 2010, pp. 31, 62)

INTERNAL COMMUNICATION

The food safety management standards require an extensive internal communication process that communicates not only the food safety policy statement, measurable objectives, and the status of the management system, but also everyone's roles in the manufacture of a safe product. It is critical that top management communicates the company's commitment to food safety to all levels of the organization and supports this communication through visual support, commitment, and actions. Examples of communication measures include plant-wide meetings, weekly staff meetings, postings, and newsletters. There must always be objective evidence (records) that a requirement has been met. Meeting minutes, including attendance lists and justification for decisions, must be identified and maintained in compliance with the record control program (Chapter 7).

As an auditor, the question has been asked: "How is management commitment measured?" This is a multitiered process starting with having a conversation with the top management team to hear their role as they see it through to interviews and process reviews. What role do they play and how do they accomplish this? Through evaluations of specific areas, we confirm that associates understand the food safety policy statement, measurable objectives, what these mean to them, and an explanation of their roles in food safety. Commitment is also evident through the evaluation of the support elements (i.e., internal audit, management review, CAPA, and continuous improvement). If a manager states how great everything is going and the management review minutes state how great the system is doing, but the majority of audits are late and findings are not addressed in a timely manner, it becomes evident that the top management team is not actually aware of the true status of the system. I did an audit once where the audits were about 75% late, and the ones that were finished had not been recorded in compliance with the company's documented requirements. The management review meetings didn't say anything about the status of the audit process. When we received the root cause for this finding, it was stated that it was the food safety team leader's fault for not making sure the audits had been done. Besides this not being a true root cause analysis, it also leaves some major questions on commitment from top management.

Comments previously in this chapter have touched on requirements for internal communication related to ensuring that associates representing all levels of the organization understand the food safety policy statement, measurable objectives, and their roles in the system and food safety overall. Additionally, communication must include food safety training presented to all associates. This training should be repeated at a frequency determined by management as necessary for effectiveness. Some organizations state that this is done annually; however, effectiveness is much enhanced when different food safety topics are presented monthly or quarterly. This communication provides strong reinforcement and the opportunity to cover more topics. Many organizations have combined this with their monthly safety (OHSAS) programs and have been very pleased with the results. Attendance must be taken at these meetings, which identifies absent associates; makeup sessions can then be scheduled which may be presented by department supervisors.

As mentioned previously, methods of internal communication may include staff meetings, postings, one-on-one conversations, newsletters, and whatever means the top management team uses to effectively communicate requirements and evaluate the performance of the food safety management system. Keep in mind that the effectiveness of this communication must be confirmed and actions applied when effectiveness cannot be confirmed. Good success has been achieved when managers and supervisors, assisted by the food safety team, create a list of 5–10 basic system questions and randomly ask these questions of associates representing all levels of the organization. A raffle for a free dinner, movie, tee shirt, etc., for those who answer the questions correctly would be a good incentive. If the company has a cafeteria or company store, a reward such as a free soda, meal coupon, or store item could possibly be given. Keep it simple, but fun. It is important to monitor the percentage of accuracy in order to track effectiveness and make adjustments to the process, so that improvements result from continual questioning. Requirements for and performance of the communication process may be recorded and tracked through management review meetings, food safety team meetings, prerequisite verification, or the internal audit process. Records must be maintained demonstrating compliance and effectiveness of the communication activities.

Before moving into the next topic, it is important to emphasize that communication is critical. Effective communication is essential to building confidence and to the long-term value-added opportunities of the management system. Understandably, management of some organizations are concerned about auditors interviewing their associates. This is a change of focus for many with the concern being twofold:

1. "What are the associates going to tell the auditors? Even if they know their responsibilities, will they be able to communicate this effectively to the auditor?"
2. "What about production? We do not want auditors interrupting the associates doing their job."

Dr. Tatiana Lorca, Manager, Food Safety Education and Training, Ecolab, Food & Beverage Division, provides the following guideline related to enhancing the confidence level of the associates and the ultimate focus (confirmation and verification) required by an external auditor through his/her interviews:

Interviews are one of many ways that an auditor collects data during an audit. Although the auditor should not let his/her audit activities interfere with production, it is important that he/she speak to [a sampling of] teammates on the floor in order to ascertain whether [the] staff understands what they are doing and why it is important. This is important for all staff positions, but those team members involved in monitoring Critical Control Points (CCPs) in a HACCP plan will get extra attention from the auditor. The auditor needs to determine whether [the organization's] system requirements are actually being implemented on the floor. If there are discrepancies, [the organization] will have a problem during the audit and this will be recorded as a non-conformity (or several depending on the situation). However, it is important that the auditor be respectful of [the organization's] associates and try not to interrupt operations. It is expected that someone being asked questions during an audit will get nervous. The ability to ask good questions in a non-threatening way is one of many personal attributes an auditor of any GFSI-recognized system should have. If a team member cannot answer a question, the auditor should ask the question in another way or move onto another person and try again. It is a sampling technique after all. One of the best ways to prepare [the organization's] staff is to conduct audit simulations. Interview associates during internal audits and GMP inspections by asking them to demonstrate key activities they perform on a regular basis, show where they record test or monitoring results and ask them to explain what they need to do in the event a control measure fails. Be sure to let them know when the auditor is coming for a visit and that they should be prepared to answer questions openly and honestly.

Additional information and guidelines related to auditor techniques and preparing for an internal audit are addressed in Chapter 5 (Internal Audit).

ASSOCIATE INVOLVEMENT

Charlie Stecher, an independent consultant and third-party auditor, Anchar, Inc. shares "that when auditing an organization with a strong, supported management system, top management empowers its associates in understanding and enjoying the opportunity to demonstrate his or her knowledge and commitment to the management system. Many associates are prepared and ready to demonstrate what they know and are disappointed when the auditor does not get to talk to them."

Frank Yiannas, VP of Food Safety for Walmart, states that it is appropriate to expect a proper food safety attitude from every associate:

> Expect employees to have a right attitude about food safety, because an employee with a right attitude will be much more likely to take right actions. Also, every single day, each employee will influence those around him or her, whether we realize it or not. If they have a negative attitude about following proper food safety and sanitation procedures, trust me – it will be evident to others by what they say and do. Instead, if they demonstrate a positive attitude toward food safety, food safety performance will increase exponentially because of their positive influence on others around them. (Yiannas 2010, p. 31)

Dr. John Surak shares his thoughts on this by stating that "the company must motivate and enable all of its employees so they can reach their full potential, thus enabling the company to achieve its goals" (Surak 2001, p. 705).

INTERNAL COMMUNICATION RELATED TO *CHANGE*

Related to internal communication, ISO 22000:2005 5.6.2 specifically addresses the requirement for the food safety team to be notified of any changes related to food safety in a timely manner, including but not limited to the following:

a. Products or new products
b. Raw materials, ingredients, and services
c. Production systems and equipment
d. Production premises, location of equipment, and surrounding environment
e. Cleaning and sanitation programs
f. Packaging, storage, and distribution systems
g. Personnel qualification levels and/or allocation of responsibilities and authorizations
h. Statutory and regulatory requirements
i. Knowledge regarding food safety hazards and control measures
j. Customer, sector, and other requirements that the organization observes
k. Relevant inquiries from interested external parties
l. Complaints indicating food safety hazards associated with the product
m. Any other conditions that have an impact on food safety

ISO 22000:2005 5.6.2 also states that the "food safety team must ensure that this information is included in the updating of the food safety management system [and that] top management must ensure that relevant information is included as input to the management review."

A significant part of food safety management system planning is ensuring that planned and potential changes (as they may affect food safety) are monitored and communicated to the food safety team. In the past, despite the best of intentions, communication and evaluations related to food safety and quality just had to wait until there was more time, which, most of the time, was after the change was put in place. Often, the waiting resulted in mistakes or having products out-of-specification that could have been avoided if extra time had just been taken to ensure everything was in place and any concerns addressed. Some of us remember the days when there was never enough time to stop and do it right, but there was always time to do it over again. No one meant any harm; it was just the way it was. As mentioned in other sections, the time for us to learn from our mistakes is gone. We cannot make mistakes. The approach must be *proactive* and *preventive,* not reactive and corrective. Top management must ensure that the system includes an effective process for managing and communicating change. The process for managing change is mentioned throughout this text and in more specific detail in a subsequent chapter. Top management has the ultimate responsibility for ensuring that the process for change management (see Chapter 15, "Change Management") is in place, it is effective, and that any specific changes are discussed at the management review meetings.

EMERGENCY PREPAREDNESS AND RESPONSE

Emergency preparedness and response is included under management responsibility as stated in FSSC 22000:2005 5.7 and also in SQF Code, edition 7 element 2.1.6 Business Continuity Planning. Most organizations have a defined crisis management program that identifies what to do during a crisis; this may include a recall, but should also include tornados, earthquakes, hurricanes, fires, floods, etc. What is not included very often is the requirement to perform a thorough evaluation after a crisis to confirm the safety of the product. Was the product affected during the crisis? If so, what needs to be done to ensure that no product currently has or could have a food safety hazard if released? Crisis management and emergency preparedness are reviewed in more detail in Chapter 14.

FOOD SAFETY PLAN

The food safety plan must be created and maintained by top management; in most instances, top management actually addresses requirements for planning through an annual specifically focused management review meeting or through an annual business planning meeting. Basically this requires top management to ensure that the organization's planning process meets both the requirements of the chosen standard and the measurable objectives for food safety and quality. Keep in mind that the organization's scope must be realistic for its operation. It is usually during these planning meetings that the top management team evaluates the previous year's goals and objectives, making the decision to revise, add, or delete them for the next period (i.e., 1-year plan, 5-year plan, etc.). It is also through these meetings that top management may choose to include quality, environmental, and other operational/business goals in the process. Keep in mind that some food safety management system standards do include quality considerations such as SQF Code, edition 7 Level 3, so these must also be addressed. However, it is encouraged that whether the food safety management system includes quality or not, that top management does include quality in its system development and maintenance. As previously stated, this just makes good business sense.

RESPONSIBILITY AND AUTHORITY

The food safety management standards require that responsibility and authority for food safety and quality are defined. (Remember whether the standard of choice includes quality or not, it is strongly recommended that the organization does include quality in the management system. It just makes good business sense and truly is much more advantageous to do so in the development process rather than having to go back to it later. This is not efficient and can be very confusing.)

Defining responsibility and authority includes the identification of a food safety team leader (SQF Practitioner) and the food safety team, which is discussed in more detail in Chapter 16. It also requires that responsibility be defined for all positions that play a role in food safety and quality. Not having a clear definition of responsibilities and authority will create an increase in variation and the likelihood of a food safety or quality event. Responsibilities and authority can be defined through a matrix, an organizational chart, procedures, and/or job descriptions. Training in responsibility and authority must be confirmed and content verified as effective. Associates must be able to demonstrate familiarity with their areas of responsibility, authority, and roles in the management system. Responsibility and authority requirements may also be defined in various prerequisite programs and other procedures depending on the focus of the function.

ISO 22000:2005 5.4 defines the following related to responsibility and authority: "All personnel must have the responsibility to report problems with the food safety management system to identified person(s). Designated personnel must have defined responsibility and authority to initiate and record actions." This is also well defined in SQF Code, Edition 7 2.1.2 requiring that management responsibility of the organizational structure be communicated throughout all levels of the organization. The BRC Global Standard for Food Safety Issue 6 requires that the most senior manager be present at the opening and closing meetings held during the certification audit.

It is recommended that responsibilities and authority include a primary and backup (relief) position for each area of responsibility related to food safety. This training must be supported by records confirming that requirements have been completed. More specific discussions related to defining, implementing, and maintaining an effective training, awareness, and competency program are discussed in more detail in Chapter 8.

FOOD SAFETY TEAM LEADER (THE SQF PRACTITIONER)

A food safety team leader or the SQF Practitioner must be identified. What the position is called depends on the standard. This leader has the ultimate responsibility for managing the food safety team, organizing its activities, and ensuring that the food safety management system has been established, implemented, and is being maintained per defined requirements. This position has the responsibility to report the effectiveness and the suitability of the food safety management system to top management. The food safety team leader ensures that the members of the food safety team are trained and have sufficient education related to food safety to be effective members of the team. Management must define the competency requirements for the food safety team leader. The standard of choice may not require a backup person, but having a person who is able to assist with these responsibilities adds confidence and a stronger foundation to the food safety management system overall.

Dr. Tatiana Lorca, Manager, Food Safety Education and Training, Ecolab, Food & Beverage Division, provides the following insight on the identification of the SQF Practitioner and his/her responsibilities as required by the SQF Code Edition 7:

> This person is the champion of SQF within the business, responsible for ensuring the program is maintained. The SQF Practitioner is responsible for validating and verifying the multiple control measures in place at the site including the food safety plan, food quality plan and the SQF program as a whole. He/she is also responsible for maintaining the integrity of the program and communicating information about the program within the business. This is a very important person because he/she is ultimately responsible for assuring that the program is maintained. As such, you want to ensure there is always a back-up for the SQF Practitioner in your facility and that the back-up is suitably trained and qualified. Many large facilities will have more than one SQF Practitioner appointed so that they can share the work and the responsibilities.

The SQF Code, Edition 7, Appendix 2: Glossary provides the following definition of the SQF Practitioner and related requirements for this position: "An individual, designated by a producer/supplier to develop, validate, verify, implement and maintain that producer's/supplier's own SQF System. The SQF practitioner details shall be verified by the SQF auditor as meeting the following requirements:

1. Be employed by the company as a permanent, full-time employee and hold a position of responsibility in regard to the management of the company's SQF System;
2. Have completed a HACCP training course and be experienced and competent to implement and maintain HACCP-based food safety plans;
3. Have an understanding of the SQF Code and the requirements to implement and maintain SQF System relevant to the company's scope of certification. Successful completion of the 'Implementing SQF System Training Course Exam' would meet this requirement."

EXTERNAL COMMUNICATION

External communication is a topic that can be a challenge to fully implement, but it is critical to ensure that top management defines these requirements carefully and extensively to ensure its completeness. ISO 22000:2005 5.6.1 requires the following related to external communication:

[External] communication must provide information on food safety aspects of the organization's products that may be relevant to other organizations in the food chain. This applies especially to known food safety hazards that need to be controlled by other organizations in the food chain. Records of communications must be maintained. Food safety requirements from statutory and regulatory authorities and customers must be available. Designated personnel must have defined responsibility and authority to communicate externally any information concerning food safety. Information obtained through external communication must be included as input to system updating and management review.

External communication requirements for SQF Code, Edition 7, are defined in elements 2.16 Business Continuity Planning, 2.3.2 Raw and Packaging Materials, 2.3.3 Contract Service Providers, 2.3.4 Contract Manufacturers, 2.3.5 Finished Product, and 2.4.1 Food Legislation.

External communication with customers, suppliers, or any other organization may affect the processes or products. This may include information, assurances, or concerns related to specifications or food safety hazards. It is up to top management or their qualified designees to evaluate and identify what must be done and then ensure that it is completed and compliant. Communication with suppliers related to their products and their food safety programs is addressed in Chapter 13, which defines requirements for purchasing, outsourcing, and supplier management. External communication related to *crisis management* is discussed in more detail in the chapter for crisis management/emergency preparedness (Chapter 14).

Responsibilities for external communication must be assigned to a company representative who is trained (competent) in food safety and understands the requirements for communicating this to an external organization. Although a crisis does not need to occur for this to take place, many companies do train key associates for this responsibility and practice through mock trials. There have been more occurrences than any of us want to think about where an event is totally blown out of proportion, with an inaccurate or wrong comment made externally. This is even more critical if the wrong information or incomplete information is communicated, increasing the severity of a crisis.

This also applies when food safety information is received from the customers, suppliers, and other external sources. Ensure that an associate (i.e., food safety team leader) or team of associates (food safety team) are trained effectively to review this for impact, internal communication requirements, reporting to management review, and updating the food safety program if deemed necessary by the food safety team.

VISITORS AND CONTRACTORS

The process for training visitors and contractors must include sufficient information related to food safety and quality based upon their purpose for being on the site. A process to ensure that food safety and quality requirements are communicated to every nonemployee to the degree necessary for working at a site must be defined, implemented, and maintained effectively and supported by records confirming compliance. Contractors may also include service suppliers such as pest control operators or a representative of the cleaning chemical company. One of the best processes experienced was a two-page (front and back) registration log maintained at both the receptionist desk and the guard shack. The blank logs were kept in an open binder. The back side of the form defined the basic requirements. The front side had the sign-in information. All visitors and contractors were asked to read the requirements and then sign in to receive the visitor's badge. The person whom the visitor or contractor was visiting or working with always questioned us to make sure we understood the rules and did not have any questions; then he or she initialed the log. These logs became the record of proof. For more complex operations, we had to view videos outlining the requirements. These have been as short as 10 min and as long as an hour, but all have been very well done, most of which included not only food safety concerns, but also basic GMP, quality, personnel safety, and environmental concerns.

It is not uncommon to find several (some serious) nonconformances caused by a visitor or contractor, such as an equipment representative working on a packaging machine without a hair net and chewing gum. When asked why that person was not wearing a hair net, the response was that he was a contractor. It is also common to find nonfood approved chemicals in processing areas that are traced back to belonging to the contractors.

MANAGEMENT REVIEW

Defining, implementing, and maintaining an effective management review process is one of the most critical aspects of an effective food safety management system and absolutely essential to the long-term health of every management system. The management review process is included as one of the legs of the three-legged stool (Chapter 1), along with internal audit and CAPA. If one or more of these processes weaken, the total effectiveness and sustainability of the system becomes at risk for self-destructing. This may sound intense, but it has been my experience to see some of the best, top-rated systems self-destruct faster than one could imagine, when one or more of the three legs of the stool faltered. The reverse is also true when one of the worst systems matures into one of the best through the strength and effectiveness of the internal audits, management review, and CAPA processes:

> As one that is involved in overall business improvement, I am amazed at the ability of the management review meetings to measure the state of our business with reasonable accuracy. When our management system was struggling, there were numerous other areas of our business also struggling. Where the three legged stool was strong and effective our management system was strong and effective. (*Anonymous*)

> It is important for all to realize that the system will only be as good as the company as a whole will allow it to be. Culture change issues, a lack of genuine management support, and other conflicting variables will undoubtedly provide uncomfortable bumps in the road early on. The management review process, under a management team that is genuinely committed to the management system, provides a voice for food safety and quality that few organizations may have. It is the driving force for continuous improvement and provides peer accountability for the completion of the measurable objectives. No one wants to attend the next management review meeting without finishing all of

the items that were assigned to him or her at the last meeting. The management review meeting should also be used by the management representative/food safety team leader to acquire assistance with system resources, implementations, and maintenance, if it is needed. The management representative/food safety team leader has the ear of the company's management and must take full advantage of the opportunities available to them. (*Russ Marchiando, Packaging Plant Manager, Wixon, Inc.*)

No matter which standard is chosen, top management must be an integral part of the food safety management system, demonstrating strong commitment, knowledge, and leadership through everyday system activities and also through the management review meetings.

Management review meetings must be performed at a defined frequency to evaluate the status of the system, its suitability, adequacy, and effectiveness overall to the operation and to ensure the adequacy of its resources. In addition, this review should also monitor the system by evaluating and identifying continuous improvement opportunities related to the food safety and quality management system.

Depending on the standard of choice the *frequency* may be up to the organization to define, but the recommendation is to be done quarterly. Documentation may state that frequency be at a minimum of every 6 months; however, in actuality, top management should make every effort to have these meetings at least quarterly. The standards are very specific in listing minimum requirements (inputs and outputs) to be reviewed at these meetings. For some items, quarterly may be too frequent to obtain effective data such as monitoring some goals and objectives and to review the adequacy of the food safety policy statement. These items may be reviewed every 6 months or even less frequently, but the quarterly time frame is most effective for monitoring and addressing any issues of concern, proposed or potential changes, food safety–related consumer complaints, follow-up on outstanding issues, confirmation of sufficient resources, or the identification of resources. This review must include assessing opportunities for improvement and the need for change to the food safety management system, including the food safety policy.

Requirements for performing management review meetings can be defined through numerous methods, such as the food safety upper-level manual, meeting minutes agenda, or a specific procedure that addresses either management responsibility overall or, specifically, the management review process. In reality, the simplest process would be to define the organization's commitment to perform the management review meetings in compliance with the upper-level system manual and then link to a management review agenda form for the requirements. The blank version of this form would be a controlled document containing all the requirements (inputs and outputs, attendance, etc.); once completed, it becomes the minutes (record), which demonstrates compliance that all requirements have been evaluated. Any specific presentations or data shared during the meeting should be attached for reference. The management review meeting minutes must also confirm attendance by the required attendees. Required attendance may be listed on the management review agenda form. However, be certain both names and positions are identified. Over time, positions may change. During an audit, if the minutes list only names, be prepared to produce an organization chart that will confirm that the required attendees per specific positions during that specific time period were in attendance. It is much easier to have this information in the minutes than to have to search archives for it during an audit.

MINIMUM MANAGEMENT REVIEW INPUT

At a minimum, the following input must be included (reviewed) during the management review meetings as defined by ISO 22000:2005 5.8.2:

> "a. Follow-up actions from previous management reviews
> b. Analysis of results of verification activities (internal audits, PRP verification, etc.)
> c. Changing circumstances that can affect food safety
> d. Emergency situations, accidents and withdrawals (i.e. recalls)
> e. Reviewing results of system-updating activities
> f. Review of communication activities, including customer feedback
> g. External audits or inspections"

ISO 22000:2005 5.8.2 goes on to state "that the data must be presented in a manner that enables top management to relate the information to stated objectives of the food safety management system." Remember, the standards do not say that these items have to be evaluated at every management review, but at a defined frequency. If the decision is to have more frequent meetings, it would be good to identify the minimum frequencies required for each item. For example, the suitability and the adequacy of the food safety policy statement are reviewed, at a minimum annually, at a management review meeting. This is an example of an item that would not have to be reviewed more frequently, whereas customer complaints related to food safety should be evaluated at every meeting. Ensure that the management review meeting minutes provide evidence that the frequencies are met per the organization's defined requirement. It is recommended that the minimum frequencies be defined on the management review agenda form to ensure compliance. However, remember to define the frequency in one document; if it is on the form, then reference the form for the defined frequencies. If defined in the food safety system manual, then the form should state that "frequencies are defined in the food safety system manual." Never define the same requirement in more than one place.

MINIMUM MANAGEMENT REVIEW OUTPUT

ISO 22000:2005 5.8.3 defines that at a minimum, the management review output must include decisions and actions related to the following:

> "a. Assurance of food safety
> b. Improvement of the effectiveness of the food safety management system
> c. Resource needs
> d. Revisions of the organization's food safety policy and related objectives"

(Note: SQF Code, Edition 7, requirements for management review are stated in element 2.1.4 Management Review)

In addition, remember that the purpose of the management review meeting is to evaluate the suitability, adequacy, and effectiveness of the food safety management system; thus, in conclusion, top management must include a statement related to the suitability, adequacy, and effectiveness of the system based on the items reviewed during the meeting. This conclusion statement must also include a comment on the status and the production of a safe product.

A well-defined and well-structured management review process supported by top management is critical to building and improving the effectiveness of the management system overall. Remember, a management system with management commitment issues will self-destruct resulting in potentially serious weaknesses and nonconformances throughout all system elements.

MORE ON *RESOURCES*

Next to the term effectiveness, the term resources is very likely the most used word in this text when discussing management system compliance and effectiveness. Ensuring that the system has the required resources for its success is the responsibility of the top management team. The following are some comments related to resource opportunities during the development stage from industry professionals who have experienced firsthand some of the trials and challenges of implementation:

I believe that the optimal situation is to have all critical parts - relative to food safety and product conformance to design – completely and seamlessly integrated into an organizations ERP (Enterprise Resource Planning) system. Most solutions now offer management system modules which can be implemented before the system goes live; or "bolted-on" over time. Implementation of a food safety and quality management system is more efficient and effective as part of any ERP. (*Tom Marchisello, Industry Development Director, Grocery Retail GS1 US*)

At the time, there were very few available resources. The only vague information available was the GFSI web page and the various web pages of the schemes. (*Anonymous*)

We predominately utilized the standard itself, as we were forging new ground in trying to get a packaging company certified to a food standard. At the time, we were in a ground breaking lead role as the current standards had not yet addressed packaging specifically As it were, we were fortunate to be able to utilize our previous Auditor as a resource, as she had recently become certified to GFSI, (SQF), standards and she was still learning as she went. We did have some lively discussions about how to comply with a food standard when we didn't have any food in the plant. (*Alan Lane, former packaging manager now an independent consultant*)

We developed many of our own yet also utilized outside resources to build a library of educational tools that have been incorporated into our overall food safety management system. (*Mike Burness, Vice President, Global Quality and Food Safety, Chiquita Brands International, Inc.*)

It would have been very helpful if we had some of the now present gap analysis tools that we found out about later in our development stage. Once certified, it becomes evident that the standards hold the key to your food safety system. We used examples of programs, procedures & charts from other companies that were certified, the more we looked at these, the better ideas that we had in development of our programs, procedures & manuals. I think the more lessons learned and best practices that you can review as long as you are not just copying and changing company names the better your system will be. (*Russ Patty, an independent consultant and former quality manager of a beverage operation*)

We found it difficult to find relative microbiological data to support the deliberations of the HACCP Team. We did use the Codex website and a consultant, but had thought there would be more relative information available. (*Anonymous*)

Although we had some for reference, it would have been very helpful if we had had more relative examples from similar operations who were ISO 22000 certified. (*Joe Hembd, Management Systems Manager, Chesterman Company*)

We would have benefited with a stronger foundation for the time management required. There never seems to be enough time to get systems exactly how you would like them to be especially when it comes to maintaining rigorous and effective internal audit programs. Also, more guidance on what is an acceptable mitigation, as related to system compliance, would have been a huge benefit for a system. The best resources that I found were tools that our consultants provided us which made my life much easier. An example of this is a matrix to aid in the control and measurement of my PRPs. (*Eric Putnam, Food Safety, Quality & Training Systems Manager, Wixon, Inc.*)

Charlie Stecher, an independent consultant and third-party auditor, Anchar, Inc. shares the following:

During a recent audit, the plant manager from a large company shared his mixed feelings about the corporation requiring that all sites attain FSSC 22000 certification. The same manager nine (9) months after certification was proud to state that he could really see an improvement and the benefit to getting certified. It was evident during my return visit that associates representing all levels of the organization were engaged wanting to now be part of the management system. This plant manager also offered and actually encouraged associates, if interested, to go through Internal Audit training so they can get a better understanding of food safety and quality management standards. It was great to see first hand such a great turnaround.

EFFECTIVE SYSTEM IS DIRECTLY PROPORTIONAL TO THE STRENGTH OF THE FOOD SAFETY CULTURE

Yiannas (2010, pp. 16–19) states that

Having a strong food safety culture is a choice. The leaders of an organization should proactively choose to have a strong food safety culture because it's the right thing to do, as opposed to reacting to a significant issue or outbreak … Creating or strengthening a food safety culture will require the intentional commitment and hard work by leaders at all levels of the organization, starting at the top.

KEY POINTS

a. Top management (individually or the top management team) must be defined.

b. Measurable food safety goals and objectives must be developed by top management and should be reviewed with the food safety team as well as during the management review process.

c. The current food safety policy statement must be defined and accepted through a management review meeting with confirmation (status) recorded as part of the management review meeting minutes.

d. The current food safety policy statement must be reviewed for continued suitability during management review meetings with results recorded in the management review meeting minutes.

e. The top management team must be intimately involved with current policies and activities to ensure continued compliance, effectiveness and available resources of the food safety management system.

f. Minutes of management review meetings, staff meetings, food safety team meetings, and food safety/HACCP team meetings must be maintained in compliance with the record control program.

g. Meeting minutes must also include attendees to the meeting. Recommend recording names and positions.

h. Top management team must be aggressive with communication activities and define a program for measuring the effectiveness of their actions.

i. The top management team must ensure effective communication with associates representing all levels of the organization.

j. Through the interview process, it must be confirmed that the top management team and associates representing all levels of the operation are able to demonstrate a commitment to and an overall understanding of the requirements of the management system and their role in food safety.

k. Top management must confirm that the food safety and quality management system planning is in compliance with the requirements of the chosen standard or standards, and that this planning includes defining measurable objectives for food safety and quality.

l. The change management process must monitor and communicate changes within the organization; this is often done through planning meetings and establishing next year's goals.

m. Organization must ensure that its management of change process is effective and that records are available to demonstrate compliance and effectiveness.

n. Required training of the food safety team leader or SQF practitioner to ensure *competency* must be defined.

o. Identification of the food safety team leader may be confirmed by an organizational chart or in the management review minutes.

p. Top management should designate, train, and record a backup or alternate food safety team leader.

q. Top management or their designee must communicate the necessity for the food management system to provide information on food safety aspects of the organization's products relevant to other organizations in the food chain.

r. Top management must ensure that site-controlled critical suppliers, including those for *outsourced* services, are identified and compliant with the supplier management program. Records must be identified and maintained to confirm this compliance.

s. To be most effective, food safety training may be a continual reinforced process of communication to all associates.

t. Top management can present and reinforce system requirements, including food safety awareness and other specific food safety requirements, through many means including one-on-one discussions, shift meetings, specific training exercises, and quarterly town hall meetings with all associates.

u. The internal audit program is an excellent tool and should include the confirmation of effective internal communication. Internal auditors should always take a sample among associates within the areas being audited, posing questions that would confirm familiarity with required documents, records, measurable objectives, food safety and quality policy and their role in the food safety quality system.

v. Top management, including the food safety team, must ensure that the effectiveness of the communication process is confirmed. Results of effectiveness, evaluations, status, and resource requirements for compliance should, at a minimum, be recorded through the management review meetings, food safety team meetings, PRP verification, and the internal audit program.

w. Management review records (meeting minutes) must provide objective evidence that all required inputs and outputs have been evaluated and results recorded.

x. It is recommended that a compliant management review meeting be held at a minimum of every 6–8 weeks until the system is established and then quarterly as the system matures.

y. The management review meeting agenda may identify and provide the basis for evaluating and recording the required inputs and outputs.

COMMON FINDINGS/NONCONFORMANCES

a. The organization had not defined and implemented a food safety policy statement.

b. The organization had not defined, implemented, and communicated the food safety objectives to all levels of the operation.

c. It could not be confirmed through interviews and review of records that associates representing all levels of the operation had a comprehensive understanding of their food safety management system and had implemented their responsibility accordingly.

d. The process was not defined for communicating statutory and regulatory requirements either internally or externally within the system.

e. Site had two separate lists of *goals*. Through interviews and review of internal records, it appeared that the goals reviewed were those of the facility rather than those identified in the food safety manual.

f. The list of measurable goals, which was posted throughout the facility, was not identified and maintained as a controlled document.

g. Data analysis results related to measurable food safety goals and objectives were neither identified and maintained as a record, nor included in the management review meeting minutes.

h. It was not clear during the review of the current food safety policy statement that it accurately addressed regulatory and statutory requirements.

i. It could not be confirmed through review of documents, records, and interviews that management had defined a process that assured that the integrity of the food safety management system was maintained when changes to the food safety management system were planned and implemented.

j. The current organizational chart did not provide information or reference to where requirements for responsibility and authority were defined as required by the organization's food safety manual.

k. Records were not available to confirm that responsibilities and authority, as related to food safety, had been defined.

l. Although it was verbally stated that the operations manager had been assigned as the food safety team leader, records were not available to confirm this or to confirm that required competency training had been defined and completed.

m. Current documentation did not define the specific required training for the food safety team leader position.

n. The maintenance manager had been identified as the backup position for the food safety team leader; however, records to confirm that the person in this position had completed the required competency training were not available.

o. Program for training visitors and contractors prior to an on-site visit provided only a limited overview of quality, GMP, and food safety concerns.

p. It was stated verbally that the organization had customers with whom they communicated directly; however, the requirements for this process as related to food safety were not formally defined and implemented.

q. Records did not confirm that the janitorial service had been trained in the procedure for "Visitor and Contractor Requirements," which included training for approved chemicals and good manufacturing practices requirements.

r. Although "Visitor Requirements" were posted in the lobby and visitors were asked to read them prior to signing the visitor log, actual records did not provide confirmation that the person signing this log had read and understood the referenced requirements.

s. Some confusion existed about the distinction between a visitor, contractor, and vendor, and the *communication requirements* for each.

t. Process had been neither developed nor implemented for communication with plant associates on issues having an impact on food safety.

u. Records were not available to confirm that the food safety team had performed a food safety evaluation of potential and existing *changes*.

v. The management review procedure did not identify required attendance for the management review meetings.

w. Management review procedure defined required attendance as a *quorum*; however, actual positions required to be a quorum were not defined.

x. It could not be confirmed through the review of the sampling of management review minutes that all required inputs and outputs were being reviewed with actions/resources applied by the top management team (i.e., CAPAs, resource status, and customer complaints related to food safety).

y. Management review meeting minutes did not provide a clear and accurate statement as to the current suitability and effectiveness of this FSMS based on the status of the system and the items reviewed during the meeting.

z. It was stated in the meeting minutes that the system was suitable and effective; however, the system to date was not in compliance.

aa. It was stated that primary communication was presented to all associates at mandatory quarterly town hall meetings; however, attendance was not taken in a manner that identified those associates who were absent, thus makeup sessions were not mandatory attendance as defined in the related documented procedures.

bb. Records did not confirm that management review meetings were being performed semiannually as required by the food safety management system manual section 5.

cc. As a result of system evaluation and the current state of nonconformances identified related to key elements such as internal audits, PRP verification, and document control, it could not be demonstrated that top management had effectively addressed and ensured the adequacy of resources for this management system.

dd. Through review of the organizational chart and related defined responsibilities and authority, it could not be demonstrated that the food safety team leader had sufficient responsibility and authority to perform required activities.

ee. Review of minutes from the management review meeting did not provide objective evidence that the management review meetings were being conducted at the frequencies defined in the procedure for management review.

ff. Management review meetings were being held, but meeting minutes did not provide evidence that one or more of the requirements (e.g., review of the results of internal audits and CAPA activities) were being evaluated at these meetings.

gg. Records of management review meetings were not identified and maintained in compliance with the record control program.

hh. The posted food safety policy was not signed by the current general manager. The current general manager had been in his position for 6 months; however, the food safety policy still represented the support of the previous manager rather than the current one.

ii. Although implied, minutes from the management review meetings did not include a statement on the current effectiveness and suitability of the system as a result of the items reviewed.

REFERENCES

Link, E., *An Audit of the System, Not of the People, An ISO 22000:2005 Pocket Guide for Every Employee,* Quality Pursuit, Inc., Fairport, NY.

Surak, J.G. 2001. International Organization for Standardization ISO 9000 and related standards. In: *Food Safety Handbook*, Schmidt, R. and Rodrick, G. (eds.), Wiley-Interscience, Hoboken, NJ.

Yiannas, F. 2010. *Food Safety Culture Creating a Behavior-Based Food Safety Management System*, Springer, New York.

---

*Additional Food Safety Management System References (not all inclusive)

FSSC 22000:2010

    ISO 22000:2005 Sections 5, 8.5

SQF 7th Edition: Section 2.1

BRC Issue 6: Sections 1.1, 1.2

IFS Version 6: Section 1

*Note: Always review the standard or visit the website for your chosen standard to ensure that all related requirements are identified and addressed.

4 Corrective Action/ Preventive Action (CAPA)
Continuous Improvement

An effective corrective action/preventive action (CAPA) program is absolutely essential to an effective food safety and quality management system. Each of the GFSI-approved schemes require that the organization has a formalized corrective action process designed to bring any nonconformances back into compliance and to prevent recurrence, but the emphasis on a formalized system overall varies depending on the standard. In my experience, having a system-wide effective formalized CAPA program is critical to the overall effectiveness and sustainability of a management system, no matter which standard is chosen. Corrections and corrective actions related to critical control points and what to do if a deviation occurs are addressed in Chapter 17. This chapter discusses the CAPA program that comprises one of the three legs of the three-legged stool (Chapter 1).

Corrective actions focus on *existing* nonconformances. Preventive actions focus on addressing *potential* nonconformances before they happen. CAPAs are applied to eliminate the causes of actual or potential nonconformities and must be, to a degree, appropriate to the magnitude of problems and proportionate to the risks. The current version of the ISO 9001 standard, although a quality management standard, provides an excellent foundation for an effective CAPA program through its defined requirements.

A compliant preventive action program evaluates product, process, and system issues to detect, analyze, and eliminate potential causes of nonconformances. A nonconformance or nonconforming situation (also referred to as a *finding*) is defined as an activity that is not in compliance with a defined requirement of the management system. The CAPA program is a tool that propels continuous improvement. A well-defined effective program not only addresses existing issues, but also is a means to capture and track continuous improvement effectively in a manner that adds value to the system overall.

Russ Marchiando, Packaging Plant Manager, Wixon, Inc., provides an excellent description of his experience with an effective CAPA program developed during Wixon's journey to certification:

> In my opinion, the most useful system/benefit that has been experienced as a result of certification is the corrective and preventive action system. I am finding that having a channel for the identification and execution of corrective actions is driving the changes necessary to improve our business systems from within and to improve our ability to meet our customers' expectations. It is a tremendous benefit having a structured

program whereby actual and potential problems can be documented and addressed. The structure of this system allows for a systematic and organized approach to issues and also requires that the root causes are addressed completely. The best things about developing a CAPA program is having a formal channel to address internal issues and customer complaints, to address potential issues before they get out of hand, and to allow all personnel the opportunity to revise the processes that affect his/her work. Every issue is handled through the same channels regardless of the source. Our system calls for a cross section of the company's management each week in order to review the week's CAPA requests and customer complaints. This program has opened channels of communication that have not previously existed providing a measure of accountability. It also provides a forum for feedback on the CAPA program on a weekly basis. I'm not sure that we would have the same success at continuous improvement without this program. Our CAPA program requires accountability to the food safety and quality system. This is a powerful tool to drive change.

UNDERSTANDING THE CAPA PROGRAM

Another excellent description is provided by Mr. Peach (1997, p. 139), which can and should be applied to every management system, no matter what its focus (i.e., food safety, quality, environmental, safety, etc.):

> Often the weakest part of [management system's] corrective action loops [activities to address the findings] is that they are frequently designed only to address the immediate problem while failing to act to avoid its recurrence. Another common problem is that they often deal only with matters of processes, products, or services while overlooking the system. [This] requires a rigorous examination of all of the data and records to detect and remove all potential, as well as, actual causes of nonconformances.

The CAPA program is considered proactive, not reactive. Organizations that apply the CAPA program only to product issues must expand their program to also address process and system issues. As an auditor, we confirm that the program is applied to product, process, and system issues, any one of which can, and do, result in nonconformance.

It is recommended that an associate with responsibility and authority within the system be the manager or coordinator of the CAPA program.

CAPA RECORDS

CAPA records must be identified and maintained in compliance with the record control program. At a minimum, the CAPA record (objective evidence) must include the following:

- Proof that the program is being applied to product, process, and system issues including those related to food safety, customer complaints, specification compliance, regulatory requirements, additional requirements of the organization, and trending from other programs (i.e., nonconforming material and products, GMP inspections, etc.)
- Results of the root cause analysis

- Identification of what is required to correct or prevent the nonconformance
- Description of follow-up evaluation confirming the effectiveness of the actions taken
- Relevant information related to the CAPA program submitted for top management evaluation during a management review meeting
- Actions as a result of CAPA process related to implementation and any changes to documents

It is recommended that records related to the CAPA program be maintained for a minimum of 3 years to provide sufficient evidence of the maintenance and effectiveness of system activities.

IDENTIFYING A CAPA

When identifying a CAPA, it is important to record the finding clearly, ensuring that it is a statement of fact. This program must never be used as a means to apply discipline or as a process to air gripes. The commitment date for completion must be directly related to the potential risk of the situation. The program itself must clearly define the criteria for identifying and issuing CAPA programs.

Most organizations find it most effective to have separate programs for addressing individual nonfood safety–related customer complaints and nonconforming product issues. It is much more effective to monitor and record these on an individual basis through a program designed especially for that purpose. However, this information must be monitored for food safety issues and trends that would then, as determined necessary by the food safety team or its designee, be addressed through the CAPA program. A favorite example of this would be one complaint related to loose bottle caps. This may be addressed on an individual basis with the customer; however, if several complaints are received on this same issue, a formal corrective action may be required. The program will struggle if a responsible manager has to respond formally through the CAPA process for every individual instance.

COMMENTS BASED ON EXPERIENCE

The following comments from professionals who have experienced this journey provide excellent insight. Unfortunately many of these comments are being reported as *anonymous* due to organization's policies restricting associates from identifying themselves and/or their companies, but no matter, comments are based on the experience of living the process and surviving that all-important learning curve:

At the onset, we were setting ourselves up for failure and struggled mightily by attempting to address every complaint through our CAPA program. It took a while to realize that everything that goes wrong in a company is not necessarily worthy of a CAPA program. It was also difficult changing a culture where the documentation of problems and solutions were nearly nonexistent. By requiring the documentation of a formal root cause analysis, its corrective action and follow up investigation, we received significant resistance from people claiming that this process was "just more paperwork" to deal with. Even as our system matures, it is difficult to keep people from taking shortcuts on the paperwork.

Finally, another problem that we have is the use of the system to air departmental gripes instead of working out minor problems of communication outside the system; some people are content to use the system to solve all of their problems for them. (*Anonymous*)

One of the most difficult aspects of maintaining the system after certification has been getting true corrective action and real change. [Prior to management system implementation], all too often the quick fix at the last minute was used. There wasn't the resolve or mandate to institute truly effective corrective action. Those who were charged [with responsibility for this program] were always on edge walking a fine line between making sure the problem appeared fixed and knowing that more could and should have been done. (*Anonymous*)

DETAILS OF AN EFFECTIVE CAPA PROGRAM

At this point, the discussion has been on the CAPA program overall. Now let's briefly discuss some specific requirements and recommendations for the development, implementation, and maintenance of an effective CAPA program.

There are many different means (i.e., forms, software programs, etc.) that may be used to capture and manage an effective CAPA program. The following recommendations may be applied to a hard copy form or incorporated into a software program that the organization uses to manage its CAPA program:

- Record the nonconforming situation, which may also be referred to as the *finding*. Findings *must* be based on the facts.
- Identify the source of the noncompliance (i.e., internal audits, management review, customer complaints, product nonconformance, employee feedback, failed effectiveness review, external audit, etc.).
- Review and assign the CAPA to the responsible department manager unless the program has a formalized meeting that is used to discuss and assign the CAPA. Note that this is an excellent example of when an organization must define its process, which in this case would state that the action is assigned as defined in the CAPA program.
- The responsible manager, depending on the organization's program, may address the finding or assign it to a designee to work on it under his/her direction.
- The *root cause* must be identified and recorded.
- The action required to correct the *root cause* with a realistic time frame based on the potential risk of the situation is identified and recorded. Every effort must be made to address the corrective action in a timely manner based on the potential risk of the situation.
- Complete the corrective action; document the action taken and the date completed.
- After the corrective action is completed, the CAPA manager generally reviews the results and assigns a date to evaluate the effectiveness of the action taken. The elapsed time period must provide the opportunity for significant evidence/activities to occur so that the results of the actions taken can be effectively evaluated.
- Effectiveness is evaluated and recorded.

- If the effectiveness cannot be confirmed, then this must be recorded with *next step actions* identified and implemented. The requirements for addressing those CAPAs that cannot be confirmed effective must be defined in the CAPA procedure. The most obvious choices are to either issue a new CAPA or reopen the existing one, but either way, if the finding is not effectively addressed, then it must be reevaluated, root cause identified, and a new action identified.

Let's look at the descriptions given earlier in more detail starting with recording the finding. The first step is to define *the finding*, which must be a statement of fact, not what went wrong or how to fix it. Let's review some examples:

Example 1: Records were not available to confirm that the certificate of analysis was being reviewed and initialed/dated by the receiving operator prior to accepting the raw ingredient, citric acid, as required by procedure REC-01 Rev. 2.

This example is a statement of fact. It identifies the nonconformance and the fact that it is an activity not being performed as defined in procedures.

Example 2: Training procedures and work instructions did not define the requirements for associates performing food safety–related packaging activities.

Example 3: Training procedures and work instructions must be written to define the requirements that need to be done for associates as part of performing food safety–related packaging responsibilities.

The difference between Examples 2 and 3 is that Example 2 states the fact whereas Example 3 directs how to fix it. The *how to fix it* must be identified by the person responsible for fixing the situation. Often, the person who identifies the situation would not be familiar enough with the operation to fix it. If the writer dictates the fix, then he/she is accepting ownership of the finding. Ownership belongs to the responsible department manager or his/her designee.

Example 4: The incoming inspection log was not identified and maintained in compliance with the record control program.

This is a statement of fact.

Example 5: Review of a sampling of records and the activity for performing inspections on the empty trailers prior to loading did not provide evidence that these activities were being recorded on the bill of lading as required by the procedure SHP-01 Rev 0.

This is a statement of fact.

Example 6: There was no documented procedure defining the requirements for truck drivers when they checked in through the guard gate.

This appears to be a statement of fact, but it also is *assuming* or *directing* that there must be a documented procedure for this activity. A finding must be based on a defined requirement. The requirement should be referenced in the finding. A finding cannot be written just because it is something an auditor likes to see. A more correct manner of writing this finding would be: There was no documented procedure defining the requirements for truck drivers when they checked in through the guard gate as required by SOP SH-02 Rev 2.

Example 7: Numerous cartons of product were being damaged/destroyed because the warehouse roof was old and needed replacement.

The decision as to *why the roof is leaking and damaging the product* should be left to the person responsible for addressing the finding.

As stated previously in this chapter, the review and assignment of the CAPA plans may be performed by an individual assigned to this responsibility or managed through a CAPA team that meets at a defined frequency to review CAPA status and make decisions on the acceptance and assignment of recently identified CAPA programs. As *Mr. Russ Marchiando, Packaging Plant Manager for Wixon, Inc.* stated previously, the team approach is very effective. The team also monitors timeliness and assignment of resources. Results and any concerns related to the CAPA program as identified by the team are reported to top management at management review meetings.

UNDERSTANDING ROOT CAUSE

The next step is to identify the root cause, which must be done by the person, group, or team *fixing* the finding. Some programs encourage internal auditors to assist with this, possibly discussing best practices with area associates. Although this can be helpful, internal (and external) auditors must be careful not to offer the fix, since the ownership for addressing the finding belongs to the department responsible for the finding.

What is required when identifying the root cause?
Is there a clear definition for *root cause*?
What exactly is a *root cause*?
What is a good *root cause*?
How do we know we are doing this correctly?
How involved must we be in evaluating the *root cause*?

There are many, many very good workshops, books, and subject matter experts available to assist an organization in answering these questions. It is strongly recommended that key personnel such as those responsible for the CAPA program attend such a workshop and then bring this knowledge in-house to share with the associate team. In house training enhances the understanding and application of the CAPA program.

Okes (2009) provides an interesting viewpoint on the foundation of root cause development:

We live in a complex world. People and organizations often don't believe they have the time to perform the in-depth analyses required to solve problems. Instead, they take remedial actions to make the problem less visible and implement a patchwork of ad hoc solutions they hope will prevent recurrence. Then when the problem returns, they get frustrated and the cycle repeats. This book provides detailed steps for how to solve problems, focusing more heavily on the analytical process involved in finding the actual causes of problems. It does so using a large number of figures, diagrams, and tools useful for helping make our thinking visible.

Mr. Okes states that the root cause evaluation process should focus "on solving repetitive problems and ... the logic of finding causes. It has sometimes been described in training workshops as Six Sigma like...problem solving without the all the heavy statistics."

One of the most common issues found related to *root cause*, which is significant, is that organizations when stating the root cause actually restate the finding.

A brief search of the Internet provided the following definitions and *root cause* explanations:

A *root cause* "is a 'cause' (harmful factor) that is 'rooted' (deep, basic, fundamental, underlying, or the like) [of why something goes wrong]." (http://en.wikipedia.org/wiki/root_cause)

Root cause analysis (RCA) "is a method of problem solving that tries to identify the root causes of faults or problems that cause operating events. RCA practice tries to solve problems by attempting to identify and correct the root causes of events, as opposed to simply addressing their symptoms. By focusing correction on root causes, problem recurrence can be prevented. RCFA (Root Cause Failure Analysis) recognizes that complete prevention of recurrence by one corrective action is not always possible." (http://en.wikipedia.org/wiki/Root_cause_analysis)

The primary aim of RCA is to identify the factors that resulted in the nature, the magnitude, the location, and the timing of the harmful outcomes (consequences) of one or more past events in order to identify what behaviors, actions, inactions, or conditions need to be changed to prevent recurrence of similar harmful outcomes and to identify the lessons to be learned to promote the achievement of better consequences. "Success" is defined as the near-certain prevention of recurrence. (http://en.wikipedia.org/wiki/Root_cause_analysis)

To be effective, RCA must be performed systematically, usually as part of an investigation, with conclusions and root causes that are identified and backed up by documented evidence. (http://en.wikipedia.org/wiki/Root_cause_analysis)

There is so much excellent information available related to effective application of *root cause analysis* to the CAPA program that any further discussion in this text cannot do it justice. Do keep in mind that a root cause can and is identified based on experience and the basic application of common sense. However, a more complex issue may require a more complex tool such as the *5 Whys*, 6 Sigma, and similar programs.

CORRECTIVE ACTION

Once the root cause has been identified and recorded, the next step is to identify what action must be taken (i.e., the corrective action) to correct the finding. This should be identified by the same individual or team that identified the *root cause*. Once the correction is identified (addressing the root cause), a commitment date to have this completed, based on the potential risk of the situation, must be identified. This commitment must be realistic and practical. Many times, making a commitment becomes an issue because the responsible associate is concerned about what will happen if the date cannot be met. This is another advantage of the team approach where the team would jointly discuss, identify, and record a realistic date based on the potential risk. Requirements for approving and documenting extension dates for completing corrective actions must also be defined. It is recommended that a request to extend the completion date be presented in writing with a clear description of why the extension is required and identifying the new date for completion. Approved extension dates must be recorded as such. Dates not achieved without clear justification are considered as *overdue*. A justified (approved) completion date extension should not be tracked as *late*.

TIMELINESS

Tracking timeliness, approved extensions, and overdue corrective actions brings us to another topic. Tracking the number of open and closed (completed) CAPA can communicate a negative message. Trying to see how quickly CAPA programs can be closed must not become a race or a fault-finding tool. It is strongly recommended that this program be monitored by the timeliness/on-time completion of the CAPA, and absolutely not on the number opened, or how fast they are closed.

One more thought on corrective action is that in some instances, there may not be a *quick* fix. Many times the responsible associate may have to define an immediate action that mitigates the risk of the finding until a time when it can be fixed completely. Capital expenditures are an excellent example of this.

It is also recommended that a CAPA is considered closed based on when the CAPA has been completed. It then remains active until the effectiveness of the action taken is evaluated.

FOLLOWING UP ON EFFECTIVENESS

Following up on the actions taken to confirm effectiveness is not only one of the toughest aspects of an effective CAPA program, but it is also one of, if not the most, advantageous aspects of an effective program. An effective CAPA program focuses on identifying and solving existing and potential problems in a timely manner with a strong focus on preventing reoccurrence. The situation is often improved through these actions, but not totally corrected and effective.

When identifying an estimated date for evaluating the effectiveness of an action taken, it is critical that the time frame is such that sufficient objective evidence is available to make this determination.

Example 1: The roof is leaking. The corrective action may be to patch the roof; however, the effectiveness cannot be truly evaluated until it actually rains.

Example 2: The finding is that "records in the blending area were not being completed as required by procedure BLD-2 Rev 2." Initially, the responsible manager may identify the root cause to be inadequate or insufficient training. The commitment for training is identified and performed. Depending on the frequency in completing the records or forms (i.e., daily), enough data after 2–3 months may be sufficient to confirm effectiveness, but if these forms are completed quarterly, then obviously a time period of 2–3 months would not provide sufficient objective evidence to confirm effectiveness.

Determining that a corrective action may not have been effective or only partially effective may be that the original root cause analysis did not address the full scope of the finding. Most times it is the most obvious root cause that is identified and addressed. This may be part of the issue, but not the complete reason for the nonconforming situation. The initial appearance is that the situation has been corrected; however, the true root cause has not been addressed, thus the situation reoccurs. This forces further review of the situation, identifying and addressing root causes until the actions taken are confirmed 100% effective.

CORRECTIVE AND PREVENTIVE ACTION PROGRAM

Many years ago, a management representative who was identified as an industry professional but due to company policy could not be identified stated that "an effective CAPA program is the 'to-do list' that doesn't go away."

Some organizations choose to have a separate corrective action–type programs for their supplier management and for their internal audit programs. Other organizations prefer one program with the sources identified on the corrective/preventive action form itself. Although multiple programs can be successful, a single program is generally easier and more efficient to document, maintain, and monitor. It is common during the implementation stage to have two or more distinct CAPA-type programs; however, as the system matures, management frequently decides to combine the reporting and documenting format into one program. Generally, associates have indicated that they are more comfortable having to learn one program, using one form or software program, and having one specific means to address existing and potential nonconformances. Keep in mind that one central system must not be expected to address every individual product nonconformance or customer complaint. These must still be monitored for trends and concerns and then, as determined appropriate by the food safety team or designee, evaluated and recorded through the formal CAPA program.

CORRECTIONS VERSUS CORRECTIVE ACTIONS

Understanding corrections vs. corrective actions can be a challenge, no matter how familiar one is with the standards themselves. ISO 22000:2005 defines these terms as follows:

Correction is the "action to eliminate a detected nonconformity." (3.13)

Corrective Action is the "action to eliminate the cause of a detected nonconformity or other undesirable situation. Corrective action includes root cause analysis and is taken to prevent recurrence." (3.14)

Typically a correction does not require a root cause and is a single or limited event such as a loose cap or a case of leaking containers. The situation is corrected without a full analysis; however, corrections must be monitored for trends. Corrections that result in trends should be recorded and tracked through the formalized CAPA program.

Corrective action as related to the management system is a formalized process that requires root cause analysis, identified actions to be taken to correct the situation in a timely manner based on the risk of the situation, actions required to minimize or prevent its recurrence, and the evaluation of effectiveness to ensure that the action or actions applied are effective.

Corrections are best identified and tracked via a spread sheet. The information recorded on the spread sheet should be used to not only confirm that corrections are made but also to be used in the trend analysis. Typically single occurrences can be addressed by just fixing them, but should a trend develop, then a complete root cause analysis may be appropriate. It is important to define the process that is best for the organization. The balance between corrections and corrective actions is important but must be a learned skill based on what is most effective for the organization. Too many insignificant corrective actions that don't require a root cause can and will strangle the CAPA process, but that said not everything is a *correction* only. Again, this must be a learned skill based on the operation and the individual findings. Customer complaints and GMP inspection results most often start out as corrections; however, if a trend is identified, then the trend should be recorded and tracked through the formalized corrective action process.

An example would be in reviewing results of a GMP inspection program: every month restrooms were reported as *dirty*. This finding was repeated month after month. When questioned, the response was that these were different restrooms. This is definitely an example of a trend that should be raised from a correction to a corrective action. A root cause and subsequent action to correct the situation are definitely warranted.

SOFTWARE PROGRAMS

There are many different software programs available that claim to be the best in recording and tracking the findings including the root cause identification, application of corrective action in a timely manner based on the potential risk of the situation, and recording effectiveness of the actions taken to ensure that the finding has been specifically addressed.

Many of these programs are very cumbersome and may actually make it more difficult because in some instances, the number of questions asked and data required appear to be more cumbersome than the basic evaluation. As an organization, be careful when choosing and/or developing a software program that initially is meant to save time and add value. There are programs that do just that, and a good software program for this purpose is invaluable to the program. The author wants to emphasize that it is critical to apply caution and ensure that the program is specific to the organization, its products, and its food sector. Also ensure that a clearly defined contract for technical support is agreed upon before the purchase.

CONTINUOUS IMPROVEMENT

The food safety management standards plus all of the other management system standards do focus on continuous improvement. They require the organization to be compliant while investigating continuous improvement opportunities. Measurable goals and objectives must be defined and tracked for compliance. These goals may also be an example of continuous improvement since these results along with the items reviewed during management review through the review process are monitored and improvement opportunities are either identified specifically or accomplished indirectly (Chapter 3).

During the early years of management standards; identifying, recording, and tracking preventive actions were one of the sole means to track and take credit for continuous improvement activities. However, as management system standards have evolved, preventive action has become one of many tools for addressing continuous improvement activities. Other examples for identifying and tracking continuous improvement are project teams, associate feedback, management of change activities, and management review output. Ongoing discussions can lead to additional opportunities depending on how an organization has defined its system.

It is common to address improvement opportunities through everyday activities without documenting and taking credit for what is being done. System records must provide ongoing and effective evidence that the food safety management system not only maintains compliance but also has an effective program for continuous improvement. Management must take credit for what they are doing. It's very common during an audit for the organization to have many different continuous improvement-type active projects; however, few, if any, have been formally recorded in a formalized process.

MORE THOUGHTS

The CAPA program is required input for the management review meetings. This may include reporting timeliness in completing actions or any other aspects of the program that communicates its status and effectiveness. The key is that if there is an issue with the program, including the fact that resources are not available to manage the program effectively, must be reported and addressed. If the organization has defined that the program is to be maintained through weekly CAPA team meetings, but due to other issues, the team has not met in a month or only 25% of the team members are able to take part in the meetings, this would be a strong indication of a resource and/or commitment issue, which top management must address during management review meetings.

> The CAPA program is a critical element of every food safety and quality management system. It is imperative that this process be well defined and maintained in an aggressive and effective manner to achieve the highest degree of value and usefulness for the management system. This program will go through many phases as the system itself matures. It is difficult to always know what should be recorded and what shouldn't. This can be achieved through effective management and basically, living the learning curve. This is where the CAPA team can be very useful. (Newslow 2001, p. 163)

Tim Sonntag, Vice President Quality & Technical, Wixon, Inc., provides us with the following insight on the advantages of an effective CAPA process that their management system experienced through its implementation stage and now long-term maturation process:

> I feel there have been numerous process/system benefits of obtaining certification that have helped move our food safety and quality management system to the next level. However, I think the single biggest benefit has been the institution of a true corrective action/preventive action based program. If properly structured and utilized, it's an unlimited vehicle for driving continuous improvement throughout the organization. It has provided a structure to address issues at every level of the organization and in every department within the organization. It has provided a structure not only for identification and documentation of current or potential problems, but it also demands a root cause determination to be done and a commitment by the person(s) responsible for proposing what corrective action is to be instituted within a targeted time frame. The progress being made within this type of structured system can easily be monitored and periodic reports generated for distribution to all levels of management. The unlimited inputs make this system a very dynamic tool for continuous improvement. The primary inputs to our CAPA process incorporates food safety concerns, customer complaints (both external and internal), product returns, internal audits, customer audits of our facility, internal trend analysis, sanitation/GMP inspection results, safety committee inspection results, HACCP program activities, SPC system results, regulatory agency audits, and supplier management opportunities.

> Corrective action can be thought of as first aid for manufacturing and service problems. What is really needed is a greater emphasis on being proactive or the use of more preventive action. Good closure means that the problem has been completely analyzed, the solution well thought out and implemented, and the effectiveness verified as required by ISO 9000..... Preventive action results in the reduction of variation. As the variation is reduced, the likelihood of or the potential for the occurrence of the addressed nonconformity is reduced. (Link 2008, pp. 102–104)

> The success of [the] organization depends on the strength of [the] CAPA process. Barriers are removed and problems solved when top management demonstrates their support and commitment to the problem solving efforts applied through an effective CAPA program. (Newslow 2001, p. 164)

> Good corrective action and root cause analysis is hard to find in most companies. It is disappointing to see corrective actions become just paperwork and not a means of improving the management system. It is firmly believed that more training in corrective action and root cause would help companies improve their systems. (*Charlie Stecher, an independent consultant and third-party auditor, Anchar, Inc.*)

Over the years, many have compared the CAPA program to Deming's well-known *plan-do-check-set* cycle. The following is an example of how this fits:

a. Plan the action to be taken after identifying and evaluating possible solutions.
b. Take the action.
c. Confirm that the action is effective.
d. Continue to monitor results to ensure that the problem has been solved.

KEY POINTS

a. Effectiveness is not verifying that the corrective action is done. Completing the corrective action is the responsibility of the person assigned to the corrective action. That person or group must take ownership for their commitment. Verification for effectiveness is just that, verifying that the corrective action is effective. The time period necessary to determine the effectiveness must be realistic, providing enough evidence to form an accurate conclusion as to whether or not the action was effective (Peach 1997, p. 230).

b. The CAPA program must not be presented as a *fear* program. Issues and commitments must be looked at as a means to strengthen the system through the identification of existing and potential nonconforming situations.

c. An excellent method to ensure that hard copy CAPA reports are not lost in a pile of paperwork is to print these forms on bright-colored stationery. Bright red or kelly green are easily identified in a pile of white papers. One organization changes the color of its forms every 3 months to help control the time frames.

d. As effective CAPA programs have evolved over the years, most organizations have developed some means of tracking the status of its actions through a CAPA log or spreadsheet that identifies each action by its unique number, responsible department, and person; commitment date for completion; actual completion date; the date to be evaluated for effectiveness; and the date effectiveness is confirmed. A numbering code can be applied to distinguish between corrective actions and preventive actions.

COMMON FINDINGS/NONCONFORMANCES

a. It could not be confirmed through records review that the effectiveness of the corrective actions taken were reviewed and recorded.

b. Review of the CAPA reports did not provide evidence that actions were being performed in a timely manner based on the potential risk of the situation.

c. Through the review of a sampling of CAPA programs, it was noted that a significant number (at least 50%) of corrective actions were not being completed on time.

d. The corrective action reports did not provide the complete record such as date to be completed and root cause analysis.

e. It was stated that corrective action reports were monitored for timeliness; however, review of a sampling of completed reports indicated that actual completion dates were not recorded.

f. Review of a sampling of management review minutes did not provide evidence that the CAPA programs were being evaluated during these meetings as required by procedure MR-02 Rev 3.

g. Records did not provide evidence that product, process, and system issues were being addressed through the organization's CAPA program. Hundred percent of CAPAs from the previous years were from external audits.

h. It could not be confirmed through the review of records that corrections were evaluated for trends and that these trends were addressed through the formalized CAPA program.

REFERENCES

Link, E. 2008. *An Audit of the System, Not of the People: An ISO 22000:2005 Pocket Guide for Every Employee*, Quality Pursuit, Inc., Fairport, NY.

Newslow, D.L. 2001. *The ISO 9000 Quality System: Applications in Food and Technology*, Wiley-Interscience, New York.

Okes, D. 2009. *Root Cause Analysis: The Core of Problem Solving and Corrective Action*, ASQ, Milwaukee, WI.

Peach, R. 1997. *The ISO 9000 Handbook*, 3rd edn., Irwin, Homewood, IL.

---

*Additional Food Safety Management System References (not all inclusive)

FSSC 22000:2010

   ISO 22000:2005 Sections 7.10, 8.5

SQF 7th Edition: Section 2.5.5

BRC Issue 7: Section 3.7

IFS Version 6: Section 5.11

*Note: Always review the standard or visit the website for your chosen standard to ensure that all related requirements are identified and addressed.

---

5 Internal Audit

The internal audit program is one of the key elements identified as the *heart* of an effective management system, no matter which standard is chosen. It is designed to focus on confirming effectiveness and fine tuning the system. A well-defined and effective audit program is essential to the growth and maturity of the management system. It is complex, with many required *inputs* that are critical to its effectiveness and success. In today's world of management systems, customer audits, third-party audits, and tougher regulations, many ask for assistance in not only setting up an effective internal audit program, but also understanding the difference between an effective program and just doing inspections. Simply put, having effective and value-added internal audits is very important for a sustainable, compliant, and value-added management system.

The chapters for management responsibility (Chapter 3) and corrective action/ preventive action (CAPA) (Chapter 4) discuss the three-legged stool (Chapter 1). The internal audit process is the third leg of that stool. An effective program is critical to the overall effectiveness of the system. Each leg of the stool must be strong and effective. Third-party auditors are trained to evaluate based on evidence confirming that the internal audit program is suitable and effective for the management system. If the auditors are not, then the system's overall compliance status may be affected.

Associates and management must understand that these audits are not meant to find every possible noncompliance, but to provide a snapshot picture through independent eyes. It is essential that management and associates maintain their areas to be compliant, addressing nonconformance issues as they arise and not waiting for the auditor to find them.

It is equally important that the audit be welcomed as a positive tool that aids in the identification of both existing and potential issues. When issues from an audit are addressed, it assists a company's overall status and opportunities for improvement. The old saying, "one can't see the forest for the trees," holds very true when applied to the management of individual areas within the system. Many times, those responsible are just too close to the situation and can't see what independent eyes can see. The audit is only as effective as the auditor and auditee combined. Top management of the area being audited must support the program and communicate through its team that the audit is fact finding, *not fault finding*.

Establishment and maintenance of a thorough, yet efficient internal audit program is the single most valuable tool that the organization has to help make sure that the management system is in compliance. These audits pinpoint areas that have fallen or could eventually fall out of compliance. They provide the framework, via the corrective action program, to support continual compliance. (*Tim Sonntag, VP Quality & Technical, Wixon Inc.*)

The internal audit program was one of our biggest challenges to manage. It was even more time and resource consuming than writing our procedures! We did not have the luxury of having dedicated auditors. We had to schedule employees' time away from their normal duties to perform internal audits. The way we managed the problem of resources was to get several people from each department trained as internal auditors. This allowed us to rotate auditors throughout the year based on the audit schedule. The benefits of this approach were many. First, everyone could find the time to perform an occasional audit. In fact, we usually looked forward to the change of pace from our normal duties. Secondly, auditors were assigned to audit a different department from the one in which they worked. The advantage to this was the auditor usually was not familiar with the procedures in the department being audited. They did not have any biases or could not make any assumptions about the procedures they audited; they had to be convinced during the audit, through evidence, that the procedures were being followed. Another benefit was that the process of having to explain what you do to someone who does not necessarily understand your procedures. The auditor may be challenging the *usual* method, finding errors or gaps that allow for improvements to be made. It's amazing what an unfamiliar pair of eyes can find! This is really the value of the internal audit process. It is continuous improvement being driven by continuous review of the system. When every finding and nonconformance from an internal audit is logged into the corrective/preventive action process, the system continues to grow and improve. (*Anonymous*) (Newslow 2001, p. 165)

When asking industry professionals if certification was worth it, many times the answer includes a specific element or program that has especially added value to the organization. The following *anonymous* contributor explains how the internal audit program has improved his organization:

Achieving certification was positively worth the effort! The certification as an entity was not the benefit; the plus was that the internal audits mandated by the system forced people into a number of good practices, which they may have otherwise let slip. (*Anonymous*) (Newslow 2001, pp. 165–166)

KEY FACTORS OF AN EFFECTIVE INTERNAL AUDIT PROGRAM

There are so many key factors that come together in an effective internal audit program. These key factors are critical to an effective audit process, and each in itself can be a challenge to maintain at the effective level that adds the best value overall to the organization:

- Top management's commitment to the entire program
- Management's role and responsibility clearly defined
- Internal audit procedures that clearly define the requirements
- A responsible individual assigned to manage the audit program, which includes scheduling, performing, and monitoring of the audit function
- Sufficient resources available and applied to effectively perform all audit activities, including providing adequate time for auditors to complete their audit responsibilities
- Audits planned that include each element and process area in the scope of the management system

- Audits scheduled at defined frequencies based on the importance and potential risk to the system and to the specific element
- Audits performed verifying the effectiveness of the management system
- An audit plan defining the scope and providing guidance to the audit team
- Effective performance of the audit
- Verification and validation of the audit program
- On-time performance of audits
- Audits performed by individuals not having direct responsibility for the area or function being audited
- Auditors not auditing their own work
- Auditor competency (skills, completion of the training requirements for the defined requirements for the internal audit program)
- A familiarity with the requirements of the standard of choice
- Commitment from responsible management to address findings in a timely manner
- Corrective actions tracked for timeliness of the actions taken
- Audit activities and results evaluated at the management review meetings with required actions implemented and resources applied, as appropriate
- Records identified and maintained to demonstrate compliance to defined requirements for both the audit program and for the record program

TOP MANAGEMENT SUPPORT

The audit function must be supported from top management through all levels of the operation. Management must clearly communicate that the internal audit is a positive activity and not a fault-finding exercise. Associates must be encouraged to use the results of the audit findings as areas of opportunities that, when corrected, will help strengthen the entire system. Audit activities and findings must not be used as a means to assess blame.

MANAGEMENT COMMITMENT LINKS TO PROVIDING THE RESOURCES

The audit program must have adequate resources. *Resources* may be defined as a sufficient number of trained auditors provided with sufficient time to complete their audit responsibilities. The resource of time and availability must also be provided to the auditee. This must not be just a paper promise. The audit manager, audit team, and the auditee must be provided adequate time and support for the effective performance of all audit activities. An accurate accountability and status of the audit program must be reported to top management at the management review meetings. This review must be conducted in a positive manner defining resource allocation and other appropriate assistance to ensure the effectiveness of audit activities.

In an effort to display management system commitment to all associates, the entire management staff of our company became internal auditors. Even the president of the company participated as an internal auditor. The results were substantial. By completing internal audits, the management group forfeited the relative isolation of the offices and was out on the "front lines" to see exactly what was going on. When opportunities were discovered, the managers were in a position to accommodate a quick resolution. The associates, seeing the manager on the floor, got an idea of the importance of the management system at the top of the organization. (*Russ Marchiando, Packaging Plant Manager, Wixon Inc.*)

Top management must assign a competent person as the audit manager or coordinator with the authority and responsibility to manage audit activities. This person is responsible to report an accurate accountability and status of the audit program to top management at the management review meetings. The discussion of the audit program at these meetings must include, at a minimum, a summary of on-time performance, any trends in the findings, resource issues for both auditor and auditee, and timeliness of the corrective actions taken.

AUDIT PROGRAM OR SCHEDULE

Requirements for the creation and maintenance of the audit program or schedule must be defined. The schedule must ensure that all system elements and operational processes are audited at specific frequencies based on the importance of and risk to the area. Some programs split the system into individual audit sections; others audit the complete system each time. Whichever option is chosen will depend on the organization's size, resources, and management system scope:

> It is important to realize that, like the management system as a whole, the internal audit program must evolve as the management system matures. The primary focus of the audits, to a newly certified organization, will likely be on compliance to procedures and work instructions. This will certainly help build the discipline necessary to maintain a compliant management system. However, as the system matures, the internal audit focus should evolve toward continuous improvement, moving from an element focus to more of a process focus. This type of audit focus is, to some extent, easier to maintain. (*Tim Sonntag, Vice President of Quality and Technical, Wixon Inc.*)

Ideally, audits of specific processes should be a combination of both management system elements and specific requirements (procedures, customer requirements, etc.) related to the area being audited. For example, multiple elements may apply to a specific process area, whereas an audit of the document control program or element focuses more on specific document control activities. To summarize, in addition to being evaluated as applicable in process areas, system-specific elements, at a minimum, include the following: management responsibility, management review, communication, document control, calibration, control of nonconforming material and product, CAPA, record control, internal audits, and training.

> The approach to performing audits from an element or process standpoint is a very good one. We began doing this approximately one year after we were certified. This was a fundamentally different auditing style from what our auditors were used to. It forced auditors to sharpen their skills and to take a "big picture" approach to the audit process.

They now were asked to manage several elements at a time and form their questions to address these rather than focus on one element of the system. In the long run, they became better auditors because of this. (*Russ Marchiando, Packaging Plant Manager, Wixon Inc.*)

It is strongly recommended that the schedule be created and maintained in a manner that tracks each audit. This can be accomplished through a matrix that identifies each area to be audited along with the specific assigned time periods for performing the audit. This matrix can also provide the record demonstrating that each element of the chosen standard has been audited (Figure 5.1).

AUDIT FREQUENCY

It is recommended that audits be performed at a minimum of once per quarter. Effective systems can be sustained on less frequent audits; however, it is much more advantageous to the system to perform smaller scope audits more frequently than to perform one large audit, annually or semiannually. A large annual audit may become burdensome in time constraints for performance, report writing, and responding to audit findings. Also, audits with a smaller scope can be extensions of each other. This provides an excellent opportunity for auditees and auditors to become more comfortable with the process. It has also proven to be a way for audit findings to contribute to continuous improvement.

The individual audit frequencies must be based on the importance and risk of the audit scope, element, or process. When identifying the time periods, be careful not to create a schedule with too many constraints. If a specific day is scheduled, then the audit must be done on that day. It is much more practical to focus on a specific month or quarter that provides flexibility in the schedule.

Scheduling audits quarterly provides ample opportunity to identify a time frame convenient for both the auditor and the department to be audited. If this audit is unable to be performed, then the responsible auditor must notify the audit manager, stating the reason for the delay. At the discretion of the audit manager, based on the importance and the risk of the area, the audit would either be rescheduled or assigned to a different audit team. These requirements must be defined in the internal audit procedure, which must also define the acceptable time frame to constitute an *on-time* performance of the audit.

PERFORMING *ON-TIME* AUDITS

Performing *on-time* audits must be taken seriously with every effort made to perform the audits as scheduled. The audit schedule must be maintained as a living document coded to identify performed, in-process, completed, and rescheduled audits. The code may also be used to identify the audit team. Requirements for rescheduling audits must be defined and identified as such on the schedule. It is essential to the effectiveness of the system to be able to track audit performance. Develop a realistic schedule that does not burden the program but contributes to the effectiveness of the audit program and the management system, adding overall

AUDIT SCHEDULE

Internal Audit elements/process areas			1st Qtr			2nd Qtr			3rd Qtr			4th Qtr			LEGEND	
Audit	Process Owner	Jan	Feb	Mar	Apr	May	June	July	Aug	Sept	Oct	Nov	Dec		Audit Scheduled	
Document Control & Record Keeping															Audit Completed	
1 ISO 22000 4.2																
Management Responsibility - ISO 22000 4.1,															Audit Rescheduled	
2 5.1, 5.2, 5.3, 5.4, 5.5, 5.8, 6.1																
Communication ISO 22000 5.6 & Product *Information/Consumer Awareness* PAS 17.0 (including State, Regulatory, & Customer Requirements, Employee Communication, Food																
3 Safety Team Communication															Rescheduled completed	
Emergency Preparedness & Response																
4 ISO 22000 5.7															Audit Not Scheduled	
5 *Human Resources* ISO 22000 6.2																
5a *HACCP Programs* - ISO 22000 7.1, 7.3, 7.4, 7.5, 7.6, 7.7																
5b *HACCP Programs* - ISO 22000 7.1, 7.3, 7.4, 7.5, 7.6, 7.7																
5c *HACCP Programs* - ISO 22000 7.1, 7.3, 7.4, 7.5, 7.6, 7.7																
5d *HACCP Programs* - ISO 22000 7.1, 7.3, 7.4, 7.5, 7.6, 7.7																
5e 7.4, 7.5, 7.6, 7.7																

AUDITORS	AREA

External Audits		Jan	Feb	Mar	Apr	May	Jun	Jul	Aug	Sep	Oct	Nov	Dec

Monthly Internal Audits: GMP/Food Safety	Jan	Feb	Mar	Apr	May	Jun	Jul	Aug	Sep	Oct	Nov	Dec
Receiving												
GMP												
Warehouse												
Laboratory												

FIGURE 5.1   Audit schedule.

value to the entire organization. Remember, the audit program is defined by the organization. Build in some flexibility. Be prepared to revise and grow as the management system matures.

EVIDENCE OF AUDIT PLANNING

The audit schedule should extend for a 12-month period. The thought patterns (the importance of and risk to the particular area) for determining frequencies for specific elements and processes may change through the course of a 12-month period. For example, in one system, the retired purchasing manager was replaced with an associate from outside the company. Although an audit of the purchasing department was not scheduled for another 4 months, the audit manager made the decision to add an audit of the area 2 months after the change to provide an opportunity to confirm that purchasing requirements had been effectively maintained through the transition.

Auditors must be independent of the area being audited. *Independent* is defined as not having direct responsibility for the areas being audited. Auditors must not audit their own work! This is important in that it provides each area an opportunity to be evaluated by a *fresh set of eyes.*

AUDIT TEAM

The audit team should consist of a diverse group of associates representing all levels of the operation. Ask for volunteers from the work force. Being part of the audit team is an excellent opportunity for an associate to not only learn about other aspects of the operation, but also contribute ideas and add extra motivation to the system. In addition, ideas and knowledge may be brought back to be applied within their own departments, which will strengthen their team. The audit manager or coordinator should hold periodic team meetings, monthly or quarterly, to discuss findings, audit program activities, refresh techniques, and receive feedback from the auditors. Many organizations plan quarterly pizza lunches for these team meetings. It is critical to remember that resource availability for both the auditors and the auditees is a must to an effective management system:

> We have had difficulty with auditors to find the opportunities in their schedules to allow time to audit. Often times when they do find time to audit, the auditors become a little overzealous and want to identify all of the company's shortcomings in one day. This inevitably results in a large amount of paperwork and a feeling of unpleasantness and distrust with the audit. This problem seems to be dissipating as the system matures and as we evolve to more frequent audits with smaller scopes. (*Russ Marchiando, Packaging Plant Manager, Wixon Inc.*)

An opportunity may arise to have the system audited by an external auditor such as an associate from a sister plant or a consultant hired to evaluate the system for compliance and improvement opportunities. These audits are invaluable to the total audit program by presenting a new viewpoint from an independent source. In order to include these audits in the internal audit program, define in the internal audit procedures or work instructions that these types of audits are acceptable as long as the external auditor or audit team meets the organization's defined auditor training criteria. The auditor/audit team must provide a record demonstrating this compliance. These audits must be identified on the audit schedule.

PLANNING THE AUDIT

The scope for each audit must be clearly defined and communicated between the audit manager, the audit team, and the auditee (the group to be audited). The planned audit basically defines the starting point, the evaluation criteria, and the projected end point for the audit. For example, an audit of a blending department would identify the area procedures, work instructions, and related ISO elements, including training, calibration, record control, and document control. When creating the scope, review the historical information relating to the area to be audited such as previous audit reports, CAPA, customer complaints, and any other information related to the element or process to be audited. The audit manager can assist with providing this information.

AUDIT CHECKLIST

Audit preparation activities should include the creation of an audit checklist. The checklist is an important tool to the audit function. It promotes a thorough, effective, and uniform audit. Generic checklists are available for almost every aspect of any audit. These are good for reference purposes; however, the audit team should prepare a checklist that focuses on the specific scope of their specific audit assignment. The checklist acts as the audit *to do list*. An audit checklist can serve many purposes such as the following:

- Clearly defines the audit objectives
- Provides guidance for keeping the audit on track through preplanned questions
- Aids auditors in the review of related documentation
- Defines the sample
- Provides an excellent training tool for new auditors
- Becomes an important part of the audit record

When preparing the checklist, it is recommended that the following be reviewed:

- Requirements of the food safety management system of choice
- Related top-level manual sections that apply to the scope of the audit
- Food safety and quality policy statement; measurable objectives
- Related procedures, work instructions, forms, external documents, etc.
- Any other sources of information such as previous audits, regulations, CAPAs, any known food safety and quality problems
- Management priorities and customer requirements

- Company information (brochures, newsletters, etc.)
- The auditor's background and experiences, which may lead to specific questions and inquiries

Keep in mind that the checklist may merely be a handwritten list of questions accompanied by an uncontrolled version of area/process procedures and documentation.

Russ Marchiando, Packaging Plant Manager, Wixon Inc., offered some interesting advice on auditing that he learned from a third-party auditor. (The third-party auditor is generally the term used to refer to the auditor working for a registrar.)

Our business is very complex, with many different things going on at any one time. As a result, our management system contains many different documents outlining each process. Sometimes our auditors feel a little overwhelmed by reading through a considerable stack of documentation. I've learned something from watching our third party auditor that I've passed on to our auditors. You don't necessarily need to read all of the documentation that is applicable to the situation. You just need to have an idea of what questions you want to ask and let the auditee provide you with the answers and be able to show the appropriate documentation. I'm not saying that you don't need to prepare, but I am saying that the onus should be put on the auditee. They, after all, should know what is contained in their documents. Finding this middle road takes some of the pressure off the auditors.

PERFORMING THE AUDIT

Actual performance of the audit brings together many key points previously discussed, such as the following:

- Management communicating their support to associates at all levels of the operation
- Trained auditors independent of the function to be audited, performing scheduled audits
- Efficient audit planning resulting in effective audit performance
- A positive audit focus that is not an effort to find things wrong and assess blame
- Verification that defined activities are in place and effective

Prior to actually starting the audit, the audit team should make arrangements to meet with the responsible management of the area to be audited. This provides the opportunity to review the scope and any other pertinent information such as associate availability, identification of the person(s) to accompany the audit team (the guide), and an agreement of the time for the closing meeting to discuss the audit findings.

AUDIT GUIDE

A responsible representative, from the area being audited, must accompany each auditor for the full duration of the audit. An audit must never be performed without an area representative accompanying the auditor. It is very important that this person experiences and sees what the auditor sees. This concept may be one that is the most challenged by department management. Keep in mind that the auditor is independent of the area being audited. Many times, if not accompanied by an area representative, then those being audited (auditee) may not truly understand the auditor, and may feel that they don't have to take much time or be concerned with the results from the audit. After all, by design, the auditors are evaluating a process unfamiliar to them. What the auditor sees provides information vital to the overall audit scope. Never leave anything to interpretation. Remember that the audit process is only as effective as the relationship between the auditor and the auditee.

Let's review some examples of various audit situations and their outcomes showing what happens when the area representative accompanies the auditor vs. what happens when they don't accompany the auditor:

> *Example 1*: An auditor asks the packaging operator to demonstrate his verification of the metal detector, which has been identified as a critical control point (CCP) in their process. As a result, the operator has three metal standards. She places one at a time on a bag of product confirming that a bag is rejected. The operator records the results on her form. The auditor states: "It appeared that the system was rejecting the bag after the standard, not the bag with the standard." The operator stated that she did not notice that, but she did run through the process again confirming that in actuality, the line is rejecting the bag prior to the standard, not the bag with the standard.
>
> a. *The auditor is accompanied by an area representative (guide)*: The area representative (guide) saw this and was very concerned stopping the line immediately and calling maintenance. The operator was instructed to follow the procedure for a CCP deviation. The area representative (guide) contacted the department manager making him aware of the serious situation. It was investigated, the product evaluated, and the situation corrected.
>
> b. *The auditor is not accompanied by an area representative (guide)*: The operator stated that she would notify her supervisor because she would get in trouble if she stopped the machine. The operator stated that a bag was rejected (not the correct bag), so the metal detector was working. The line kept running and the auditor made a note of the finding and moved on. The auditor did notify his audit program manager who stated that he would tell the department manager. At the closing meeting, the auditor read his finding. The department manager stated that he sent maintenance to check it out, but maintenance could not find anything wrong. *So, what do you think happens next? How many serious situations could evolve from this?*

> *Example 2*: An auditor is doing a good manufacturing practice (GMP) audit of a warehouse. The scope of this audit is to review the shipping process. The auditor interviews four different associates, asking each of them what the process is for loading the trucks and for knowing what needs to be shipped. The auditees were able to explain the warehouse software system to ensure the product that is not approved or that is on hold is not

pulled for shipments. In the process of auditing, the auditor noted an empty transport backed up to the warehouse. He asked what this transport was for; the auditee stated that it was waiting to be loaded. The auditor noted that the transport had two holes (daylight coming through) on the bottom of the transport and that there was a strange odor coming from inside the transport. The auditor questioned the inspection process, and the auditee stated that it would be inspected by the forklift driver assigned to load it and that the result of this inspection would be documented on the bill of lading. The auditor asked to see a record of the load that the auditee was working on, and he provided a copy of the bill of lading with the inspection stamp present and signed as acceptable for the current load.

a. *The auditor is accompanied by an area representative (guide)*: As a result of this observation, the area representative (guide) also noted the holes in the transport and the foul odor. The representative immediately contacted the area lead person who stated that they had turned down the transport the day before and that the transport had been cleaned. The transport company stated that holes on the bottom were not of concern. The representative asked that the transport be returned to the supplier stating that it did not meet their requirements.

b. *The auditor is not accompanied by an area representative (guide)*: The auditor stated that, as defined in the warehouse Standard Operations Procedure, this transport was not in compliance and asked if it would be used. The auditee again stated that it was up to the person loading to make that decision, but that he would leave a note in the office for the lead person to check it out. The auditor moved on. However, later in the day, he returned to review the paperwork and found a copy of the bill of lading for the load that went into the questionable transport. It was stamped with an acceptable evaluation of condition, loaded, and released. When noted in the report, the department manager's only comment was that since the transport was gone, he had no way to follow up on this issue and felt that a corrective action identifying this would be useless. *Wow—not even a true assessment of the situation* or *root cause or solution suggested for the future!*

Example 3: The audit team is auditing the receiving area. The scope of this audit is to review the incoming inspection process to confirm if all required evaluations are being completed. The auditee demonstrated positive knowledge of the required activities, including being able to access the document control software for the most current version of the specifications and requirements for acceptance. The auditor asked the associate if he used any calibrated equipment. He stated that his company did have a calibrated thermometer. The auditor asked the status of this calibration and the auditee stated she did not know, but thought that the quality department checked it once a week. The auditor also asked how she was trained and if there were records maintained on the training. The auditee stated that she learned from the previous person and that she did not know of any records.

a. *The auditor is accompanied by an area representative (guide)*: The guide stated that the quality department did perform the calibration process and recommended that they go there to find the record of this thermometer. The guide noted that they would need to review the calibration Standard Operating Procedures to confirm that defined requirements for the identification of the calibration status were met. The guide also confirmed that records of training were kept in the supervisor's office, leading the auditor there to confirm this, prior to visiting the quality lab to complete the calibration trail. The auditor was able to review all related information to identify findings and observations. Results were documented and reported at the closing meeting. The audit was completed on time.

b. *The auditor is not accompanied by an area representative (guide)*: The auditor was unaware of required trails to follow, so she completed the audit writing nonconformances for calibration (one for the thermometer not being calibrated and one for not having the status clearly identified) and training (one for lack of training related to calibration and one for no training records confirming that the operator had been trained). The auditor presented findings to the department manager who was quick to argue each finding. He stated that these were worthless and did not add any value because he was sure that they had all the required information. The auditor just didn't know where to look or whom to ask.

These are only three simple examples. There are examples like this for just about every audit activity comparing the effectiveness with and without a department representative (guide). Having a guide greatly increases the effectiveness of the audit. As stated previously, it is essential that the department representative sees what the auditor sees. Additional benefits include being able to explain the process to the auditor; answer questions that relate outside of the responsibility of the auditee; provide guidance on trails that may answer the questions; and, where possible, provide evidence of compliance.

Another resource issue is the availability of the department management for an opening or preaudit planning meeting. It is essential that prior to actually beginning the interview process that the audit team makes arrangements to meet with the responsible management of the area to be audited. This provides the opportunity to review the scope and any other pertinent information. Examples of pertinent information would be associate availability, identification of the person(s) who will accompany the auditor (department representative or guide), and agreement of a time for the closing meeting to discuss the audit findings.

PROCESS APPROACH

A process is defined as a "set of interrelated activities that transform inputs into outputs." ISO makes extensive use of this definition. First, a process model is used to define the relation of the various elements of a ... management system. The process model is also used to further define the relationship between inputs, value-added activities, procedures,

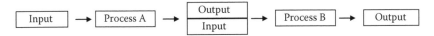

FIGURE 5.2 Process relationship. (From Surak, J., International Organization for Standardization ISO 9000 and related standards, in Schmidt, R. and Rodrick, G., eds., *The Food Safety Handbook*, Wiley, New York, p. 707, 2001.)

measurements, and outputs. The inputs include materials (ingredients or components), machines, time and finances. Procedures may or may not be documented (Chapter 6). Monitoring and measurement opportunities are used to provide feedback and feed-forward information on the process and product quality (Surak 2001).

Figure 5.2 describes the relationship between process inputs and outputs.

PROCESS AUDIT

Process audits vs. element-specific audits were mentioned briefly previously in relation to scheduling. In scheduling the audit and maintaining the audit schedule or program, the combination of process or area audits and specific element audits should be discussed.

Process audits have proven to be much more efficient and value-added than element-specific audits, especially in a mature system. Think of the system as having processes within a process. An organization is defined as a collection of processes, all of which work together to transform *inputs* into *outputs*. The key point to remember is that all processes have inputs that must be met in order for the process to successfully meet its outputs (goals and objectives). The output of every process is the input to another process. The output must meet its objectives and be successful for it to be a successful input to another process. When preparing for a process audit, it is best to divide the inputs into six different focused parameters: people, equipment, environment, measures, methods, and materials. Russell (2003) when applying these parameters, ask the following question. This question can and does lead to a successful gathering of information and the identification of related trails that may not have surfaced otherwise:

> What are the outputs of the process or area of responsibilities? In other words, what are the objectives, and what measures are being used to know if this has been accomplished successfully? Once the goals/objectives and an idea of the measure of success are determined, then ask about related inputs. Figure 5.3 provides a sampling of *process audit* questions.

Process auditing is an excellent audit tool. It is recommended that every organization move into the process audit approach as its management system matures. The parameters, such as those mentioned earlier, are just one of many methods applicable for learning process auditing.

There are many other good ways to approach process auditing. One very popular tool that is good for internal auditors is the *turtle diagram*. The use of the *turtle diagram*

a. Are *people* involved?
b. Is specific *equipment* required to complete the process?
c. Are there any *environmental* considerations?
d. Are *materials* needed?
e. Are there *measures* or standards?
f. Is a specific *method* required?

FIGURE 5.3 Process audit questions.

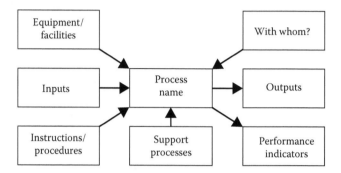

FIGURE 5.4 A basic example of a *turtle diagram*. (From Micklewright, M., Auditors, turtle diagrams and waste, *Quality Digest*, 1/8/2007. http://www.qualitydigest.com/inside/quality-insider-article/auditors-turtle-diagrams-and-waste.html.)

creates a *picture* of the process which includes related system elements along with related process inputs. Figure 5.4 is an example of the turtle diagram that is then completed during the audit preparation stage. It can also be maintained as a reference document for future audits although content may vary depending on the scope of the audit.

As with every tool, there are advantages and disadvantages; thus, it is recommended that the audit program owner review and learn its application, and then share with the audit team to determine its role in the organization's audit program.

It is recommended that management also review other sources such as textbooks and workshops focusing specifically on process auditing. Once the process auditing concept and the best way for the organization to apply it to its internal audit program are understood, then it is recommended that training and practice audits be performed. It takes practice to get used to the concept, but this type of approach used long term, is invaluable to the management system overall.

Keep in mind that the questions that may relate best to a specific area, process, or element are going to vary depending on the system itself, the organization, the auditors, and the scope of a particular audit. It is important that the auditors, in their preparation stage, do not worry about *getting the questions right* or *missing something critical*, but focus on preparation, experience, and common sense. Once this confidence or comfort level is achieved, then effective audit activities fall into place for the audit. Each audit is different. It is the combination of all of the audits that result in an effective and value-added internal audit program.

ADDITIONAL SAMPLE QUESTIONS

The following are sample questions that may be applied to various processes and elements:

1. Are there procedures and work instructions present to comply with the chosen standard and the management system?
2. Are the procedures and work instructions under document control? Review the area and identify/question any posted documents to confirm whether they are controlled or uncontrolled. If controlled, are they the current version? Keep in mind that any posted document that provides information on how to do something must be a controlled document. For example, if removed, the outcome of the task or process would be different and not completed as defined. Posted controlled documents must be identified as to their location, so that when the document is revised, the process knows to replace the posted document with the revised version. Review Chapter 6 for more information on document control.
3. Do the procedures and work instructions reflect what is actually occurring?
4. Are the associates who are responsible for the defined requirements familiar with the procedures and where they are located?
5. Are the associates familiar with the food safety policy, customer requirements, objectives, and goals of the organization and their role in successfully achieving them? (Chapter 3)
6. Is there objective evidence (records demonstrating compliance) that proves that what is stated in the procedures or work instructions has actually occurred? (Chapter 7)
7. Do identified nonconformances follow the related procedures?
8. Are records identified and maintained according to the record control program? (Chapter 7)
9. Are any external documents used or referenced? Are they defined and maintained under document control? (Chapter 6)

Although easier said than done, it is important not to worry about what questions to ask and even more important to be careful with getting in the habit of asking *canned* questions. These may work at first, but become repetitive very quickly. It is more important to focus on asking questions with such phrases as *Show me* and *How do you know*. Whenever possible, do not ask questions that can be responded to by a simple *yes* or *no*. Investigative questions may be used to clarify the information and also encourages auditees to continue talking. Questions should have a distinct focus. Remember that the ability to discover relevant information is directly dependent on the ability to ask the right questions.

ISO 19011:2011

ISO 19011:2011: Guidelines for Auditing Management Systems is an ISO guideline developed to provide guidance when auditing management systems. It includes principles of auditing, managing the audit program, and basic audit techniques for auditing management system audits. This standard also defines its recommendations for the audit program owner and audit team member competencies. It is a good document providing guidance for the audit of any management system whether being an external or internal audit. It is recommended that the audit program owner obtain a copy of this ISO standard and use it as additional guidance in managing the audit program. It is a guideline and not an auditable standard, but does contain excellent information that should be reviewed and applied to the organization's program as identified appropriate by the audit program owner.

AUDITORS MUST BE OBJECTIVE

The auditor must remain objective and remember to collect the essential facts. The audit function is not a fault-finding exercise. Facts must be listed clearly. Conclusions must be based on facts. The audit is a sample, and sampling activities do have limitations. Auditors must keep a positive attitude, creating an atmosphere that promotes good communication. Thank the auditees for their assistance and time regardless of the outcome. Auditees must never feel threatened by the auditor. An auditor must treat the auditee as he/she would want to be treated.

Audit activities must focus on the scope, with the auditors making clear audit notes as to the questions asked, the location of the related materials, any observations and concerns. However, many times an audit trail must be investigated after it surfaces during an audit. It depends on the situation and the discretion of the audit team whether time is spent investigating the audit trails during the current audit or during a subsequent audit. For example, the auditors may identify a potential training issue in the receiving area. Depending on the situation, they may follow that trail during this audit or provide the necessary information so that it can be investigated during a subsequent audit.

These situations are, of course, unforeseen in the planning stage, but must be given appropriate attention during the audit depending on the potential risk of the situation. The audit report must note such findings and the ultimate decision. By noting this, the record provides subsequent auditors the opportunity to follow up on an area originally planned, but preempted in order to address unplanned activities essential to the audit outcome.

Auditee comments, opinions, and hearsay information may be noted informally and investigated, but only included in the actual report if the auditor can substantiate by fact. Do not mention names or include he said/she said comments. Focus on the facts and keep in mind that assessing blame on the individual responsible for the process is almost never the true root cause of the situation. Management responsible for addressing audit findings must be cautious not to assess blame or identify individuals as the root cause of a finding.

AUDIT NOTES

Audit notes may not only be part of the audit record, but also are likely to be used in the planning stage of the next audit. Related to audit notes, the following applies:

- Record all relevant facts, seen and heard.
- Be patient developing the technique.

- Record sufficient facts to make an informed judgment.
- Clearly identify/reference pertinent documents, batch numbers, order numbers, etc.
- Take the time to ensure notes are legible and understandable.
- Identify recommended relevant areas/trails for the next audit.

AUDIT FINDINGS

Some organizations insist that auditors grade their findings. Grades may include *major, minor,* and *observations.* This can be effective, but be sure that criteria for grading is clearly defined and understood. Unfortunately, this type of grading system frequently becomes subjective, causing confusion within the system. Most systems, as they mature, realize that it is more effective not to grade the finding, but to apply the time frame commitment for completion based on the potential risk as the *priority* guideline.

A nonconformance or finding is defined as a situation that violates a requirement of the standard or a defined requirement of the system. Examples of frequent findings or nonconformances are included at the end of each chapter related to the element or process discussed. An observation may be defined as either a situation that, if not addressed, would likely result in a nonconformance. An example of an observation would be the following:

> Although audits are now being performed on time, the recent decrease in the number of auditors may result in difficulty scheduling and adhering to on-time performance during subsequent audit periods.

It is very important to the strength and effectiveness of the audit program and the system overall that the responsible area managers investigate both findings and observations. An observation may be a warning of a potential problem. Requirements for an effective CAPA program are discussed in Chapter 4.

CLOSING MEETING

There should not be any surprises at the closing meeting or in the audit report. Auditor's concerns must be discussed on the spot, in a positive manner, with the auditee and the area guide. Many times answers can be provided; at other times, the situation may not be as it was initially observed. The guide or area representative can provide the information required. As mentioned previously, it is important that management representing the scope of the audit be present at the closing meeting. This is a positive exhibit of commitment.

The closing meeting must be held as close to the completion of the audit as possible. The purpose of this meeting is to discuss the findings with the responsible area management and as many area representatives as possible. Ideally it is best to submit the written findings at the closing meeting; however, depending on the particular situation, this may not be possible. If this is not possible, the written report must be completed as soon as possible after the completion of the audit and no longer than one week from the closing meeting. Taking any longer than this decreases the focus and the effectiveness of the audit. The audit findings are areas of opportunity to strengthen the system. As stated many times in this chapter, findings must be based on fact and not assessing blame. Do not mention names in the report. Findings must be reported and recorded constructively in a predetermined format. Responsible area management must ensure that root cause analysis and corrective actions are performed in a timely manner based on the potential risk of the finding.

NOTES FOR THE AUDITEE

Being an auditee may present some challenges. The following guidelines apply to the auditee:

- Always be prepared and take the time to answer a question. If the auditee doesn't know the answer to a question, he/she should say so. The auditee should not make up answers or try to give the auditor what the auditee thinks the auditor wants to hear. Be honest; don't try to adjust the answer. Experienced auditors will detect dishonesty.
- Be careful not to answer questions that are not asked; however, be helpful.
- Be confident; after all, no one knows the job better than the auditee.
- Do not argue with the auditor. There is a difference between arguing and challenging. The auditee may challenge the auditor if the challenge is based on fact, but be careful challenging everything.
- Do not argue just to argue or waste time.
- Know the job, related customer (internal and external) requirements, and his/her responsibilities.
- Know related procedures and work instructions.
- Be familiar with how to access those procedures/work instructions related to his/her area(s) of responsibility.
- Know what records are within his/her responsibilities.
- Know the retention times and retention locations for those records within your responsibility or know where to find the requirements.
- Know your training procedures.
- Know the food safety and quality policy and how it relates to your responsibility. (*Doing the best you possibly can.*)
- Always think about the answer before responding. Be honest.

FOLLOW-UP FOR EFFECTIVENESS

The audit program should include a follow-up evaluation to ensure that corrective actions taken are effective. Many programs actually schedule specific audits at defined frequencies, such as monthly or quarterly, to provide the opportunity for trained auditors to complete this evaluation. Be sure that enough time and evidence are available to confirm effectiveness. For example, if the root cause is identified as insufficient training and the corrective action is to train, then significant evidence must be available to confirm that the training was effective. Confirmation of effectiveness is not merely the completion of the training exercise, but a review of the activity, records, and any other appropriate information necessary to confirm that the issue had been addressed and is currently in compliance.

Some organizations define audit corrective actions as being incorporated and tracked through the CAPA program. In this particular situation, the audit may be considered closed after the report is issued. Although some organizations successfully manage the process with the audit corrective actions handled separately, over time, most systems generally find that it is best to have one CAPA program. Responsible associates find it more effective to manage, track, and monitor through a centralized system. It may actually be best to handle them separately in the beginning, combining the activities as the system matures. This is a management decision. The CAPA program is discussed in more detail in Chapter 4.

AUDITOR TRAINING

The training that the internal auditors receive will likely be directly proportionate to the overall effectiveness of the audit program. Auditors should be trained in the standard of choice per requirements (current version) and also in basic auditor techniques. Auditors must also be trained in the positive attributes of being an auditor. The auditor must understand that his/her role is to act as an independent set of eyes to confirm and reinforce the effectiveness of the management system. An on-site internal auditor training course for the audit team, which combines classroom training and actual audit performance, can be very beneficial. Each auditor's training certificate confirming successful completion of his/her training must be maintained as a record (Chapter 7).

Auditors must focus on a positive performance style. Auditors must always treat auditees as they would want to be treated. Nothing causes the program to fail quicker than to have an auditor that presents himself/herself as an overpowering force. Interpersonal skills of the auditor are as important, if not more important, as the knowledge of the standard. One doesn't have to memorize a dictionary to know how to use it. Using the standard works on the same philosophy. Even professional auditors that use the standard every day do not specifically memorize it, but know how to reference it.

Auditor attributes should include being:

- A good communicator, good listener, good organizer, and exhibit a professional image
- Patient, tolerant, polite, and confident
- Objective, persevering, and have a positive attitude
- Precise, but practical, and practice interpersonal skills
- Nonthreatening
- Nonjudgmental and not opinionated
- Empathetic (not sympathetic)
- Able to validate *tip-offs*

Auditor training must include an emphasis on these attributes and also on the characteristics already discussed earlier in this chapter. An auditor must approach the auditee in a positive, respectable manner. After all, no one knows the process better than the person performing the activity (i.e., the operator). An auditor's focus is to confirm compliance. The success of the audit will be greatly hampered if the auditor acts like a *bull in a china shop*. It must be emphasized that the audit is a fact-finding, not a fault-finding program. The auditor must always treat the auditee with the same respect that he/she would want to be treated.

AUDIT RECORDS

Audit records should, at a minimum, include audit notes, the checklist, a summary of the findings, and the report including observations and noncompliances. This information is invaluable to the planning phase of subsequent audits. Auditor training records may be maintained as part of the internal audit program or with the other system training records. Internal audit records must be maintained in compliance with the record control program (Chapter 7).

EFFECTIVE INTERNAL AUDIT PROGRAM

We took an unusual approach at internal audits and started them very early in the quality system implementation effort. By starting the audits so early in the process, we felt that we could take advantage of a good opportunity to identify potential implementation weaknesses as well as allow our auditors to gain confidence and become seasoned by the time we were ready for certification. (*Russ Marchiando, Packaging Plant Manager, Wixon Inc.*)

Our auditing process was the most useful because it broke down so many barriers and led to enormous improvements throughout the organization. This occurred in ways it was not even designed to do. Besides the obvious benefits of auditing, our process led to a large amount of associate cross-training, team-based action plans, and an increase in intra-department communication and consistency. Our auditing program progressed to auditing for effectiveness and efficiency rather than auditing to meet requirements. (*Anonymous*) (Newslow 2001, p. 180)

After certification, one of the toughest aspects of the system was to keep the auditing function challenging and growing as the system matured. Although the audits overall continued to improve, the auditors found it more and more difficult to find the time to perform the audits as scheduled. Prioritizing and communicating the importance of timely audits, as well as the timely response to audit findings, is an important management review opportunity. (*Anonymous*) (Newslow 2001, p. 180)

The audit program provides a disciplined, proactive approach to system improvement. Audits are an excellent tool for identifying weaknesses before they manifest themselves as costly failures. (*Anonymous*) (Newslow 2001, p. 180)

TEAM APPROACH

This chapter provides a generic overview for establishing and maintaining the audit function of an effective management system, no matter which standard is chosen. It is generic in nature and can be applied to many different processes.

An organization must develop its own program, specifically designed for its needs, backed by top management support and commitment, while focused on team involvement. A baseball pitcher could not pitch a no-hitter without eight players on the field and a full roster of trained and competent players on the bench ready to do what they do best. Each department must take ownership for compliance rather than waiting for an audit to find out that something is wrong. The audit team is meant to be an independent set of eyes that sees what those close to the process miss, not because of neglect or uncaring, but because they are just too close to the situation.

In discussing effective internal audits, there are a few key principles that stand out. The first, of course, is having a trained team that understands the concept of auditing and their role in the process. This team must have a leader that plans the audits, leads the team, and acts as a technical resource with a strong understanding of the organization, its processes, and the focused standards. This leader, known most frequently as the audit program owner or as mentioned previously, the audit manager, must have the resources available to create and manage an audit schedule that is appropriate to the organization. The audit manager must also have the responsibility and authority to report to the top management team (may be a member of the team) on the status and effectiveness of the audit program. Another resource issue is the availability of the department management for an opening or preaudit planning meeting. As stated previously, it is essential that prior to actually beginning the interview process, the audit team make arrangements to meet with the responsible management of the area to be audited. It is recommended that a process map be developed for each identified audit process. From this information, identify all related elements for the process. This can be done in the process map or in a related matrix (Figure 5.1). Once this is done, then divide the total process into four- six *components* to be audited throughout the 12-month period.

Edward Link (2008) in An ISO 22000 Pocket Guide for Every Employee sums up the thoughts on internal auditing very well in the following quotation:

Management's Role—The investment made in the food safety management system (FSMS) will provide significant payback if the system is kept in a healthy state. The internal audit subsystem is the self-check necessary to maintain management system health. The effort, in conjunction with the third party audit conducted by the registrar, virtually assures a highly effective management system, both in the short term and the long term. Recognizing the benefit and supporting the effort with the necessary resources is required management.

Employee's Role—In this area, everyone is involved as an auditee and/or an auditor. As an auditee, be prepared to be totally open with your internal auditors so that the measurement being made on the health of the management system is as accurate, as can be, despite its often subjective nature. Internal auditors should be mindful of their training that was provided to become an internal auditor. If unsure about the exact interpretation of the requirement, seek help from more knowledgeable and experienced users of the standard. Declare yourself ineligible to audit an area where you think there is a possibility that you could not be both fair and thorough. Do everything that you can to make the measurement you are making as accurate as possible. (pp. 99–100).

MANAGEMENT COMMITMENT

It is not a coincidence that the subject matter returns to the fact that top management must be committed. Management commitment must be evident throughout the process, ensuring the provision of required resources for an effective internal audit program. It is critical that top management communicate through its ranks, that the internal audit program is important to the system, and that findings are based on facts, not a tool for discipline. The audit is an audit of the process, not the people.

KEY POINTS

a. Audit schedule must demonstrate flexibility and assignment based on risks, concerns, changes, etc.
b. Each element from each standard should be audited within 3 months of the certification audit. All elements must be audited prior to the registration audit.
c. An audit program must be planned, taking into consideration the importance of the processes and areas to be audited, as well as any updating actions resulting from previous audits.
d. The author recommends that GMP audits be included and tracked through the internal audit schedule.
e. The author recommends larger audit teams, which will provide the opportunity for more frequent audits, but with condensed scopes. Note that it is recommended that the organization train and manage an audit program that includes 20%–25% of the total associate base with representatives from each department. Related to individual audit teams, teams of two auditors have proven to be successful. Having two minds is always better than one, plus having an extra person to follow-up on trails is also valuable, but even without trails, it is much more fun to do these in teams of two sharing the responsibilities and activities.
f. Some thoughts to remember about the audit:
   • The audit is a sampling exercise.
   • Never perform an internal audit without a representative of the area present as a *guide*.
   • The audit is not meant to find everything that is wrong.
   • The audit is a snapshot in time.
   • Findings are based on factual evidence.
   • Focus is on the system.
   • The audit is not a fault-finding exercise.
g. The following are some benefits that have been identified as a result of a successful audit:
   • An increase in system awareness among personnel
   • Enhanced continuous improvement opportunities
   • Objective evidence of the suitability and effectiveness of the management system

- Objective feedback based on facts
- Information for effective allocation of resources
- Early detection of potential problem areas
- Compliance with the standard of choice
- Confirmation that customer requirements are being met

h. The benefits from an extra pair of eyes, independent of the area being audited, are one of the most valuable sources of information available to an effective management system, no matter which standard is chosen.

COMMON FINDINGS/NONCONFORMANCES

a. Records did not confirm that audits were being performed as scheduled.

b. The majority of audits performed were late; an explanation as to why they were late was not available.

c. Training criteria for auditors were either not defined or records were not available to confirm that auditors had met the defined criteria.

d. Records of auditor training did not provide evidence of training in the standard of choice.

e. The audit schedule was not maintained in a manner that provided evidence that audits were planned based on the potential risk of the area.

f. The required frequency for performing audits was not defined.

g. Training criteria for internal auditors did not identify training on the standard of choice as a requirement.

h. There was no evidence that all the elements and processes were being audited within the defined time frame identified on the audit program.

i. Review of the corrective actions issued from the audits did not provide evidence that findings were being reported to the responsible management and addressed in a timely manner based on the potential risk of the finding.

j. Records were not available to confirm that the effectiveness of the actions taken after completion were being monitored and recorded.

k. Audit record requirements were defined, but records were not being maintained as required by the record control program.

l. The organization had not formally implemented its internal audit program.

m. The audit schedule did not provide a revision history, thus it could not be confirmed that audits were scheduled and performed based on the potential risk of the area.

n. The audit schedules were not maintained in compliance with the procedure for document control and for record control.

o. The current audit schedule did not include all the elements of the chosen food safety standard.

p. The current internal audit program did not provide sufficient objective evidence that an effective audit program had been established and was effective and suitable for this system.

REFERENCES

Link, E. 2008. *An Audit of the System, Not of the People: An ISO 22000:2005 Pocket Guide for Every Employee*, Quality Pursuit, Inc., Fairport, NY.

Micklewright, M. 2007. Auditors, turtle diagrams and waste, *Quality Digest*, http://www.qualitydigest.com/inside/quality-insider-article/auditors-turtle-diagrams-and-waste.html, accessed September 5, 2013.

Newslow, D.L. 2001. *The ISO 9000 Quality System: Applications in Food and Technology*, Wiley-Interscience, New York.

Russell, J.P. 2003. *The Process Auditing Technique Guide*, ASQ Quality Press, Milwaukee, WI.

Surak, J. 2001. International Organization for standardization ISO 9000 and related standards. In: *Food Safety Handbook*, Schmidt, R. and Rodrick, G. (eds.), Wiley-Interscience, Hoboken, NJ.

---

*Additional Food Safety Management System References (not all inclusive)

FSSC 22000:2010

    ISO 22000:2005 Section 8.4.1

SQF 7th Edition: Section 2.5.7

BRC Issue 6: Section 3.4

IFS Version 6: Sections 5.1, 5.2

*Note: Always review the standard or visit the website for your chosen standard to ensure that all related requirements are identified and addressed.

---

6 Document Control

Document control and record control are actually two distinctly different processes that overlap each other, causing much confusion when trying to develop the processes in a manner that fits the organization. Let's first discuss *document control.*

One of the most frequent causes of things going wrong is either not having a defined procedure to follow or following an obsolete procedure or outdated specification. One of the most difficult processes to harness is document control. An operation that has been functioning for many years will definitely have many different versions of requirements, specifications, and/or operating procedures that are referenced by associates. Additionally, associates may use operating procedures based on their memories and *how it has always been done.* A system for identifying the documents, including current versions and distribution locations, must be created. This can be accomplished through a formalized purchased document control software program or developed using basic office software programs. The latter will work; however, a formalized program can absolutely save time and provide an efficient method that does not require extra people to manually manage the process. The choice becomes a management resource issue which will be discussed more in detail.

REQUIREMENTS

Victor Muliyil (technical manager for North America Food Safety Services, SGS) and Supreeya Sansawat (global food business manager, SGS) in their white paper Comparing Global Food Safety Initiative (GFSI) Recognized Standards October 2012, provides the following summary of documentation requirements between the approved schemes.

> All schemes require documented procedures to demonstrate conformance with the specified scheme requirements and records to demonstrate the effective control of processes and food safety management. BRC requires that electronic records be backed up, and both the BRC and SQF schemes require that documents are in a language or languages spoken by the organization's staff and that they are sufficiently detailed.

As stated, each food safety standard identifies specific documents required for the system; however, in addition to these specific requirements, common sense and experience must be applied to identify the documents that are necessary to ensure "the effective development, implementation, and updating of the food safety management system" (ISO 22000:2005 4.2.1c). An essential task for management is to ensure that associates understand and are as familiar with the documents as if they had written them themselves. Communication is discussed in more detail in Chapter 3.

Remember the team approach; it is strongly felt that associates should assist with writing and reviewing documents that relate to their areas of responsibility. No one knows and understands the processes better than those doing the job. This not only results in accurate documentation, but also enhances ownership and adds resources to the implementation team. Associates must also understand where to find and how to access the most current versions of the documents required for their areas of responsibility.

REQUIRED DOCUMENTATION

Figure 6.1 provides some examples of documents that are required by various food safety management system standards. Each standard has its specific requirements, but these are very similar between the standards.

Keep in mind that this is only a partial list. Review the standard of choice, and wherever it states that a *documented* procedure is required, this means that the subject matter must be defined in a written procedure. However, procedures can use many different forms such as pictures, videos, and flow diagrams. Flow diagrams are becoming more and more popular because these provide a *picture* of the process. Remember the adage, "A picture is worth a thousand words." Management, in identifying procedures required by the operation, many times goes overboard. There must be a good balance between documented requirements, experience, and training. Requiring a written procedure for a professional second baseman would be ludicrous. This is a learned skill, and stopping to read a procedure before covering the base would be ridiculous. When one gets his driver's license, studying a handbook of rules and regulations is the initial requirement along with demonstrating the driving skills; however, once knowledge of this information and demonstration of the skills is completed, one rarely needs to review the book again except possibly for reference.

WHEN IS A PROCEDURE REQUIRED?

Not every process or activity requires a written procedure. Management must accept that fact. A good rule of thumb is to apply common sense by documenting as little as possible, but as much as necessary. Management or those individuals responsible for identifying which processes require a documented procedure must objectively make this determination. Another good rule of thumb (which was actually learned in

---

- Document control program
- Written statements of food safety policies
- Procedures prescribed by the standard
- Work instructions for operations
- Record control program
- Corrective action and preventive action program
- Internal audit program
- Hazard assessment methods and results
- Control of nonconforming materials and products

FIGURE 6.1   Examples of required documentation.

the early ISO 9001 days) is that documented procedures are required wherever their absence would have an adverse effect on food safety and/or the manufacture of a safe product. This, of course, can and should be applied to the entire management system. If a documented process is required to ensure the safety of the product, then the need for a document must be evaluated and written, if confirmed necessary.

Again, keep in mind that required information may be more user friendly if presented in a flow diagram or video. The range and detail of the procedures depend on the complexity of the process or activity, the methods used, and the skills and training needed by personnel involved in carrying out the activity (ISO 9001:2008 4.2.1). It is very common to find a large number of documents required for a laboratory or quality assurance department, whereas less would be required for operating machinery, which is specifically skill based. Laboratory testing is also skill based; however, with the number of different tests usually required, having reference information is important in performing these tests on a less frequent basis. Procedures, work instructions, work aids, standard operating procedures (SOPs), or whatever documentation is referenced in a specific system must be written in a manner that is user friendly and useful to the system overall. Requirements must be defined in a manner that ensure activities are performed consistently, in compliance, and effective.

In the past, many thought that the answer to every problem was to write a procedure. As time has evolved, it is evident that there are needs for defined requirements where consistency is critical. Experience and training, however, must also play a key role in what is actually written. One of the most common findings through external audits is that while an organization or department is in compliance with the standards, they are not in compliance with their own documented requirements.

Three phrases to remember when addressing the need for and the preparation of documentation:

Phrase 1: "Do not write *War and Peace*." This means do not write more than needed.

Phrase 2: "Do not shoot yourself in the foot" by including too much detail or by defining a *wish list* rather than the actual activities being performed. (Keep in mind, this does not mean to avoid defining specific requirements; just refrain from including too much minute or unnecessary detail.) Define the necessary activities that must be done consistently by everyone without going overboard with extra activities that management might feel would be good to do. In most processes, it doesn't matter if an associate records data with his/her right hand or left hand, but it does matter which form the data is recorded on and what is done if the process is nonconforming.

Phrase 3: "Document as much as necessary but as little as possible."

This leads to another thought on documentation. A written document is not a *wish list* as to what management wants to be done, but a document that reflects what is actually being done. If improvements are possible, make the improvement and reflect this in the document. However, many times management will define what they want done, knowing that the process may not be able to do this. For example, clean-in-place requirements may be defined at 185°F when the operation can only achieve 170°F. Thus, the operation is noncompliant and stays noncompliant when the document could have identified the achievable temperature and any related parameters (i.e., cleaner concentration, time, velocity) being applied to ensure effective cleaning.

DOCUMENT FORMAT

Many times, the question is asked, "What format is required?" The format is entirely up to the organization, but keep in mind that the key is what the document says. Management should identify a document control team who evaluates the current needs of the organization, including any existing documentation, and identifies the most user-friendly format available for the associate team. It is possible to purchase templates, buy software that includes templates, or hire someone to write the documents; however, caution is critical. Remember, this is the organization's system, and the resulting documents must truly reflect the operation and be user friendly for the associate team. Even large companies that have multiple locations find rather quickly that the documents must be unique for each operation. It is very important that assistance from an outside source, whether a company associate or an outside consultant, function as a coach assisting the team in creating the documents. If someone writes the documents, then leaves or retires, the system has their documents, not the organization's. This type of issue will surface quickly, especially during an external audit when the auditee doesn't understand the document because he or she was not involved in the process of creating it, and the document may not reflect what is actually done.

When writing procedures, make every possible effort to avoid writing the same requirement in multiple documents; state it once and then reference (or hyperlink if documents are maintained electronically) to this document if the information is required. For example, there may be a procedure on sampling requirements. The receiving document that defines the requirements for the receiving technician would state, "Sampling is performed as defined in the procedure for 'Sampling and Testing' (add document number)." This is very important because as the system matures, certain requirements may be revised, and if that requirement is stated in multiple documents, some of those documents may be overlooked, creating potentially serious contradictions within the requirements.

PRINTED CONTROLLED DOCUMENTS

The goal to go paperless is becoming more and more popular. It is much more efficient to have current controlled documents accessible through an electronic system. In years past, many stressed document coordinators spent most of their time printing, stamping, distributing, and updating hard copy controlled document manuals. Some organizations may have had as many as 20 manuals throughout the facility. Electronic control is recommended wherever possible, but associates must have *ready* access to the documents required for their areas of responsibility. In some environments, it is not practical to have a computer station for each operational area, thus the need for some hard copy controlled documents will most likely exist.

Identification and distribution of a controlled hard copy document must be defined. This is critical so that when a document is revised, the document system or coordinator knows where all hard copies of the documents are located so that they can be removed and replaced with the current version. This can be done by one document coordinator. It's more advantageous to distribute the revised documents to the area management to update (and train, if necessary) on the revised version. The latter scenario enhances ownership by area management.

Hard copies of uncontrolled documents are something that always challenge even the best processes. Realistically, hard copies are helpful and many times required for auditing, training, and basic reference. It is best to identify all printed hard copies with an expiration date. Associates are trained to know that the most current version of a document is always wherever it is defined to be, such as electronically on a specific drive (i.e., J drive, etc.). The printed document is used and then destroyed avoiding compliance issues. In a system that also has hard copy controlled documents, it is best to define a specific identifying factor that identifies it as *controlled*. A successful means of control involves printing on special paper that turns a certain color or shows a watermark if copied; however, this can be very expensive. Another method is to stamp in red *controlled copy* for hard copy controlled documents. Remember that the person responsible for document control must ensure that the stamp or the special paper is under control, so it is not accessible to unauthorized individuals.

DOCUMENTATION REQUIREMENTS

Requirements to be included in the document control procedure may vary depending on the specific food safety or management standard; however, for an effective program, at a minimum, the following must be included:

- *Process for approving documents prior to issue.* Keep it simple. Some companies identify five or more associates to read and approve, which not only delays the process, but also adds additional tasks. Some management documents may require review and approval by several departmental managers; other documents may require only one or two approvals.
- *Process for reviewing, updating, and reapproving documents.* It is recommended that upper-level documents, such as the food safety manual, be reviewed annually, at a minimum. Work instructions, work aids, and other specifically focused documents, a minimum frequency of every 2 years may be sufficient. The review process should be defined as 1 year past the last revision date; however, if this is done, make certain that the entire document is reviewed during a revision. Again, these frequencies should be identified based on what is best for the organization. Ensure that the review is listed in the revision history even if no changes are made to the document. This is the objective evidence that the reviews had been completed as scheduled.

- *Process for making changes to controlled documents.* Handwritten changes must not be allowed on a controlled document. Some systems successfully deal with this option; however, each copy of the controlled document must have the same handwritten change. This will catch up to even the best of processes eventually. Required changes should be made to a document and the document reissued. Associates reviewing or identifying a needed change should print out an uncontrolled hard copy, make changes, and then submit the document to a supervisor or document coordinator. Controls related to revising or changing a controlled document must ensure that the proposed changes are reviewed prior to reissuing. Remember, rights to make *official* changes to a controlled document must be managed effectively and controlled per defined requirements.
- *Process for identifying changes to a document.* This can be done in many different ways; however, through experience the most efficient way is to both highlight the changes in a document (i.e., yellow highlight or shading) and then state the change in a "revision history" section. Some include the revision history as the last section of a document. Many formalized document control software programs actually state the changes in a section of the informational portion of the document. The latter may require special access, whereas having a specific section becomes part of the document text. The revision history section can also be where required training on the revision is identified. In some instances, such as fixing typos or changing a noncritical content item, training may not be required, which would be so noted in that section.
- *Process to ensure that associates have applicable documents readily available.* This is a tricky requirement. What is meant by *readily available*? The key here is that if an associate must have a document to know how to perform a required responsibility, he/she must have that document available where it is needed. An example may be a document that tells the operator what code to put in each day. This document should be readily available in the area where the code is set. Having to walk a distance to find a key and/or a password for accessing a document takes time and effort and may not be done. If the activity is taught or learned like covering second base when a ground ball is hit to the shortstop, a trained second baseman, like a trained associate does not have to read a document to know when it is his responsibility to perform a special task.
- *Process for ensuring documents is legible and able to be identified.* This really refers to handwritten directions or changes to an existing document that may be posted. Handwritten directions are rare, but finding posted documents in areas is fairly common. If a memo is making a change or giving a direction, then that information must be stated in a formally controlled document. It is acceptable to post a hard copy controlled document as long as the location of this document is identified (such as on the master list of documents). When the document is revised, the obsolete document can be replaced with the most current version. However, in some instances, temporary process adjustments may be required. This can be handled successfully

by requiring that any posted document such as a directional dated memo has a specific expiration date such as 30 days. If the information is still required by the process after that period, then the information will become part of a controlled document.

EXTERNAL DOCUMENTS

There must be a defined process for identifying and controlling the ever popular and misunderstood external documents. Depending on the type of system, identifying and controlling external documents can be simple or monumental. An external document is defined as a document required by the system, but not created and controlled by the system:

Example 1: If a calibration document states that the pH meter is calibrated according to the owner's manual, that manual must be controlled. It is recommended, wherever possible, that the information needed is included in an internal document to eliminate the added burden of keeping up with the owner's manual. If identified as a controlled external document, the document must be identified as *controlled* (given an ID number) and distribution (location) defined on a master list of external documents. External documents that are accessible on the Internet or through an intercompany web system may have the current version hyperlinked. A good example of this may be the food safety standard or a regulatory requirement with a link to FDA.gov for 21 CFR 110 for the most current version of the GMP requirements.

Example 2: A food safety management system compliant with the requirements of SQF 2000 Version 7 must have at least one copy of this standard identified and controlled as an external document. Without the document, how can the system be compliant with the standard?

Example 3: It is very important to understand that not every textbook and outside document is an external document. Owner's manuals, the Internet, a dictionary, or college chemistry book most likely are reference documents. However, if the college chemistry book is referenced as the source for directions/requirements for running a specific test, and that test cannot be performed without it, then it must be identified and controlled as an external document.

Example 4: Examples of external documents may include standards, legal regulations, customer specifications, validation data, corporate manuals, supplier contracts, testing procedures, and any other document identified within the operation as required by the system but not created in the system.

UNINTENDED USE OF OBSOLETE DOCUMENTS

The process for preventing the unintended use of an obsolete document requires that if an obsolete document is maintained for reference, it must be properly identified to prevent unintended use. Systems that are composed mainly of hard copy documents will normally identify and file one obsolete copy of each document version. All other copies must be shredded or otherwise destroyed. Electronic document control processes usually automatically archive the previous version of a document, but allow access by the document coordinator should it be needed for reference. Many times, obsolete versions are retrieved and reviewed should a change be questioned; however, the change may also be listed in the revision history. Whether or not obsolete versions are maintained becomes an individual decision for each organization.

ELECTRONIC BACKUP

The process for electronic backup of all software programs used within the system or required by the system must be clearly defined and supported by objective evidence (records) demonstrating compliance. This aspect of document control is often overlooked. If the documentation (and also record process) is maintained electronically and crashes, then processes to ensure data is protected must be defined. Do not ever assume that the information is backed up! The process must be defined with records maintained to confirm compliance and effectiveness. It must also be challenged at a defined frequency, such as semi-annually, to ensure that it is truly effective. This includes all software programs, even those individual programs, such as those on the lab computer to provide statistical data or the maintenance program for preventive maintenance. If this is handled off-site through another company location or through an external service supplier, then this requirement must be addressed accordingly to ensure that it is defined and effective, with records maintained to demonstrate compliance. Responsibility and process for monitoring may be defined in the document control procedure. The external service supplier must be managed in compliance with the requirements for a critical food safety and quality supplier.

MASTER LIST OF DOCUMENTS

The master list of controlled documents has been referenced several times in this chapter. Some standards require a master list of controlled documents; however even if it's not required, it is an excellent tool for an effective document control program. The master list of documents can be a formalized document that the software program produces or a manually created list or matrix that is created to uniquely identify (list) every controlled document, its current revision, and the location of any hard copies. The current revision can be identified by a date, such as the effective date or a revision number. Be cautious about having too many dates on a document; more often than not, these dates can become very confusing. Some programs actually have the revision number as part of the document location number such as "LAB – 23-xx"; "xx" is the version. When one document makes reference to another document, it is recommended using the "xx," which is a symbol for the current revision. If a true revision number is used, then every time that number changes, all documents referencing that document revision number must be revised. This quickly becomes a nuisance and almost impossible to manage effectively long term.

HIERARCHY AND NUMBERING

The hierarchy and numbering protocol of the document control process and its complexity is totally up to the organization. It is recommended that this be kept as simple and logical as possible. Upper-level documents such as the food safety manual (management system manual) and procedures may address policies and the basic response to the requirements (who, what, when, where, and a simple how) of the standard. More specific directions (complex directions on how to do something) may be identified as a work instruction, work aid, or SOP. However, some operations identify an SOP as a higher-level procedure. The important aspect of this is that the process is clearly defined, associates are trained in understanding the protocol, and that it is effective for the operation. Remember, what works great for one operation may be a black hole for another.

Surak (2001, p. 709) explains the most popular hierarchy arrangement for an effective document control program:

> The standards require that all documented procedures must be controlled to ensure that employees have the correct and the most up-to-date procedures. In addition, there are requirements that ensure that all records be controlled and retained for a specified amount of time. To achieve the requirements of the standard, most companies use a three-tier system to document the processes. This system is supported by a fourth layer that consists of various quality records (Figure 6.2).

WRITING PROTOCOL

When writing procedures, it is best to use present tense such as *management reviews are held quarterly* rather than *management reviews must be held a minimum of quarterly*. The latter is appropriate if it is a corporate document with directives to

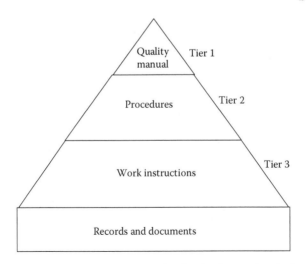

FIGURE 6.2 Document hierarchy. (From Surak, J., International Organization for Standardization ISO 9000 and related standards, Schmidt, R. and Rodrick, G. (eds.), *The Food Safety Handbook*, Wiley, New York, p. 709, 2001.)

the individual sites. The use of *shall* in an organization's documents is not appropriate. When *shall* is used in a compliance standard, this is an absolute requirement compared to the term *should*, which is more of a recommendation or an encouraged requirement. The term *shall* is stated in the standard; however, this may be answered in an organization's food safety manual by saying that top management assesses and ensures that the food safety management system is compliant by using the present tense in defining procedures, meetings, and other critical factors. *Example*: Top management reviews the organization's food safety management system at a minimum of quarterly to ensure its continuing suitability, adequacy, and effectiveness.

When writing documents, be careful when defining a *frequency*. Terms like *periodically, as necessary,* and *when appropriate* do not define a frequency. *Periodically* has different meanings to each person. Defining activities such as *as necessary* or *when appropriate* again translate differently for different individuals. However, these terms also result in questions, "who determines what is necessary, when it is appropriate, and on what basis?" Also, be careful with terms like *biweekly* or *biannually*. Officially, the use of the term *bi* could mean twice a week or twice a year or every other week or every other year. The differences can be very significant. For example, related to calibration frequency, if someone is doing it every 2 years instead of twice a year, it could have adverse results. Also would the frequency be twice a year, any time in that year, or every 6 months? Does it need to be done *only* every 6 months, a minimum of every 6 months, or more frequently if deemed necessary by the responsible person? Writing procedures (application of words and directions) in a manner that truly reflects the intentions of the writer can be challenging. This is why it is very important that written documents are reviewed by others responsible for the task, which promotes clarity, accuracy, and understanding.

When referencing other documents, keep in mind that this is a reader's roadmap or link to the *rest of the story*. Electronic systems have become very effective through hyperlinking between documents. Some formalized procedures are written with a *reference* section that actually identifies all material referenced in the text of the procedure with a hyperlink to that document.

USE OF SOFTWARE PROGRAMS: GOING PAPERLESS

There are many different software programs available for use in the management and control of documents. However, the true effectiveness of any software program depends on the users and its application in the system. It is a good idea to contact several representatives of software suppliers and request actual demonstrations of their software and how it should work within the operation. Ask for references from the companies who have purchased their software and follow up to see how the software is actually working in their operations. It is not possible to ask too many questions: What is their technical support and assistance? What is the response time from technical support when the company has a problem or an issue? What is the training protocol, and how long does it take for associates to be adequately trained on the software? Would this software be a better way to manage document control instead of using Microsoft programs, such as Word or Excel, that the operator may be familiar with and already using? Make certain that all questions are answered thoroughly before making a decision. This author had one client who experienced the ultimate frustration; every time she called technical support, they'd either not answer or put her on hold until the line would eventually disconnect.

Although more and more operations are going to a *paperless* system, which can be very effective, it must be confirmed that all associates are fully knowledgeable, understand, and are able to access the procedures and work instructions. Printed documents may still be needed for training, auditing, etc. As previously stated, it is recommended that a footer be added to the document, which identifies the date and makes a statement that the document expires xx days from the printed date.

Any software program used for controlling information and maintaining records must be designed to ensure that the content always is current and accessible. Documents that are accessible through a database must have *read only* access to users, access to make revisions must be controlled (such as *password protected*) and available only to those associates who have defined authority to make changes. The management and location of the controlled documents must be clearly communicated and defined in the appropriate procedure and/or work instruction. *John Yarborough*, CEO, Solutions Managed LLP shares the following comparison and trend toward paperless systems:

> Typically, initial investment for the software can require a capital expense; however, significant reduction in labor costs does occur after program is implemented, training completed, and associates become used to accessing and managing their documents through a controlled software process. The effective use of and the long term advantage of a formalized software system for document control depends on many factors. There are many programs available. It is essential that if this type of option is being considered that the organization create an implementation team that evaluates several different programs and then choose the program that best fits the needs of the organization. Be careful with promises made by sales folks. Request to talk to and possibly even visit the program in operation. Ensure that effective computer support is available through the organization's IT department or from the software provider on a continual program. Be careful choosing the least expensive because long term, this program must add value to the operation providing a more structured discipline to the software program. *You get what you pay for* may or may not apply. There is some excellent software available in a variety of price ranges; however, functionality for the organization must be the first consideration. Having an effective, user friendly, supported program is a definite advantage to every management system. However, this is a purchase that requires a structured and efficient approach to ensure the organization is getting the best program to fit its needs and that is user friendly for the associates.

In Figures 6.3 through 6.5, Mr. Yarborough compared hard copy controlled documents to a combination of hard copy and electronic documents, to a totally automated document control process.

- Initially inexpensive
- Labor intensive
- Simple
- Error prone
- Process is slow and can require considerable time to ensure manuals are kept current

FIGURE 6.3   Hard copy controlled documents.

Semiautomated (limited paper copies) (i.e., Excel spreadsheets—Access) document control process

- Better organization
- May or may not need organization's computer management department (i.e., IT involved)
- Improved efficiencies
- Decrease in number of errors

FIGURE 6.4   Combination (electronic and hard copies).

- Collaborative
- Systemized processes promote compliance
- Reduced errors
- Replaces the manual activities

FIGURE 6.5   Totally automated document control process.

WRITING PROCEDURES

Over the years, many have stated that "writing the procedures was the hardest part of implementation." However, many of the same people realize that the success of the procedures is directly related to the fact that they are written by the people who are actually doing the job. Having associates within the same areas of responsibility team up and write procedures, work instructions, flow diagrams, etc., which reflect what they actually do, not only enhances the consistency of the process, but also results in a document that responsible associates understand because it was written by those who must use it. It was mentioned earlier that we must reduce variation and build consistency. What better documentation method is there than to build a team empowering those who know their work. The result is a document that is consistent and user friendly and accurately describes the *must do* requirements.

The hardest part of the system implementation effort was getting everyone to agree on how we do something. It is amazing on how we think we do something the same and then putting it down in a procedure or work instruction identifies so many interesting variations. The most difficult aspect of maintaining the system after certification is remembering when you revise a procedure, that you need to look at all the other procedures or documents that the revision might affect. (*Charlie Stecher, an independent consultant and third party auditor, Anchar, Inc.*)

Since we know that following food safety is an issue of quality, it is diminished by increased variation. The use of documentation reduces variation. Management must be a proponent of strong document control systems. During management review sessions, top management must be especially attentive to the document control related issues that are reported through internal or external audits. (Link 2008, pp. 13–14)

KEY POINTS

a. Requirements for document control must be defined in a procedure.

b. Existing documents must be reviewed and, where necessary, revised to ensure accuracy.

c. Associates must understand, be compliant with, and able to demonstrate familiarity with defined requirements.

d. Management must ensure that associates have ready access to documentation related to their areas of responsibility.

e. External documents must be identified and controlled.

f. Software backup processes must be defined and supported by objective evidence (records).

g. The software backup process must be challenged to *prove* the effectiveness of the process with records maintained to provide objective evidence.

h. Associates not responsible for the revision of a document must have *read only* access.

i. Documentation revision must be controlled. For example, documents can be revised (changed and/or edited) by only the responsible department manager or the document control coordinator.

j. Control of documents and control of records are two different processes. Records are a specific kind of documentation.

k. An internal audit of the document control process must confirm that

- relevant versions of applicable documents are available at points of use.
- Obsolete documents are suitably identified and stored in a manner to protect against unintended use.
- Any posted documents are properly identified and controlled in compliance with the procedure for document control.
- External documents are identified in compliance with defined requirements.
- Documents are legible, readily identifiable, and do not include any handwritten changes.
- Document control, identified in existing or potential nonconformances must be addressed in a timely manner based on the potential risk through the organization's formalized corrective action and preventive action programs (Chapter 4).

COMMON FINDINGS/NONCONFORMANCES

a. It could not be confirmed that all processes had been evaluated to identify required food safety documents.

b. Although it was stated verbally that the document control process had been defined and implemented, through the review of process areas and interviews with operators and document coordinators, it could not be confirmed that implementation was completed and that the system was in compliance with the defined requirements of the document control procedure and specific management system standards.

c. Document review did not confirm that documentation had been reviewed for accuracy.

d. Requirements for back-up of the software program used for document control were not defined.

e. Although it was stated that procedures and work instructions were maintained on the database, associates could not demonstrate ability to access the related documents for their areas.

f. Several posted controlled documents were of a different version from that identified on the master list.

g. Several controlled documents had handwritten changes, which were not permitted according to work instruction DC-01 Rev.2.

h. Although the work instruction stated that handwritten changes were allowed by the person authorizing the document, the handwritten changes in several instances did not match the content of the master document.

i. Documentation stated that a master list of documents would be maintained in each department; however, several departments either did not have a master list or the list was outdated.

j. A master list of documents was being maintained in each department as required by Document Control Procedure DC-01 Rev. 2. The revision status and/or the distribution were not identified on the master list.

k. Evidence was not available to confirm that documents such as 21 CFR 110 and equipment manuals for the refractometer (referenced as required by the system) were being controlled as external documents.

l. An obsolete version of form PCF-02 Rev 2 was in use in the process areas.

m. Obsolete versions of controlled documents were referenced by operators in several process areas.

n. It could not be confirmed through review of the documentation process that all documents required by the food safety management standard had been identified and were being controlled through a formalized process for document control.

o. During the evaluation of the receiving department, there were several uncontrolled food safety related forms in use.

p. During the evaluation of the mixing department, there were eight uncontrolled documents posted on the bulletin board.

q. During the evaluation of the shipping department, four documents related to transport loading and inspection were not of the current version as stated on the master list of controlled documents.

r. Documents referenced in the maintenance department but external to the system were not controlled. Through the interview process, an understanding of external documents and the requirements for identifying and controlling these in compliance with the procedure for document control could not be clearly demonstrated.

s. A master list of documents was available; however, some information listed was not current, nor did it match the electronic version, which was defined as the *most current* version.

t. Records were not available to confirm that the food safety team reviewed all proposed changes to specific (SOP) documents for food safety prior to implementation to determine their effects and impact on food safety.

u. The document control procedure did not define the requirements for controlled documents that were printed for use.

v. Records were not available to confirm that challenges of the electronic backup had been performed to confirm effectiveness as required by the document control procedure and also the record control procedure.

w. Statutory and regulatory requirements specific for this operation were not identified and controlled as external documents.

x. It could not be confirmed through the interview process that associates were familiar with related procedures and work instructions for their areas of responsibility.

y. Some controlled documents did not have identification and pagination, thus the complete hard copy document was not formally controlled (pages could be changed or removed inadvertently).

z. Procedure for performing management review meetings stated that its frequency was to be "randomly, on a periodic basis," which did not reflect an actual defined frequency.

REFERENCES

Link, E. 2008. *An Audit of the System, Not of the People: An ISO 22000:2005 Pocket Guide for Every Employee*, Quality Pursuit, Inc., Fairport, NY.

Surak, J. 2001. International Organization for standardization ISO 9000 and related standards. In: *Food Safety Handbook*, Schmidt, R. and Rodrick, G. (eds.), Wiley-Interscience, Hoboken, NJ.

---

*Additional Food Safety Management System References (not all inclusive)

FSSC 22000:2010

    ISO 22000:2005 Sections 4.2, 4.2.2

SQF 7th Edition: Section 2.2

BRC Issue 6: Sections 2.13, 3.2

IFS Version 6: 2.1.1

*Note: Always review the standard or visit the website for your chosen standard to ensure that all related requirements are identified and addressed.

7 Record Control

The identification and maintenance of *records* to demonstrate compliance to defined requirements and the effective operation of the food safety management system (FSMS) are critical. Records that must be identified and maintained are required by the

a. Specific standard of choice
b. Organization's (specific site and corporate) defined procedures
c. Regulatory and statutory requirements including federal, state, and county
d. Customers

When defining the system, it is best to first review the standard closely and ensure that everywhere the word *record* is stated that this record either exists or is created within the system. After this is completed, identification of any additional required records, under the guidelines mentioned earlier, must be completed. It has been found to be most effective if responsible department managers, teams, or designees identify the records that they must maintain in order to demonstrate compliance and/or confirm the effectiveness of a process or activity.

The following is a summary of records that, at a minimum, must be formally maintained. However, keep in mind that *required records* may vary between standards:

- Internal audit reports
- Internal audit schedule
- Management review meeting minutes
- Training records
- Supplier management, approval, and evaluation
- External and internal communication
- PRP verification
- Required Hazard Analysis and Critical Control Point (HACCP) program records such as those related to critical control points, etc.
- Verified flow diagrams
- Pest control and other specific PRP records
- Individual verification results
- Analysis of results of verification activities
- Corrective action and preventive action
- Data analysis related to measurable goals and objectives

- Change management
- Control of nonconforming product
- Receiving, storage, and shipping records
- Lot traceability and recall (including mock recalls)
- Crisis management
- Food safety team meeting minutes
- HACCP team meeting minutes (if specific HACCP teams exist)
- Others, depending on the specific standard, legal requirements, and organi-
  zation's specific defined requirements

RECORD CONTROL PROCEDURE

A documented procedure defining the organization's requirements for record control must be developed and implemented. This procedure must define requirements
for the "identification, storage, protection, retrieval, retention time, and disposition
of the record" (ISO 22000:2005 4.2.3). Some organizations choose to identify the
requirements for specific records in the controlled document wherever the record
is referenced. This can be done; but should retention time or some other requirement be changed, every document or record referenced must also be changed. This
can become very difficult to manage effectively and compliantly. The most efficient method is to create one master list of records. The master list defines requirements such as responsibility, storage locations, and retention times for each record.
Associates responsible for records must be trained in the record program. If the
master list defines requirements for a record or records for a specific responsibility,
the associate must have ready access to the master list in order to understand his/her
level of responsibility. The records must be specifically identified. Just stating *internal audit records* or *training records* does not adequately explain or define the actual
required record.

MASTER LIST OF RECORDS

At a minimum, the master list of records should include columns for the following
information:

- Record name
- Responsibility
- Format and location (electronic or hard copy)
- Retention time

RETENTION TIME

Most standards do not specifically state how long records must be kept. This must
be an internal decision based on the duration required to demonstrate compliance to
specified requirements. For example, depending on the record, 3–6 months of daily

records may be sufficient to demonstrate compliance; however, monthly records may have to be kept for at least 12 months. Many registrars require that system maintenance items such as internal audit reports, management reviews, customer complaints, corrective actions, and preventive actions be maintained for 3 years at a minimum. This may be the time period required between assessments and/or to demonstrate that the system is being maintained in an effective manner. When defining retention times, consider requirements related to the FSMS standard of choice, regulatory requirements, company legal policies, shelf life of the product, and customer requirements.

Be careful when defining the retention time. Records must be maintained at a frequency that provides evidence of compliance. Issues have developed when the retention time for a training record was stated as *duration of employment*. If it is stated that audit records are maintained for 3 years, but the trained auditor leaves employment during that period, then his/her record may be destroyed; therefore, confirmation of training is lost. If calibration is performed annually but records are kept only for a year, then available records do not confirm that the activity is being done annually.

ELECTRONIC OR HARD COPY

Records may be maintained electronically. However, as discussed in regard to document control (Chapter 6), the process for controlling the software databases (i.e., backup) to ensure that they are protected and contain the current information must be defined and supported by records demonstrating compliance.

RECORD OR CONTROLLED DOCUMENT?

Confusion often develops in understanding the difference between a controlled document and a record. Document control is the process used to ensure that the most current version of instructions, requirements, and procedures are available to those who must have the information to perform the required activities. A record provides objective evidence that an activity has been performed to demonstrate conformance to a defined requirement. A blank form is a controlled document that ensures that the current version is in use; however, once information is recorded on that form, it becomes a record confirming that an activity was performed. In some instances, a document may be both a record and a controlled document. An example would be a production schedule. The person responsible for creating the production schedule maintains this as a record to demonstrate compliance to the defined requirements. Since the schedule includes information required for specific activities, it may be distributed as a controlled document for specifically identified areas of responsibilities.

KEY POINTS

1. Ensure that all required records are identified and maintained according to defined requirements.
2. Use one master list to identify individual records that are shared and referenced by all.

3. Defined requirements must include instructions for how to destroy records.
4. Define retention times by a *minimum* time frame so not to be held to an exact time frame.
5. Ensuring defined retention times make sense; keeping all records forever is not an effective method.
6. Ensure that associates understand the difference between a record and a controlled document.
7. Records must be identified correctly and readily retrievable.
8. Records must be stored so that they are protected from loss, damage, and deterioration.
9. Records must be legible and written with permanent ink:
    a. Must never be completed in pencil or corrected with *correction fluid*
    b. Must not contain erasures
    c. Must be initialed and dated by a person with approved authority for making changes and/or corrections
    d. Must have explanations for blank spaces (i.e., process not running, etc.)
10. Records must be in a form that is appropriate to the system (i.e., hard copy or electronic data).
11. Be selective in saving notations; it is not necessary to save every notation.

COMMON FINDINGS/NONCONFORMANCES

a. A formalized process for identifying and controlling records had not been defined and established.
b. Requirements for the controls needed for the identification, storage, protection, retrieval, retention time, and disposition of records had not been formally defined and implemented.
c. Although they were completed, the data recorded was not legible in 10 out of 15 records sampled.
d. Not all records required by the standard of choice were identified and maintained in compliance with the record control procedure.
e. The master list of records was not identified with a current date or with the last date it was updated (not a controlled document).
f. Master list of records did not identify specific training records.
g. Pest control records were not identified on the master list of records as required by the record control procedure.
h. Review of a sampling of completed records showed blank spaces with no explanation of why the required information was not recorded. It was stated verbally that the operation was not running during this time, but this could not be confirmed.
i. Five out of ten records reviewed in shipping showed evidence of changes (cross-outs, erasures, etc.) with no identification or approval for these changes as required by the record control procedure.
j. Master list of records defined the retention for training records as *ongoing*. The actual duration that was required could not be determined.

k. Ten out of fifteen records reviewed did not have legible signatures, thus it could not be confirmed that the associate performing the activity had been trained and confirmed competent.

l. It could not be confirmed through the interview process and review of objective evidence that all associates were familiar with defined requirements for record control in their areas of responsibility.

m. Through the interview process, associates could not demonstrate familiarity with the master list of records, which was their source for related record control requirements.

n. Defined requirements stated that records were kept for 2 years; however, 4 years of records were still being maintained. (*Would this actually be a nonconformance?*)

o. Records were not available to confirm that visitors and contractors were trained per the documented procedure for *visitor and contractor training.*

---

*Additional Food Safety Management System References (not all inclusive)

FSSC 22000:2010

     ISO 22000:2005 Section 4.2.3

SQF 7th Edition: Section 2.2

BRC Issue 6: Sections 2.3, 3.2

IFS Version 6: 2.1.2

*Note: Always review the standard or visit the website for your chosen standard to ensure that all related requirements are identified and addressed.

8 Training, Awareness, and Competency

The organization must determine and provide the resources needed to implement, maintain, and improve the processes of the food safety management system, addressing any issues or potential issues in a timely manner. The system must have the resources needed to be effective, as well as to apply improvements that ensure food safety. Resource needs must be identified, monitored, and recorded. As stated in Chapter 3 (Management Responsibility), this must be evaluated and recorded during management review meetings, but may also be evaluated during other activities such as business planning meetings and staff meetings. Meeting minutes must be maintained through the record control process, providing objective evidence of compliance (Chapter 7).

Training requirements and the definition for *qualification* place a strong emphasis on the evaluation of competency, effectiveness of training, and making sure that today's associates are aware of the importance of their actions and how they contribute to the achievement of measurable food safety and quality objectives. Evaluation should include training records that document both the positive and negative results of the training. Although follow-up and documentation may take time, the overall positive impact on the business and success of the organization is well worth the extra efforts required.

It is imperative that associates understand their roles in meeting the food safety and quality policy statement and measurable objectives of the management system. Generally this occurs through communication of the policy statement, what it actually means to the company, and what roles associates play in achieving success. Providing a work environment that ensures the production of a safe product, achieving *conformity of the product*, keeping good records, and communicating the management system requirements throughout all levels of the operation are key to an effective management system (Chapter 3).

The assignment of trained personnel, identification of training needs, and provision for providing training for all associates responsible for any aspect of food safety and quality are required. It is critical that management applies the proactive approach to education and training rather than being reactive when something goes wrong. Without a structured, proactive, and effective program to communicate and educate, the organization is putting its products at risk not only related to food safety but also quality and brand protection:

> The best thing about certification and our quality system was that it defined the ability to train operators on a consistent basis and slowly develop the concept that only in-specification product was shipped every time. No cutting of corners, any back

room decisions or short cuts. Compliance made us identify trouble spots, pay attention to them, and take actions before they become critical. It also promoted a true sense of teamwork and a common purpose, particularly in the earlier days of seeking and obtaining registration. Most employees ultimately understood and supported the system as it provided a degree of stability to their job. The system "worked" for them, not against them! (Newslow 2001, pp. 186–187)

DEFINING COMPETENCY REQUIREMENTS

When defining the necessary competencies, it is critical to approach this task by thinking about how every position affects food safety and quality. The following requirements are necessary to aid in identifying what it means to be competent and qualified to perform specific responsibilities related to food safety and quality:

- Define criteria for each area of responsibility as it affects food safety and quality.
- Define criteria based on related documentation (procedures and work instructions).
- Provide new hire orientation training.
- Provide specific training on the compliant management system.
- Define requirements for identifying training needs.
- Identify records required to maintain and demonstrate that technical training requirements are met and effective.
- Ensure that *competency* of associates is evaluated, recorded, effective, and defined in a manner that is appropriate for the specific organization.

EDUCATION VERSUS TRAINING

The training, competence, and awareness program must focus on educating the associates. *Education* is a critical term that communicates sharing the knowledge and reasons why an activity must be done in the manner presented. Training is communicating the required actions without necessarily stating why these are important:

"Education" is the act or process of imparting or acquiring general knowledge, developing the powers of reasoning and judgment.... the act or process of imparting or acquiring particular knowledge or skills. (http://dictionary.reference.com)

"Training" is to develop or form the habits, thoughts, or behavior of by discipline and instruction and to make proficient by instruction and practice, as in some art, profession, or work. (http://dictionary.reference.com)

A favorite example when comparing these two terms is that Pavlov's dog was trained (a reflex) to come to eat when the bell rang. Education is communicating an understanding of why it is necessary to perform an action in a specific manner:

The most difficult is always the same, keeping the interest of the people channeled and informed. From what I have seen there has been a shortage of good education on a continual basis to keep all persons informed and to keep interest in the system at a high level. One does not take graphs and numbers to a meeting to show to the employee. One must present a program that tells the employee how he or she will benefit. Informative meetings with drinks and snacks are received as a reward and, therefore, gain attention (*Jon Porter, President, J. Porter and Associates, Ltd.*).

Let's emphasize that the training, awareness, and competency program must be developed and presented in a manner that effectively educates our associates. This must include an understanding of the organization's food safety and quality policy statement and an understanding of each individual's role in producing a safe product that meets customer requirements. All associates must at least have an overview of what compliance to the requirements of the management system means and how it impacts the measurable food safety and quality goals of the organization.

Training criteria must include associates at all levels of operation, including those in supervision and management positions, who perform responsibilities that affect food safety and quality.

A matrix designed to identify the relationship between a responsible position and the related procedures and work instructions is an effective tool. Documentation is listed across the top and job responsibilities down the side. This matrix would be a controlled document that provides the defined requirements needed to qualify for a specific position or responsibility. To provide a *record* of the training, the same type of matrix with the same documents listed across the top and the associate's name (title) listed in the vertical column should be created. The date training is completed can be entered into the *field* that corresponds to the position/document for which the associate is trained. Remember, associates who perform relief activities must be qualified to perform these activities, and with records maintained confirming this. An example of a training matrix is presented in Figure 8.1. Blacked out sections are these areas that don't apply to position or area of responsibility.

CLARIFICATION OF THE REQUIREMENTS

ISO 22000:2005 6.2.1 states that "The food safety team and the other personnel carrying out activities having an impact on food safety must be competent and must have appropriate education, training, skills and experience." ISO 22000:2005 6.2.2 states that "the organization must

a. identify the necessary competencies for personnel whose activities have an impact on food safety,
b. provide training or take other action to ensure personnel have the necessary competencies."

BRC Global Standard for Food Safety Issue 6 7.1 states the "the company shall ensure that all personnel performing work that affects product safety, legality, and quality are demonstrably competent to carry out their activity, through training, work experience, and qualification." The other standards have similar statements. It is clear that the organization must determine what competencies or *skills* are

Production Associate Training Identification and Records

	Job Descriptions				Work Instructions		
Name	Mixing Instructions Rev 0	Operator Responsibilities Rev 01	Shipping Rev 12	Warehouse and Storage Rev 1	Ware 01 Rev 2	Nonconforming Product Rev 0	Rec InspRev 0
Line workers							
Jona Hall	■■■■■■■				■■■■■■■■■■		
Derrick Scott	12/31/11	03/01/12		03/01/12	02/01/12		
Warehouse							
Maggie Leigh	03/23/12				09/01/12	03/01/12	03/01/12
Lilee Berry	05/02/12				09/01/12	03/01/12	03/01/12
Lucky Charm	05/02/12				09/01/12	03/01/12	03/01/12
Receiving Inspections							
Sophie Rose						03/01/12	03/01/12
Skip Macfee							
Tester							
Andrea Newman	■■■■■■■				■■■■■■■■■■		
Shipping							
Gary Harris			■■■■■				
Quality lab				■■■■■			
Judy Sanders							
Patty Troni							

FIGURE 8.1   Sample training matrix.

required and ensure that these are met and effective with records maintained to confirm compliance (Chapter 7).

Cynthia Weber, president of Vinca, LLC, provides the following summary:

Employees must also be trained on the standard requirements and knowledgeable on prerequisite programs or GMPS. Plan training for your employees as you complete the design and documentation of your system. Make sure that each employee is aware of the prerequisite programs or GMPs that they are responsible for complying with as well as details on each Critical Control Point or Operational Control Point in their processes. (http://www.22000-tools.com)

MEASURING EFFECTIVENESS

Associates must be educated and trained in all aspects of food safety including their roles in ensuring that the prerequisite programs are effective. In addition to confirming the training happened, records must also provide the objective evidence that the training was effective. This can be accomplished through associates demonstrating the skill sets, through audits, through interviews, and record review. The best way to approach the measurement of effectiveness is to ask "How do we know it was effective?" "What are the results that we need to see to determine if the training was effective?" There is really no *cook book* answer to this question. It is going to depend on the activity and the required performance against its defined requirements.

TEMPORARY ASSOCIATES

Some organizations contract with outside agencies to provide temporary associates to perform various duties within the system. Qualification or competency requirements must be defined for temporary associates performing responsibilities related to food safety, quality, regulatory, or customer requirements. Records must be maintained in compliance with the record control program (Chapter 7). This may be accomplished in many ways. Examples are listed in the following:

Example 1: Provide a controlled procedure, work instruction, or pamphlet that defines all the basic requirements for training. It is recommended that specific process training be performed in the same manner as training for permanent associates. It is most efficient if the training records are maintained in the department for which the associate is working. Records should be available any time, day or night. Records secured in the human resource office or the temporary service central offices may not be readily available at 2:00 a.m. when the supervisor must confirm if an associate was trained in a specific task prior to assignment.

Example 2: Train each temporary associate in exactly the same manner as the permanent associate, maintaining all records as required by the record control program.

Example 3: Maintain a checklist with duties defined for that particular area of responsibility. The area leader must review responsibilities with the temporary associate, recording completion on the checklist. This works well in systems that have a large turnover of temporary associates. It also serves as a quick reference during off production hours.

Whatever program is defined and implemented must be supported by records confirming that the training has been completed and effective. Training records must be maintained in compliance with the record control program (Chapter 7).

The external temporary service agency must be identified and maintained in compliance with the critical food safety and quality supplier program (Chapter 13).

ROLE OF *SOFTWARE* PROGRAMS IN YOUR TRAINING PROGRAMS

Many times, an organization chooses to maintain training records in a software program designed specifically for this purpose. There are many available. It is important when choosing a software system that, prior to the final decision, an outline is created defining exactly what is needed by the organization. Contact the various suppliers and request demonstrations and other appropriate literature. Be sure that the program has technical support available for the user after purchase.

Some programs maintain hard copies while others use software programs to maintain training information and related compliance records. Software programs used for controlling information and for maintenance of records must be in compliance with the document control program (Chapter 6) and the record control program (Chapter 7).

Laura Dunn Nelson, Industry Relations Director for Alchemy Systems, provides the following description of the relationship between effective training and an effective food safety management system:

> It is not an exaggeration to conclude that effective training, or the lack of it, poses a significant risk for the food industry. Experts say that "deficient employee training" is the top food safety problem in the food processing industry (Sertkaya et al. 2006, pp. 310–315). A survey of global auditors, representing thousands of audits, noted employee training deficiencies in almost twenty-five percent (25%) of their audits including findings of inadequate document control, lack of training (i.e., seasonal and contractors), illegible paperwork, and little or no verification of learning comprehension.
>
> Experience with a structured software system that focuses on employee food safety training has successfully applied developed technology for training, addressing and preventing typical training program deficiencies. This training technology consists of industry-specific courseware that incorporates visual, attention getting material that directly relates to the employee's on-the-job environment.
>
> Employees and trainers have found that this technology is easy-to-use and especially valuable for verifying comprehension. Responses to questions during each training course are keyed into a remote location. Every question has to be answered accurately. When a question is missed, the instructor knows immediately who submitted the wrong answers, and the training platform provides automatic remediation training to assure that the question has been understood. Once the course is complete, a permanent electronic record is established of each employee's training, completely eliminating time consuming and inefficient sign-in sheets.
>
> One of the biggest challenges for Quality Managers is taking time from their busy schedule to refresh their food safety training to ensure the training reflects the current regulatory compliance requirements and continues to engage the employees. Many managers continue to use food safety training materials they have collected over the years regardless of the changing demographics of their workforce and customer requirements. As managers struggle to make time for important employee food safety training, it's no wonder that most managers never measure the effectiveness of their training and link their training efforts to measureable operational goals. The technology leaders utilize industry experts to continuously refresh their food safety training and provide easy reports to measure training compliance per assigned lesson plan.
>
> As the regulatory climate increases the compliance demands, food companies must go beyond their usual approach to employee training and adopt more comprehensive and verifiable training platforms. Utilizing technology provides a conduit for meeting the compliance demands and establishing a strong food safety culture. Once achieved, a culture of food safety will provide a significant return on investment for investments in training technology and resources.

KEY POINTS

a. Training records and contracts with consultants must be maintained in compliance with the record control program.

b. Top management, the food safety team, and responsible managers must continually review all tasks to ensure that competencies related to food safety responsibilities are identified and effective with records maintained in compliance with the record control process.

c. Creating a *training matrix* that defines the food safety and quality requirements per department and area of responsibility within a department has proven to be very successful for many organizations. An added advantage is that these records can transfer with the associate when transferred to a different department or area of responsibility.

d. The training matrix is an excellent tool to identify required training for each area of responsibility that has an impact on food safety and quality.

e. The organization's training program must, at a minimum, ensure that associates are *educated* in the food safety and quality policy statement, Hazard Analysis and Critical Control Point (HACCP), Good Manufacturing Practices (GMP), and product/facility safety during orientation. Each associate should attend a refresher course at least annually.

f. Communication related to the associate's responsibility for food safety may be accomplished through several means such as team meetings, shift meetings, awareness meetings, management review meetings, bulletin boards, puzzles, e-mails, and prizes for solving puzzles. Communication may be reinforced through daily activities and ongoing education. More information on communication, its requirements, and options for compliance are discussed in Chapter 3.

g. Management must ensure that records confirm the effectiveness of training on food safety related responsibilities.

h. Evaluations confirming successful implementation and the effectiveness of competency training may be performed by department managers and immediate supervisors. Other means of competency evaluations may include visual observations, routine GMP inspections, and internal audits.

i. Some very good software programs are available. Management should evaluate the needs of the organization and then request further information on a sampling of these programs for review. It is critical that the training protocol, delivery, and record-keeping abilities meet the needs of the organization.

COMMON FINDINGS/NONCONFORMANCES

a. It was evident that management had identified training as a resource issue; however, at this time, there wasn't any evidence (records) that a solution (corrective action/preventative action) had been identified.

b. Requirements for defining competency for the positions that affect food safety and quality had not been defined and implemented.

    c. Requirements for confirming training effectiveness had not been identified and implemented.

    d. It was stated in the management review meeting minutes that training matrixes were being created for each department; however, after 6 months of elapsed time, these had not been completed nor was there evidence that a plan to complete these had been defined and implemented (see Chapter 4 for corrective actions and preventive actions).

    e. The process for using the training software program appeared to be well managed; however, content had not been revised to include the current revisions to six operating procedures released 3 months ago. Records confirmed that associates continued to be trained for the past 3 months on the obsolete versions of the documents.

    f. Not all required training (task specific) to confirm competency for personnel carrying out activities having an impact on food safety had been formally defined and implemented.

    g. Specific training requirements and records required to demonstrate that new hires had been trained in food safety and quality had not been defined. In addition, there wasn't any formalized means to confirm and record that the training had been effective.

    h. Task-specific training requirements had not been identified for each food safety prerequisite program (i.e., chemical control, pest control, preventive maintenance, etc.).

    i. Training requirements for temporary associates in the areas of food safety, GMP, and health and safety (i.e., maintenance) had not been defined.

    j. Site had not implemented a formalized process to define competency requirements for those areas of responsibility related to food safety.

    k. Records were not available to confirm that the relief person performing food safety related responsibilities had been trained and deemed competent based on the defined requirements.

    l. Records were not available to confirm that defined training requirements were actually being performed.

    m. It was not evident through the interview process that associates had been trained in documents and document revisions related to their areas of responsibility.

    n. Temporary associates were not able to demonstrate familiarity with the organization's food safety programs and their roles within these programs.

    o. Records were not available to confirm that the temporary associate agency had been identified and was being maintained in compliance with the critical food safety and quality supplier program.

    p. Although it was defined that training needs were to be evaluated, records were not available to confirm this.

    q. Food safety related competency requirements were defined for process hourly associates, but not for management and supervisory associates.

    r. Records were not available to confirm that the software system used to maintain training records had been *backed-up* to protect the data as required by the document control program (Chapter 6).

s. It was stated verbally that food safety related responsibilities were defined in job descriptions; however, review of a sampling of job descriptions did not provide objective evidence of compliance.

t. Job descriptions were referenced as the source of training requirements for specific food safety responsibilities, but these were not identified and maintained in compliance with the document control program (Chapter 6).

REFERENCES

Newslow, D.L. 2001. *The ISO 9000 Quality System: Applications in Food and Technology*, Wiley-Interscience, New York.

Sertkaya, A. et al. 2006. Top ten food safety problems in the United States food processing industry, *Food Protection Trends*, 26(5), 310–315.

---

*Additional Food Safety Management System References (not all inclusive)

FSSC 22000:2010

    ISO 22000:2005 Sections 6.2

SQF 7th Edition: Section 2.9

BRC Issue 6: Section 7.1

IFS Version 6: Section 3.3

*Note: Always review the standard or visit the website for your chosen standard to ensure that all related requirements are identified and addressed.

---

9 Control of Nonconforming Materials and Products

Nonconforming materials and products may be considered any material or product that does not meet specifications or that, for whatever reason (damage, leaking, code date, etc.), is not acceptable for use or shipment to the customer. These items must be clearly labeled as *hold* and stored in a segregated area that protects them from inadvertent use or shipment.

The reference to nonconforming materials and products may apply to raw ingredients, packaging, in-process product, and finished products relating to customer complaints. In the food industry, operations have been addressing this subject for many years through formalized *hold* procedures. Having a documented procedure to define the requirements for identifying, evaluating, and disposing nonconforming materials and products, in such a manner as to protect against their inadvertent or unintended use, is a requirement of the food safety management standards. Every effort must be made to identify and segregate the nonconforming situation as soon after it occurs as possible. An operation must have either a designated hold area to ensure its segregation from good product or an effective process such as using *hold* tape to identify and segregate the items. This special tape is very similar to police crime tape but labeled *hold* or *do not use*. The procedure for the control of nonconforming materials and products plays a critical role in the food safety program as a means to make certain that any deviation in a critical control point (CCP) and/or an operational prerequisite program (OPRP), if applicable, prevents an unsafe product from being released for consumption. This process must also make reference or link to the crisis management and/or recall procedure for materials and/or products that may have been released before being identified as potentially unsafe or otherwise nonconforming (Chapter 14).

Through the food safety awareness training and internal communication programs, the message imparted to associates at all levels of the operation must be that each has an individual responsibility to report any concern or potential concern related to nonconforming materials or products. This must be conveyed to supervisors and management as well. It is every associate's responsibility to ensure that unsafe or potentially unsafe materials and products are identified and segregated in a manner that prevents them from reaching customers and/or consumers. This responsibility may be as simple as notifying a supervisor or designee, but whatever the process, each associate must understand his/her responsibility and respond accordingly if such a situation occurs. Depending on the process and activity, every associate must understand that he/she has the authority to stop the process should an unsafe or potentially unsafe situation be identified. Management must communicate and encourage this responsibility clearly so that the associate understands that it is better to stop the process immediately than to take the chance of continuing to produce a product that may be unsafe or otherwise nonconforming.

A documented procedure for the *control of nonconforming materials and products* must, at a minimum, define requirements for the following:

- Identification of the nonconforming product or material
- Investigation of the event including the root cause
- Process for segregation
- Required notifications (i.e., interdepartmental, shipping locations if any of the product had been shipped)
- Responsibility for the evaluation of the product and authority for disposition
- Validations of actions required to be put into place

What is done with the product (the disposition) may include such actions as reworking, repackaging, destroying, or donating. Some systems may have a waiver release process, which allows a responsible individual or designee, after evaluating the nonconformance, to issue a waiver for release of the product. If it is not a food safety issue, but a customer specification issue, a waiver or concession by the customer may be possible. Whatever the issue and justification, actions must be recorded. All records must be identified and maintained in compliance with the record control program (Chapter 7).

ISO 22000:2005 7.10.1 states that "a documented procedure must be established and maintained defining:

a. The identification and assessment of affected end products to determine their proper handling;
b. A review of the corrections carried out."

ISO 22000:2005 7.10.3.1 states that "the organization must handle nonconforming products by taking action(s) to prevent the nonconforming product from entering the food chain unless it is possible to ensure that

a. The food safety hazard(s) of concern has(ve) been reduced to the defined acceptable levels;
b. The food safety hazard(s) of concern will be reduced to identified, acceptable levels prior to entering into the food chain; or
c. The product still meets the defined acceptable level(s) of the food safety hazard(s) of concern despite the nonconformity;

d. All lots of product that may have been affected by a nonconforming situation must be held under control of the organization until they have been evaluated;

e. If products that have left the control of the organization are subsequently determined to be unsafe, the organization must notify relevant interested parties and initiate a withdrawal."

Records of all actions and decisions must be identified and maintained in compliance with the record control program (Chapter 7). Most organizations track nonconforming situations through either a physical log book or a software program. All requirements are recorded on the log or in the software program. This information is monitored not only for individual event compliance, but also for any trends. If an event or a group of events (trend) result in a formal corrective action/preventive action (CAPA), then there must be a trail that leads into the system for tracking, corrections, and measurements of effectiveness (Chapter 4).

ISO 22000:2005 7.10.3.2 addresses requirements for the evaluation of release by stating that "each lot of product affected by the nonconformity must only be released as safe when any of the following conditions apply:

a. Evidence other than the monitoring system demonstrates that the control measures have been effective;

b. Evidence shows that the combined effect of the control measures for that particular product complies with the performance intended;

c. The results of sampling, analysis, and/or other verification activities demonstrate that the affected lot of product complies with the identified acceptable levels for the food safety hazard(s) concerned."

ISO 22000:2005 7.10.3.3 addresses the disposition of nonconforming product by stating that "following evaluation, if the lot of product is unacceptable for release, it must be handled by one of the following activities:

a. Reprocessing or further processing within or outside the organization to ensure that the food safety hazard is eliminated or reduced to acceptable levels;

b. Destruction and/or disposal as waste."

ISO 22000:2005 links the next step from the control of nonconforming materials and products to the withdrawal process. The recall program, crisis management, and traceability programs are defined as individual prerequisite programs (Chapter 14). Although most (hopefully all) organizations have had effective programs for addressing nonconformances, being forced to evaluate these programs internally puts the focus on the effective handling of potentially unsafe product in a structured, disciplined, and tested manner that strengthens already existing programs.

KEY POINTS

a. A documented procedure with specific parameters for the control of non-conformance materials and products is required. Review the food safety management standard of choice to ensure compliance.

b. It is the responsibility of associates representing all levels of the organization to report any existing or potential nonconforming situation per the organization's defined requirements.

c. All records related to the nonconforming materials and products program must be identified and maintained in compliance with the record control program (Chapter 17).

d. Unsafe or potentially unsafe product that have already left the control of the organization (i.e., shipped) must be managed through the crisis management and/or product recall programs (Chapter 14).

e. Records must also be kept if the product has to be destroyed. If this is performed by an external service provider, the provider must be identified and maintained in compliance with the critical food safety and quality supplier program.

f. It is the responsibility of the organization to ensure that any material or product that is destroyed is also defaced in a manner to prevent reuse of the container or product per the organization's defined procedures.

g. Nonconforming materials and products may come from many sources including from raw ingredients, packaging, deviations from CCPs, loss of control of OPRPs (if applicable), out-of-date items, damaged items, product (in process, finished product, etc.) that do not meet specification, product not labeled (such as to meet traceability program requirements), and any other issue that may compromise the safety of the product, brand name, regulatory, quality, or customer requirements.

h. Software programs for identifying and tracking nonconforming materials and products are very popular; however, it must be assured that the software of choice meets the needs of the organization. Also, software programs must be managed in compliance with both the document control program (Chapter 6) and the record control program (Chapter 7).

COMMON FINDINGS/NONCONFORMANCES

a. Records were not available to provide evidence of the disposition for product identified as nonconforming on finished product testing logs.

b. Pallets of product were observed stored in the finished product warehouse storage area with both a *hold* sticker and an *available for shipment* label.

c. Review of 6 of 15 micro testing records for the week of September 5 indicated test results that did not meet specification requirements; however, there was no evidence that this product was placed on hold, evaluated, and dispositioned as required by the procedure for the control of nonconforming materials and products.

d. Review of the nonconforming log indicated several trends in product non-conformances; however, there was no evidence that these occurrences were reviewed, trended, and addressed through the formal CAPA program.

e. The organization had defined its requirements for the control of nonconforming products in a documented procedure; however, records were not available to confirm that the requirements of this procedure had been implemented and that activities were in compliance with the FSMS of choice.

f. A documented procedure was not available that defined requirements for corrections when critical limits defined for the CCPs were exceeded or when there was a loss of control of an OPRP (Chapter 17).

g. Current procedure for the control of nonconforming product did not include handling of food safety issues or out-of-specification test results for raw ingredients.

h. The current procedure for the control of nonconforming materials and products did not address product that was rejected by the metal detector.

---

*Additional Food Safety Management System References (not all inclusive)

FSSC 22000:2010

  ISO 22000:2005 Section 7.10

SQF 7th Edition: Section 2.4.6

BRC Issue 6: Section 3.8

IFS Version 6: Sections 5.7, 5.10

*Note: Always review the standard or visit the website for your chosen standard to ensure that all related requirements are identified and addressed.

10 Prerequisite Programs (PRPs)

Part One

UNDERSTANDING PREREQUISITE PROGRAMS (PRPS) AND THEIR ROLE IN THE FOOD SAFETY PROGRAM

Effective prerequisite programs (PRPs) are critical to the overall success of a food safety management system (FSMS) program. The individual FSMS standards are similar in their requirements for PRPs stating that PRPs must be in place, monitored, actions taken if nonconformances are identified, and that each program is verified for effectiveness. Each standard identifies the PRPs that must be defined and effective. The basic programs are similar; however, depending on the food sector, some may define more specific requirements for a select group of PRPs. An example would be the BRC requirement for physical separation and uniform control. It is critical that the organization knows its standard of choice and then applies the best practices and compliance guidelines from this text to those programs.

In addition to the requirements of the standards, some food segments are required by Food and Drug Administration (FDA) or U.S. Department of Agriculture (USDA) to have a Hazard Analysis and Critical Control Point (HACCP) program that include specific required PRPs. Also, an organization's food safety/HACCP programs may identify internal required PRPs that have been identified by the food safety/HACCP teams necessary to control or eliminate a hazard.

PRPs are activities defined and managed through operational-type programs that effectively eliminate or reduce the likelihood of a food safety occurrence. Many existing or potential hazards may be addressed through PRPs. The role of the PRP is to prevent, eliminate, or control, existing and/or potential hazards. It is critical that during the hazard analysis, the PRPs responsible for controlling or eliminating the identified hazard are referenced. These programs must be active and effective. The overall safety of the product and effectiveness of the food safety/HACCP program depend on this (Chapter 17).

As we proceed to review specific PRPs in detail, it is critical that top management, the food safety/HACCP team, and every associate be trained in related requirements of his/her area of responsibility and that everyone understands their roles and responsibilities for the production of a safe product (Chapter 8).

Let's emphasize that the production of safe products requires that the food safety/HACCP program be built on a foundation of effective PRPs. PRPs provide the basic environmental and operating conditions necessary for the production of safe, wholesome products.

ISO 22000:2005 7.2 states that "the organization must establish, implement, and maintain PRPs to assist in controlling:

- The likelihood of introducing food safety hazards into the product through the work environment;
- Biological, chemical, and physical contamination of the product(s), including cross contamination between products;
- Food safety hazard levels in the product and product processing environment.

PRPs must be

- Appropriate to the organization's needs as related to food safety
- Appropriate to the size and type of the specific process and the nature of the products being manufactured and/or handled
- Implemented across the entire production system
- Approved by the food safety team"

As stated previously, the FSMS standards are very consistent in their requirements for PRP programs. The standard of choice, use of the guidelines, and best practices shared in this chapter are meant to help to develop, implement, and maintain effective and compliant PRPs that add value to the organization overall.

PRPs must include and be in compliance with local, state, and federal statutory and regulatory requirements. ISO 22000:2005 7.2 requires that when defining and implementing PRP programs, the organization must consider and use all pertinent sources of information in its research. Requirements for each individual PRP must be defined, implemented, maintained, and effective. The requirement for creating a *documented* or *written* procedure varies per standard, but this author wants to again add a word of caution: "Document as little as possible and as much as necessary." Many times a flow diagram, checklist, or form can provide the basis for requirements without having the *war and peace* number of pages (Chapter 6). ISO 22000:2005 7.2.3 also states clearly that "verification of PRPs [must] be planned and PRPs [must] be modified as necessary" with documents defining "how activities included in the PRPs are managed."

It is absolutely critical that PRP requirements are defined in a manner that meets the needs of the organization and that objective evidence (records) are available that confirm the effectiveness of each PRP. Verification is proof of the effectiveness of each program.

It is recommended that a PRP verification chart be created that lists each PRP, the requirements, the responsibilities for verification, the outcome, and conclusions.

The PRP verification information can be effectively recorded on a one- to two-page chart that becomes an excellent tool for tracking, ensuring timeliness, and a record of effectiveness. Following is a list of the information that should be included on this verification chart:

- PRP and a brief statement of its requirements
- Control measures and the responsibility for the PRP

- Results of the food safety team's analysis of effectiveness including linkage to any required actions as a result of the analysis
- Frequency of the effectiveness evaluation, date completed, and next due date

As a record, this chart provides an excellent summary that can be used for reporting the effective analysis at management review meetings. The internal audit and GMP inspection programs must also be applied as a tool for the verification of the effectiveness of the PRPs.

Every associate is responsible for food safety. This is a message that must be communicated from top management and filtered throughout the operation. Without effective PRPs, then it is virtually impossible to have an effective food safety/ HACCP program. Is it possible to consistently produce a safe product without effective PRPs providing the foundation for the food safety program? An effective food safety program must be proactive in identifying and controlling hazards that could result in an unsafe product being consumed.

Specific requirements for Principles 1–7 related to as defined by Codex Alimentarius; (2003) Hazard Analysis and Critical Control Point [HACCP] system and guidelines for its application (www.codexalimentarius.org)) are applied by the food safety management standards as the building blocks for the food safety/HACCP programs (Chapter 17). The requirements for the five presteps are addressed in Chapter 16.

Since PRPs are the foundation for an effective food safety/HACCP program, then it is a must to review these before reviewing the development and maintenance of the food safety/HACCP programs (Chapters 16 and 17). It is important to note that these programs are not listed or presented in any special order. Many are included as part of this chapter, some have their own chapters, and others are addressed in Prerequisite Programs Part 2 (Chapter 15). Again, these programs are not presented in any specific order; each has its own critical role in food safety. There are other programs that may not be addressed in this text but may be identified in an organization's hazard analysis as a PRP. Do not be shy in identifying and giving credit to those programs that are supporting the production of safe products. Ensure that requirements for the PRP program are defined, activities are in compliance with these requirements, and records are maintained confirming compliance along with the effectiveness of the program.

A recommendation for educating associates about the importance of the PRPs is to review a sampling of current recalls in the news. As the group discusses the recalls and the statements of cause, it is safe to say that at least 90% of recalls are caused from ineffective or lack of specific PRPs. The food safety program is only as strong as its weakest link. A weak and/or ineffective PRP is an accident or recall waiting to happen.

MORE ON VERIFICATION

Be careful in identifying a metal physical (foreign object) hazard and then state that it is not a concern because the PRP program for metal control identifies the hazard through the placement and care of in-process magnets however, the program in reality is not being maintained as defined. That program states that magnets are strategically placed throughout the system, cleaned every day, and effectiveness confirmed through daily magnet inspections. Sounds great; but let's review the records to confirm that these activities are actually being performed. What was actually found were records indicating that the magnets had not been removed and cleaned in 3 months. The magnets were then removed for inspection. Each magnet was full of metal fragments with limited to no surface space available. Several nicks were noted on the magnet surface. It was this equipment that was referenced in the hazard analysis as effectively controlling an identified hazard. Imagine the possible outcome related to the identified hazard(s) or potential hazards that were not controlled at the magnet because obviously its effectiveness was questionable at best at the time of the magnet inspection. Do not assume that the testing requirements are always being done correctly. Verification is critical, and if records are not identified and maintained in compliance with the record control program, then the verification was not done.

The FSMS standards may require that the organization ensures that resources "for the establishment and maintenance of infrastructure" and also work environment as needed for compliance to the requirements of the standard (ISO 22000:2005 6.3, 6.4) are implemented and effective for the operation. Infrastructure may include buildings, workspace, associates, utilities, process equipment, supporting services, transportation, and communication. Effective infrastructure compliance also relates to information and requirements defined in the food safety/HACCP programs, PRPs, and the specific FSMS standard of choice. These requirements must be addressed to ensure compliance through the PRP programs. It is critical that the requirements for infrastructure and work environment needed to achieve food safety and conformity to product requirements in a compliant FSMS are defined and confirmed effective.

Records of verifications and modifications must be maintained in compliance with the record control program (Chapter 7).

KEY POINTS

a. PRPs are the foundation of an effective food safety/HACCP program.
b. PRPs must be defined, implemented, and effective.
c. A PRP verification spreadsheet listing each PRP, its requirements, responsibilities, findings, status, etc., is an excellent tool to track and record the effectiveness of the PRP programs. PRP effectiveness must also be monitored through the GMP inspection program and internal audit programs.
d. Every associate is responsible for food safety.
e. Associates must be educated in his/her responsibilities for food safety.
f. Be careful to not *over document*. Apply the rule "document as little as possible, but as much as necessary."
g. Records must be available to demonstrate compliance to defined requirements and the effectiveness of the PRPs.

h. Records must be identified and maintained in compliance with the record control program (Chapter 7).
i. The organization must ensure that statutory and regulatory requirements are addressed

COMMON FINDINGS/NONCONFORMANCES

a. The organization had not established and implemented PRP programs as required by the FSMS standard of choice.
b. The organization could not show evidence that the implementation of processes required for the realization of safe products had been completed with evidence available to confirm its effectiveness.
c. The organization's PRP verification programs had not been defined and implemented.
d. It could not be confirmed that statutory and regulatory requirements as required by the standard of choice had been identified, implemented, and confirmed in compliance.
e. Records were not available to confirm that PRP documents had been reviewed and accepted by the food safety team.

---

*Additional Food Safety Management System References (not all inclusive)

FSSC 22000:2010

　　ISO 22000:2005 Section 7.3

　　ISO 22002-1:2009 whole standard addresses various PRPs

SQF 7th Edition: Module 11 addresses various PRPs

BRC Issue 6: Sections 2.2, 4

IFS Version 6: Section 4 includes requirements for various PRPs

*Note: Always review the standard or visit the website for your chosen standard to ensure that all related requirements are identified and addressed.

CONSTRUCTION AND LAYOUT OF BUILDINGS

When first looking at an operation to determine if there are hazards, everything usually begins with the building or structure. How old is the building? How is it constructed? What is the layout? What was the scope of the original design? Hearing management comments such as "it is just an old building" is very common, but no matter what the comment, the focus must be on identifying hazards or potential hazards and either eliminating or reducing these to an acceptable limit. An old building with many exterior holes increases the likelihood of pests entering the building. Management may dream about a new exterior wall that does not have the funding. In the meantime, the holes must be plugged. This author remembers auditing an older facility that had major structural issues where the walls and floor came together and were not sealed. This area was a trap for dirty water and organic matter that collected and seeped under the floor area. The odor was horrific. This situation had to be corrected.

An organization must have a well-defined and implemented Good Manufacturing Practices (GMP) program. This must be primary and based on strong management commitment. Cleanliness and organization are critical. It is recommended that each organization have an ongoing GMP inspection program that includes a structured schedule, effective attention to findings, and evaluations to identify trends that may need more attention to permanently correct (Chapter 4). Good Housekeeping Practices (GHPs) and effective adherence to GMP requirements must become a way of life, not just a practice when *company* is coming.

This PRP addresses basic construction and internal and external layouts. While basic good housekeeping habits and adherence to GMPs are definitely important to compliance, this PRP also addresses some other key concerns such as hazards that may come from the environment, neighborhood, and traffic patterns.

The following is a list of some areas that organizations have incorporated as part of their GMP inspection program and also, in some instances, their preventive maintenance (PM) program:

- Roof inspections
- Inspections of openings into the plant, vents, etc., to ensure that protective screens remain intact
- Parking lot, yard areas, and driveways in good repair, which help to prevent standing water
- External areas clean and well kept with no vegetative growth growing next to the building
- External wall parameters with a 12–18 in. rock border to protect against weeds growing next to the building which aid with external pest control

When evaluating possible hazards in the environment, it is important to check the neighbors. Are there any possible hazards that could enter the process due to activities performed by the neighbors? Could the meat-processing plant next door contaminate the water? Is there a warehouse that may be infested? A good tool for this activity is to use Google Earth to check the neighborhood. This evaluation must be recorded and maintained as a record providing proof that the activity was done and its results recorded (Chapter 7). Any identified potential or existing hazards must be addressed, either by correcting, eliminating, or mitigating the risk.

The evaluation of external parameters also addresses access to the grounds, bulk storage, etc. Although there's a separate chapter on security, this is an important aspect to keep in mind. Not all organizations are able to have their facility protected by a fence, but the site boundaries must be identified, if not physically, then on a site schematic. It is necessary to ensure that all external access openings to the facility are in some manner locked or secured with key access. It is not uncommon during an audit to find doors propped open or just left open allowing direct access to the facility; many times the product and/or the process area is exposed. This also relates to the PRP for security (Chapter 14).

When considering compliance and hazard analysis, ensure that any regulations (federal, state, local, building codes, etc.) are known and compliant. It is up to the organization and its management to ensure that its operation is designed and maintained in a manner suitable to its operation and that it doesn't introduce any hazards or potential hazards into the process or product. If a hazard or potential hazard exists, it is the responsibility of the organization to ensure that the hazard is either eliminated or controlled. It is everyone's responsibility to ensure that no aspect of the process takes place where potential or existing food safety risks exist.

KEY POINTS

a. A GMP procedure can be the source for defining requirements for the external plant and grounds. A special procedure for this PRP does not have to be written. An inspection checklist can be created and used as both the training tool and the inspection record.

b. When performing a GMP Inspection, it is good to be able to make this type of notation: "The site is maintained in good order with vegetation controlled or removed. Exterior of facility is inspected in monthly GMP audits, with evidence provided that findings are addressed effectively and in a timely manner."

COMMON FINDINGS/NONCONFORMANCES

a. Site had not performed an *environmental* analysis of its neighbors for potential contaminates from the local environment.

b. Weeds were growing very close to the loading docks, creating a pest harborage.

c. There were several areas of standing water in the loading areas, which created mud slush (an excellent source for insect breeding), adjacent to the finished goods warehouse.

d. Access to boiler room along east roadside was open during processing, allowing access by unauthorized personnel.

e. Excessive trash and cigarette butts were spotted in several areas on the external grounds, including the main entrance of the shipping department.

f. External door to shipping office was not secured, allowing transport drivers ready access to entire facility.

g. Organization did not have a separate secure break area for transport drivers. Drivers were allowed to walk unescorted through the warehouse to access restrooms and break areas.

> ***Additional Food Safety Management System References (not all inclusive)**
>
> FSSC 22000:2010
>
> > ISO 22000:2005 Sections 6.3, 6.4, 7.3
> >
> > ISO 22002-1:2009 Section 4
>
> SQF 7th Edition: Section 11
>
> BRC Issue 6: Section 4.1
>
> IFS Version 6: Sections 4.6, 4.7
>
> **Note: Always review the standard or visit the website for your chosen standard to ensure that all related requirements are identified and addressed.*

LAYOUT OF PREMISES AND WORKSPACE

ISO 22002-1:2009 5.1 states that

> Internal layouts must be designed, constructed and maintained to facilitate good hygiene and manufacturing practices. The movement patterns of materials, products and people and the layout of equipment must protect against potential contamination sources. The building must provide adequate space; with a logical flow of materials, products and personnel, and the physical separation of raw from processed areas. Examples of physical separation may include walls, barriers or partitions, with sufficient distance to minimize risk.

It is critical that traffic patterns are identified, implemented, and associates trained on food safety, possible cross contamination, and quality. Many companies have had to reroute sales personnel and other associates, to eliminate passing through the processing and packaging areas to get to various internal areas. If the building design makes passage difficult, then, at a minimum, create a walkway along the wall parameter to control unauthorized traffic in the process areas.

Related to the internal layout of premises and workspace, the following are some common GMP concerns and recommendations collected from the food safety standards and basic GMP requirements. These may be monitored through the GMP inspection program and, in some instances, through the preventive maintenance (PM) program:

- Process area walls and floors must be washable or cleanable, as appropriate for the process or product risk. Materials must be resistant to the cleaning system applied.
- Wall floor junctions and corners must be designed to facilitate cleaning.
- Floors must be designed to avoid standing water.
- In wet process areas, floors must be sealed and adequately drained. Drains must be trapped and covered.

- Ceilings and overhead fixtures must be designed to minimize buildup of dirt and condensation.
- External opening windows, roof vents, or fan ducts must be insect screened and in good repair.
- External opening doors must be closed or screened when not in use (ISO 22002-1:2009 5.3).
- Sufficient space on both the interior and exterior walls must provide access to the processing area for cleaning and maintenance.
- An 18-in. perimeter is required around outside walls to ensure access for pest control. This area must be kept clean and free of storage.
- Dock doors (all doors to the outside) must be kept closed. Only authorized personnel should have access from the outside.
- Ceiling, overhead lights, and floor cleaning should be performed as defined on the master cleaning schedule (MCS).
- Cobweb removal may be tracked through preventive maintenance (PM) and work orders.
- No standing water (i.e., puddles, broken floors, plugged drains, etc.) is allowed.
- Water lines should be insulated to prevent condensation.
- Insulation should be sealed and of a cleanable surface.
- Overhead fans should be cleaned per the MCS.

This list, of course, is not complete. GMP compliance touches many and all aspects of the process. Complete GMP compliance should be monitored through the GMP training and inspection program with any identified concerns addressed in a timely manner. Ongoing items such as cobweb removal and light replacement should also be identified and tracked through the PM program.

During a warehouse audit, some overhead lights were burned out. When this was pointed out to the warehouse manager during our system audit, he stated that the problem would be identified and fixed on the next monthly GMP audit. That answer did not come across as taking ownership and responsibility for compliance, nor did it take into account how much darker the areas were because of the burned-out light bulbs.

Management must ensure that new installations are designed so that the equipment is used for its intended purpose and located in a manner that facilitates good hygiene practices and effective monitoring. Every effort must be made to ensure that equipment is located in a manner that permits access for operation, cleaning, and maintenance. Existing equipment must also meet these requirements with required resources applied to ensure compliance.

Microbiology laboratories must be designed, located, and operated so as to prevent contamination of people, plant, and products. They must not open directly onto a production area.

Temporary structures must be designed, located, and constructed to keep pests out and avoid potential contamination of products. A temporary structure could be a trailer, storage shed, temporary office on pallets, or additional equipment such as an auxiliary cooling unit that was required during a remodeling activity. A vending machine may be considered a temporary structure. Existing or potential hazards associated with temporary structures and vending machines must be assessed by the food safety team (i.e., hazard analysis) and controlled or eliminated.

KEY POINTS

a. Evaluate and address existing hazards related to internal design and layout as each relates to food safety.
b. Some organizations apply their food safety program's flow diagrams to map out traffic and other related patterns providing a clear *picture* of any existing or potential hazards that must be controlled.
c. Inline/online test facilities must be controlled to minimize risk of product contamination.

COMMON FINDINGS/NONCONFORMANCES

a. Not all issues related to internal layout including GMP compliance and *people* traffic patterns had been identified. Issues that had been identified had not been evaluated in order to correct, eliminate, or control the existing or potential hazard.
b. Review of records from monthly GMP inspections did not provide evidence that repeat items were trended and addressed through either the formalized corrective/preventive action process or the correction process.
c. There was limited evidence during this evaluation that *zones* related to food safety and process controls had been identified.
d. Overhead insulation was very dirty and in poor repair (torn in some areas and, missing entirely or torn with insulation falling out).
e. Vents and fans to the outside were either not screened or screen enclosures in poor repair.
f. Temporary storage sheds being used for current construction project next to ingredient receiving were in poor repair with trash and water collecting under the shed.
g. Floor under vending machines was very dirty and in very poor repair.
h. Tops of vending machines, refrigerators, and storage bins in the break room were being used for storage (i.e., supplies, lunch boxes, etc.).
i. Storage requirement stated that the storage temperatures must be between 40°F and 70°F with a relative humidity of less than 40%; however, at this time, there was no means to measure, monitor, and record this requirement.

---

*Additional Food Safety Management System References (not all inclusive)

FSSC 22000:2010

     ISO 22002-1:2009 Section 5

SQF 7th Edition: Section 11.2

BRC Issue 6: Section 4.3

IFS Version 6: Section 4.9

*Note: Always review the standard or visit the website for your chosen standard to ensure that all related requirements are identified and addressed.

UTILITIES: AIR, WATER, AND ENERGY

Depending on the FSMS standard, supplies of air, water, and energy may be broken down separately or included in one group such as *control of utilities*. The food safety standards want to ensure that:

> The provision and distribution routes for utilities, to and around processing and storage areas, are designed to minimize the risk of product contamination. The quality of utilities must be monitored to minimize product contamination risk. (ISO 22002-1:2009 6.0)

It is up to each organization to evaluate its *sources of utilities* for hazards or potential hazards that could affect food safety or the quality of its products or processes. This must also be addressed in the hazard analysis of the food safety HACCP program with a formalized risk assessment performed to provide accurate data on the process (Chapter 17). Generally, electric and water suppliers are the local city municipality, but if well water is used, additional concerns are likely to be present that would be unlikely if water is coming from a municipal (or city) source. Also, could there be a hazard if electricity is lost? Everything in the coolers and freezers could be lost if there is not an alternate plan to ensure against this occurrence.

In addressing the water supply, the organization must define what is required to confirm that the water supply is free of hazards and approved for use in the process. Each process is different. Some companies filter the water (i.e., 5 μm or 10 μm filter); some confirm chlorine content and then remove the chlorine so as not to add off-flavors; and some have very sophisticated water treatment systems (i.e., reverse osmosis). These requirements may also be different depending on whether the water is used as an ingredient, a processing aid (i.e., cleaning), or both. This relationship with the water supply must be evaluated and addressed based on the needs of the individual processes with records maintained to confirm compliance (Chapter 7). ISO 22002-1:2009 6.2 states that "the supply of potable water must be sufficient to meet the needs of the production processes. Facilities for storage, where the distribution is needed, and temperature control of the water, must be designed to meet specified water quality requirements."

As stated previously, the evaluation of the water supply or source must be part of the hazard analysis depending on the use of water in the process (i.e., as an ingredient, processing aid, etc.). ISO 22002-1:2009 6.2 states the following:

- "Water used as a product ingredient or in contact with products or product surfaces [must] meet specified quality and microbiological requirements relevant to the product including ice or steam (i.e., culinary steam).
- Water for cleaning or applications where there is a risk of indirect product contact (e.g., jacketed vessels, heat exchangers) must meet specific quality and microbiological requirements relevant to the application.
- Where water supplies are chlorinated, checks must ensure that the residual chlorine level at the point of use remains within defined limits.
- Non-potable water must have a separate supply system that is labeled, not connected to a potable water system, and prevented from refluxing into the potable system.
- It is recommended that water which comes into contact with the product should only flow through pipes that can be disinfected."

It is important that an effective water treatment and testing program that confirms water is free of *hazards* meets potable water standards and relates specifically to the processes and products must be defined. Objective evidence (records) must be available to confirm that testing results are being reviewed and confirmed compliant. Records must be identified and maintained in compliance with the record control program (Chapter 7).

Air quality and ventilation are often forgotten sources of hazards. Requirements for evaluating how air quality and building ventilation relate to existing or potential hazards must be defined and addressed in the hazard analysis (Chapter 17). Any existing issues or potential issues must be identified, then either eliminated or corrected. If necessary, a risk assessment should be performed that provides data to confirm the mitigation of the risk (Chapter 15).

ISO 22002-1:2009 6.4 is very specific stating that

- "The organization must establish requirements for filtration, humidity (RH%), and microbiology of air used as an ingredient or for direct product contact. Where temperature and/or humidity are deemed critical by the organization, a control system must be put in place and monitored.
- Ventilation (natural or mechanical) must be provided to remove excess or unwanted steam, dust and odors, and to facilitate drying after wet cleaning.
- Room air supply quality must be controlled to minimize risk from airborne microbiological contamination. Protocols for monitoring and control must be established in areas where products that may support growth or survival of airborne contamination are exposed.
- Ventilation systems must be designed and constructed so that air does not flow from contaminated or raw areas into clean areas. Specified air pressure

differentials must be maintained. Systems must be accessible for cleaning, filter changing, and maintenance.
- Exterior air intake ports must be examined periodically for physical integrity."

If temperature and humidity are not a concern for the product, this must be stated with the decision justified. If room air supply, due to the nature of the product, is acceptable for the process, explain how this is known and the justification for the decision.

Many times, compressed air and other gases, their existence and role in the process are also included in the category for utilities. The primary location for addressing these items is through the hazard analysis as a product contact surface, processing aid, or ingredient. For example, CO_2 in a carbonated soft drink would be considered a compressed gas. Keep in mind, depending on the products and processes used, the potential for a hazard may be through either indirect or direct contact with either the product or a product contact surface. Do not ignore this requirement. There must be evidence that it has been evaluated.

Lighting (or lack of) must be evaluated. It is recommended that each area to be evaluated ensures that sufficient lighting for the activity being performed is available. A hazard or potential hazard could exist because lighting is insufficient. Most companies do an analysis of the various areas to measure the light (candling) intensity. This, along with visual evaluations and management decisions, is used to determine what is necessary, then ensuring that the required level is obtained and maintained. There are guidelines for recommended candling levels on the FDA website. It is recommended that management do their own evaluation, setting the best parameters for the individual operation. Justification for decisions can be recorded in the food safety program or in the food safety team meeting minutes. Linking this to the glass and brittle plastic program (see "Glass and Brittle Plastic" section), all light bulbs must be either enclosed to protect against breakage or be shatterproof. Confirm whether light bulbs do or do not contain mercury. Mercury-containing bulbs are surfacing for some large warehouses. Is this a potential hazard and, if so, how is it controlled? Proof that the bulbs are shatterproof must be available. The source must be identified and managed through the critical food safety and quality supplier program (Chapter 13).

KEY POINTS

a. Steam and ice, if directly contacting the product or product contact surface (equipment), must be evaluated for existing and potential hazards with results recorded and actions taken (Chapter 17).
b. When identifying the use of water, remember the janitorial cleaning activities, which most frequently use potable city water.
c. Make a statement in the food safety program as to the source and use of water. An example may be no nonpotable water in use at this site; potable city water is used for cleaning floors and external equipment. Water used as an ingredient and for cleaning product contact surfaces passes through the reverse osmosis water treatment unit. This water is monitored and tested as

defined in the water treatment program with records maintained in compliance with the record control program (Chapter 7).

d. Boiler chemicals must be approved for use in food manufacturing. Many times, the approvals are listed on the label; however, if the label does not state this, a letter from the manufacturer must be on file and maintained as a record (Chapter 7).

e. Water treatment chemicals including boiler chemicals are considered chemicals, and all chemicals must be kept secured or locked in a separate area protecting against cross contamination (see "Measures for Prevention of Cross-Contamination" section).

f. Ensure that the nonpotable water supply has a separate delivery system that in no way could cross contaminate the potable or water-treated systems.

g. Maintaining the water treatment system (i.e., testing frequencies, filter replacement, etc.) can be managed effectively through the PM program.

COMMON FINDINGS/NONCONFORMANCES

a. Records did not confirm that an analysis of the utilities for possible hazards had been completed.

b. It was stated that the incoming water supply was filtered; however, the requirements for the filtering process (i.e., filter size, replacement frequency, etc.) were not defined, nor were there records available to confirm that any activities were taking place.

c. Requirements that water treatment chemicals be approved by the American Water Works Association were not formally defined, nor could the associate responsible for ordering this material demonstrate familiarity with the purchasing requirements.

d. The external service supplier who was responsible for the water treatment program was not identified and maintained as a critical food safety and quality supplier (Chapter 13).

e. Requirements to protect against cross contamination through indirect means (i.e., jacketed tanks, boiler, heat exchangers, etc.) had not been evaluated and recorded through the hazard analysis.

f. Site had not identified its water sources, hazards, and methods of control. Water was being used as an ingredient and a processing aid.

g. Although records were available to confirm that water testing was being performed at the defined minimum frequency of once per year, these records were not identified and maintained in compliance with the record control program.

h. Specific food safety requirements for water used for cleaning, which had been identified as a potential food safety hazard (i.e., cross contamination with primary food packaging material), were not defined.

i. Criteria for water treatment chemicals approved for manufacturing, although stated as a possible indirect product contact surface or processing aid, had not been defined.

j. Records were not available to confirm that boiler chemicals were acceptable for food manufacturing operations, nor was the person responsible for ordering these chemicals familiar with this requirement.

k. Boiler chemicals stored in the boiler room were not secure. The boiler room as designed could not be locked or secured.

l. Organization did not have data related to a current evaluation of air quality for existing or potential hazards such as records confirming effectiveness of the current *filter* program.

m. Site did not have a formalized procedure for ensuring that air that had direct product contact was free of food safety hazards.

n. Maintenance stated that only oil-free compressors were in use; however, physical review of the compressors indicated that they were not oil free. Records confirming that this had been evaluated and that the compressors were maintained per the PM program were not available.

o. Records were not available to confirm that a light intensity evaluation had been performed in the processing areas.

p. There was not a defined program based on the risk of air and the role of the air filters used in the compressed air system. For example, the purpose of the 5 μ air filter was to remove particulates; however, what particulates and whether the particulates posed a food safety hazard was not defined.

---

*Additional Food Safety Management System References (not all inclusive)

FSSC 22000:2010

    ISO 22002-1:2009 Section 6

SQF 7th Edition: Section 11.5

BRC Issue 6: Section 4.5

IFS Version 6: Section 4.9

*Note: Always review the standard or visit the website for your chosen standard to ensure that all related requirements are identified and addressed.

WASTE DISPOSAL

Waste flow, the waste disposal process and related contaminants, must be evaluated for existing or potential hazards and addressed through the food safety program. A risk assessment must be performed for any concerns that cannot be corrected or eliminated, with records maintained to justify the outcome. Cross contamination from mishandled or improperly labeled waste is of real concern, but this can be controlled through proactive and preventive actions.

Trash and waste containers must be covered and properly labeled. Trash and waste containers (including those for hazardous materials) must be managed in a manner that protects against cross contamination. This can be accomplished by developing a color coding system for trash containers. It is not uncommon for an organization to use identical trash containers throughout the operation. A trash container from the raw materials area, which has not been properly cleaned and sanitized that ends up in the finished product area, can be a *disaster waiting to happen.* Another *problem* would be a trash container used externally for dirt, cigarette butts, and various other trash on the grounds. That trash container, identical to the ones in the finished product area, could end up in a process area creating a significant opportunity for cross contamination. Color coding the containers is simple. Once associates become used to the program, the coding becomes *user friendly* and effective.

A process must be defined that addresses the segregation, storage, and removal of waste. The food safety standards combined with GMPs and basic common sense makes it clear that waste must not be allowed to accumulate, especially in process and storage areas where cross contamination can occur. The waste removal program must include the storage and removal of hazardous waste in a manner to protect against cross contamination and to ensure that it does not include existing or potential hazards. ISO 22002-1:2009 7.2 defines the following process requirements for waste and inedible or hazardous substance containers:

- "Must be clearly identified for the intended purpose;
- Located in a designated area;
- Constructed of suitable, impervious material which can be readily cleaned and sanitized;
- Closed when not in immediate use; locked, if hazardous;
- Locked where the waste may pose a risk to the product."

A process must be in place to ensure that trademark-sensitive waste (labels, labeled products, or printed packaging, etc.) is disfigured or destroyed in a manner that ensures that it cannot be reused. Removal and destruction must be performed by a service supplier identified and managed as a critical food safety and quality supplier (Chapter 13). Depending on the company policy, management may also want to ensure that destruction is witnessed and recorded by a company associate(s). Records of destruction (item, date, amount, process) must be identified and maintained in compliance with the record control program (Chapter 7).

Drains and drainage, which can be linked to engineering and process designs, are many times included in the PRPs that address waste and waste removal. Drains must be designed, constructed, and located to eliminate or minimize the likelihood of cross contamination; however, issues do occur. The first step is to ensure that drains are designed properly to fit the process and the process flow. Second, the *drain flow* must be traced or mapped. How the drains actually flow and the potential for a drain from a raw materials area flowing through the finished product rooms, creating cross contamination, must be evaluated. Drain lines must not flow over processing areas, ingredients, packaging materials, and/or finished product storage, or through any area that could introduce a risk to the product. If this is found during the evaluation, the situation must either be corrected or eliminated. If this is not possible, then a risk assessment must be performed to determine the actual risk, mitigation opportunity, and any other action that may be required to protect the safety of the product.

The drain flow report, justifications, and next steps must be recorded and maintained in compliance with the record control program (Chapter 7).

KEY POINTS

a. Recommend, as a daily task, emptying and cleaning of trash cans as listed on the MCS.
b. The suppliers used for waste handling, removal, recycling, etc., must be identified and managed in compliance with the food safety and quality critical suppliers program.
c. Recommend establishing a color coding program for handling waste receptacles.

COMMON FINDINGS/NONCONFORMANCES

a. Records were not available to confirm that an analysis of the *waste flow* had been completed.
b. Records were not available to confirm that a formalized program for managing waste to protect against cross contamination had been defined and implemented.
c. Gray trash containers used in production areas were not labeled as trash nor were they covered.
d. Outside dumpsters next to the shipping and loading entrance and next to the building were not covered. There was very strong odor in the area. Trash, debris, and organic matter collected under the dumpsters. Flies and other insects noted in the area.
e. Open top dumpsters/trash containers were not covered when not in immediate use.
f. A program was not in place to destroy/deface packaging material bags before they were discarded.
g. Records did not confirm that the organization had performed an evaluation of the drain system to ensure that there were no issues that would result in a food safety hazard.

---

*Additional Food Safety Management System References (not all inclusive)

FSSC 22000:2010

     ISO 22002-1:2009 Section 6

SQF 7th Edition: Section 11.5

BRC Issue 6: Section 4.5

IFS Version 6: Section 4.9

*Note: Always review the standard or visit the website for your chosen standard to ensure that all related requirements are identified and addressed.

EQUIPMENT SUITABILITY, CLEANING, AND MAINTENANCE

This PRP defines the basic requirements to ensure that equipment suitability, cleaning, and maintenance are designed and constructed to facilitate appropriate cleaning, disinfection, and maintenance. The focus is to ensure that equipment, cleaning, and/or maintenance do not introduce hazards or potential hazards (potential contaminants) including required activities and the effectiveness of measures taken to prevent contamination. Equipment must be designed for its specific use, and the operation must adhere to that use. Changes or revisions must be evaluated by the food safety team as part of the change management program, which may also be called *commercialization*. Larger companies often assign equipment design to the corporate engineering department; however, this does not absolve the specific site of the responsibility of ensuring that the equipment functions properly and does not contribute to food safety hazards. Hazard analysis must be done for all new equipment. The food safety team is responsible for performing the hazard analysis for all new equipment should that equipment be considered a product contact surface, which becomes part of the food safety program. If there is a concern, the hazard must either be corrected or eliminated (Chapter 17). If this is not possible, then a risk assessment must be completed that provides justification and mitigation information that ensures that the product or process is not at risk due to food safety–related hazards (Chapter 15).

Whether an organization is part of a corporate structure or a stand-alone operation, management must ensure a firm understanding of its role in regard to equipment suitability and hygienic design. It is critical that the existing structure and design be evaluated to ensure that the process for identifying and installing new equipment is best for the operation and that standards used for purchasing are understood and available. Just because equipment works in one location, at a sister company, or because a sales person states that the equipment performed great for similar operations, it may not be the best choice for another similar operation. There are many variables with equipment design and installation. Equipment or other new installations (i.e., a new lubricant) that meet the criteria of a food contact surface must be evaluated and documented through the food safety program hazard analysis (Chapter 17).

ISO 22002-1:2009 8.2 clearly defines what is expected relating to hygienic design:

- "Smooth, accessible, cleanable surfaces; self draining in wet processing areas;
- Use of materials compatible with intended products and cleaning or flushing agents;
- Framework not penetrated by holes or nuts and bolts;
- Piping and ductwork cleanable, drainable, and with no dead ends;
- Equipment designed to minimize contact between the operator's hands and the products."

One time early in this author's career, she remembers having difficulty convincing an operator in the tanker receiving area that the lines had to be capped. In those days many of the caps were a thick, see-through plastic. He did cap the line, and about ten minutes later there was a frog in the line trying to get out.

When evaluating equipment design, it is recommended that the 3-A Sanitary Standards, Inc. (www.3-a.org) be visited. This is an independent not-for-profit corporation dedicated to advancing hygienic equipment design for the food, beverage, and pharmaceutical industries. 3-A SSI represents the interests of three stakeholder groups with a common commitment to the promotion of food safety. This organization got its start in the 1920s with a strong focus on the dairy industry. In addition, HACCP international (www.haccp-international.com) and NSF (www.nsf.org) are two very good sources for food contact equipment information and guidance related to food safety. It is recommended that food industry professionals become familiar with these equipment standards when considering key decisions that must be made during installations and renovations.

Product contact surfaces must be constructed from materials designed for food use. They must be impermeable and rust and corrosion free. A hazard analysis must be completed for all product contact surfaces. This is addressed in more detail in the chapters on the development of the food safety program (Chapter 17). This is an example of how programs overlap. So many aspects relate to other programs or activities that many times it can be confusing. It is critical that the food safety team has the resources and time to ensure that gaps and connections between programs, processes, requirements, and actions are connected. It is important to acknowledge this relationship. There are many Internet sites that provide information on product contact equipment related to potential hazards or justification for not having a hazard identified. One Internet site is www.lawserver.com. This website is very comprehensive, difficult to maneuver, but worth the effort. First, find the food industry reference and then experiment with the different links. While providing some good data, it is not the only source. Always reference or link to the source for substantiation should the decision be challenged or the situation change. From a maintenance and engineering focal point, it is critical that product contact equipment be designed for food processing. Any concerns related to existing or potential food safety hazards must be identified and then addressed in a manner that ensures the production of a safe product.

ISO 22002-1:2009 8.4 specifically addresses equipment used for thermal processes stating that the "temperature gradient and holding conditions" must be defined in relevant product specifications. It goes on to state that this "equipment must provide for the monitoring and control of the temperature." This would also be addressed through the hazard analysis, depending on the role of the thermal process, with any significant hazards identified and controlled through the measure for the specific product as determined by the food safety team. Requirements to confirm accuracy of the equipment are further discussed in Chapter 11.

ISO 22002-1:2009 8.5 defines the requirement that all plant utensils and equipment be cleaned at defined frequencies. ISO 22002-1:2009 8.5 also states that "the programs [must] specify what should be cleaned (including drains), the responsibility, the method of cleaning (e.g., clean in place [CIP]) the use of dedicated cleaning tools, removal or disassembly requirements, and methods for verifying the effectiveness of cleaning." Requirements related to specific cleaning and sanitizing programs are addressed in the section for Sanitation: Cleaning and Sanitizing. Confirmation of effectiveness is discussed through verification activities. This focus must be on the equipment so that it is designed to be cleaned and sanitized in a manner that prevents or eliminates potential food safety hazards. Like most PRPs, this is everyones responsibility!

ISO 22002-1:2009 8.6 specifically addresses preventive and corrective maintenance by requiring that a PM program be in place that includes "all devices used to monitor and control food safety hazards." Examples of these include, but are not limited to, screens, filters (including air filters), magnets, metal detectors, x-ray detectors, automatic samplers, air control, CIP systems, and conveyors.

This requirement also states that

- "Maintenance requests that impact product safety [must] be given suitable priority;
- Corrective maintenance [must] be carried out in such a way that production on adjoining lines or equipment is not at risk of contamination;
- Temporary fixes must not put product safety at risk." Items such as string, tape, wire, and rubber bands should not be used as temporary fixes. If temporary repairs are allowed, it must be ensured that these repairs do not result in a food safety hazard. If breakdown affects food safety, then equipment must be shut down and repaired immediately.
- "Lubricants and heat transfer fluids [must] be food grade where there is risk of direct or indirect contact with the product." The storage and handling of both food grade and non-food grade maintenance chemicals are managed through the prerequisite program for chemical control.

An aspect of this element, which has resulted in vast improvement overall, is the requirement that a program be put in place that formalizes the release of equipment back to production after maintenance has completed their work. This relates specifically to product contact surfaces, but many organizations have expanded it to all equipment. There must be a program defined and supported by records confirming that maintenance is communicating to a designee (i.e., operations quality) once work is completed to ensure that the equipment is cleaned and sanitized per defined requirements prior to use. This release back to production includes cleanup, sanitizing (where specified in process sanitation procedures), and preuse inspection.

Requirements defined for PRPs must apply to maintenance areas and maintenance activities. Maintenance associates must be trained in the food safety hazards associated with their activities. Maintenance associates have a very specific critical role in food safety and quality, with this responsibility overlapping throughout the entire FSMS. Existing or potential hazards related to cross contamination can be traced to maintenance tools, uniforms, shoes, and just a general misunderstanding of their roles. As food safety program requirements have expanded, organizations have provided more intensive food safety education to the maintenance teams. This has included, in some instances, significant financial investment to prevent the potential of cross contamination. It is common for quality managers or operators to seriously ask a maintenance technician where a specific tool has been prior to using it on a product contact surface. Dedicated tools (recommend applying color coding to the tools) have been a positive step forward for many operations.

As previously stated, maintenance associates play a key role in the management system. It is critical that system requirements be incorporated into this department's goals and activities.

The following are system elements and activities that also may relate to maintenance department responsibilities:

- Controlled documents (internal and external) in compliance with the document control program (Chapter 6).
- Records demonstrating compliance and maintained as defined with the record control program (Chapter 7).
- Program for releasing equipment to production after maintenance repairs is defined to ensure that equipment is clean and free of any hazards prior to production.
- Defined criteria for confirming competencies (skills), both initially, and as ongoing training needs (Chapter 8).
- Measurable department goals and objectives that relate not only to measurable food safety and quality system objectives, but also to the management system overall (Chapter 3).
- Training related to roles and responsibilities for food safety, quality, and the effectiveness of the management system (Chapter 8).
- Application of supplier management requirements for critical food safety and quality supplier management program (Chapter 13).

Management must apply the necessary resources to ensure that gaps related to maintenance department compliance are identified and addressed in an effective and compliant manner. Initially the most common gaps are those related to cleanliness of tools and carts; potential for cross contamination (i.e., tools, uniforms, etc); use and storage of food grade vs. nonfood grade chemicals; and ensuring that equipments cleaned and inspected prior to resuming production. Records confirming compliance must be identified and maintained per the record control program (Chapter 7).

Food-grade chemicals and non-food grade chemicals must be stored separately. If stored together, then the food grade should be stored above the non-food chemicals. Every effort must be made to protect against cross contamination. This includes when chemicals are transported to the work areas. It is common to see a mixture of food-grade chemicals and non-food grade chemicals scattered on maintenance carts and in the equipment areas. At that point, the opportunity for cross contamination becomes a concern. In addition to previously mentioned concerns and recommendations, consider the following:

- Evaluate PMs to determine which are related to food safety. If possible, group together in the PM program under a food safety category. If defined that an associate determines food safety–related items and assigns them accordingly, evidence must be available showing that the associate is trained and knows how to identify food safety items. This process, however, leaves much opportunity for errors or omissions and is not recommended. It is best to prepare a list of food safety critical PM requirements with the assistance of the food safety team and use this list for reference and assignments.

- Define the ordering process so that references, catalogs, and other specifics provide the purchasing associate ready access to food safety–related supplies, including chemicals (food-grade lubricants) that may have a *specific* requirement. It is not uncommon to find the wrong chemical in stock just because it was ordered by mistake and not caught either through the ordering process or upon receipt.
- At a minimum, depending on the areas of responsibility, train maintenance associates in the same basic food safety as everyone else in the operation (i.e., GMP, role in food safety, what to do if any issues are observed, etc.).
- Ensure that all chemicals and lubricants used by the maintenance associates meet defined requirements for chemical control (See section on Sanitation).
- Storage areas must be maintained in compliance with GMPs and GHPs. Storage must be organized, off the floor, and pest free. Floors must be kept clean.

KEY POINTS

a. Recommend that the master cleaning schedule (MCS) include drain cleaning.
b. Color coding of cleaning equipment and maintenance tools should be defined and implemented to protect against cross contamination.
c. Maintenance manuals required by the system, but not created within the organization, must be identified and managed as external controlled documents (Chapter 6).
d. Food safety–related PM items must be given top priority. Recommend that 100% food safety–related PMs are identified as a food safety measurable goal.
e. Temporary fixes must not put product safety at risk. Items such as string, tape, wire, and rubber bands should not be used as temporary fixes. If *temporary repairs* are allowed, it must be ensured that these *repairs* do not result in a food safety issue.
f. The maintenance associates must be trained in food safety–related requirements including those for GMP and PRPs.
g. There must be a program put in place that formalizes the release of equipment back to production after maintenance has completed their work.
h. Contractors and other external service companies must be identified and maintained in compliance with the critical food safety and quality supplier management program.
i. Contractors and visitors working with the maintenance department must be trained in related food safety requirements.
j. Records must be maintained to demonstrate compliance with defined requirements as related to maintenance department FSMS responsibilities and maintained in compliance with the record control program (Chapter 7).

COMMON FINDINGS/NONCONFORMANCES

a. Although it was stated verbally that there was a change management program in place, records were not available to confirm that the recent equipment installation had gone through this process prior to installation.

b. Current maintenance program used to identify and track completion of PM activities did not have the capability to prioritize food safety–related items.

c. It was stated that a food safety measurable objective existed that required a minimum of 98% of food safety–related PM items be completed on time. However, records were not available to provide this data. It was also stated that this was done manually, then estimated and reported verbally to management during staff meetings. Review of a sampling of records indicated that approximately 50% of those scheduled for the previous quarter were either incomplete or completed after the due date with no action identified to address this situation.

d. The food safety programs were still being developed, thus it could not be confirmed if the food safety team had effectively identified and performed a hazard analysis on product contact surfaces and processing aids used by maintenance associates.

e. It was observed that maintenance was using tools on the finished product lines that had earlier been used to adjust the conveyor for raw product. There was no evidence that these tools had been cleaned and sanitized prior to use on the finished product line.

f. It was stated that a program for color coding maintenance tools to protect against cross contamination had been developed and implemented; however, through interviews and review of records, implementation could not be confirmed.

g. Although it was stated that daily drain cleaning was on the MCS, records related to completed MCS tasks did not provide objective evidence confirming compliance with the requirements.

h. Site had not implemented a program for managing *tools* to protect against cross contamination.

i. It was noted during the audit that air was used to form the package; however, records were not available to confirm that the air supply had been evaluated through the hazard analysis for product contact surfaces.

j. Records were not available to confirm that the PM checklists were completed as scheduled. Note: it was stated that paper copies were destroyed once completed; the current PM system kept limited historical data.

k. The external service company for contracted maintenance management was not identified and managed as a critical food safety and quality supplier.

l. Records and documentation were not available to confirm that maintenance related training requirements related to specific skills, food safety, and PRP management had been defined, implemented, and maintained.

m. Maintenance manual used to calibrate the flow meters was referenced as required to complete the task in the work order system, but manual was not available, nor was it listed as an external controlled document.

n. Although stated verbally that it was done in practice, requirements for (including training) performing corrective maintenance in a manner that did not contaminate materials and products were not defined.

o. Requirements for temporary fixes (i.e., string, tape, and wire) were not addressed.

p. Screen replacement was not being managed through the PM program. Records related to replacement were not available.

q. There was no verification program defined through the PM program for completed PMs. Current process was for PMs to stay open indefinitely until completed, thus on-time completion could not be tracked.

---

*Additional Food Safety Management System References (not all inclusive)

FSSC 22000:2010

     ISO 22002-1:2009 Section 8

SQF 7th Edition: Section 11.2

BRC Issue 6: Sections 4.6, 4.7

IFS Version 6: Sections 4.16, 4.17

*Note: Always review the standard or visit the website for your chosen standard to ensure that all related requirements are identified and addressed.

MEASURES FOR PREVENTION OF CROSS CONTAMINATION

The hazard analysis is used to identify existing or potential food safety hazards that may occur through cross contamination. The hazard analysis must make reference to the PRPs that are in place to control or eliminate these hazards. There are many PRPs that relate to the prevention of cross contamination. Examples of these programs include

- Basic GMP
- Building design, traffic patterns (people, trash, equipment, etc.)
- Employee practices
- Approved chemicals list
- Chemical control program
- Glass and brittle plastic program
- Warehouse, shipping, distribution program
- Security program
- Pest control program
- Maintenance programs (i.e., PM, corrective maintenance, etc.)
- Storage practices (raw ingredients, packaging materials, finished product, etc.)
- Allergen control
- Incoming receipt and inspection program

- Contractor and visitor training program
- Trailer security and seal inspection program
- Utility control program (air quality, water treatment, etc.)

Typically, when reviewing a process for potential or existing cross contamination, the focus is divided into the three categories of hazards: microbial (pathogenic), chemical, and physical. Potential hazards must be identified through the hazard analysis portion of the food safety program, but this author has found through experience that it reinforces the food safety program overall to apply common sense and preventive thinking in looking for and addressing cross contamination (Chapter 17).

Microbial pathogenic cross contamination could be in the design of the equipment, such as allowing raw product to enter pasteurized product. It can also be caused by cross contamination from associates that do not wash their hands or use dirty gloves. Cross contamination can occur if equipment is not cleaned properly; this could result in either microbial and/or chemical contamination if cleaning chemicals are not sufficiently rinsed from the system. Historically, there was a microbial pathogen hazard created in a dairy when the cleaning and sanitizing agents in the cleaning system for milk crates ran out. The crates were being rinsed with hot water, which was not sufficient enough to kill the pathogens on the crates. The crates contaminated the exterior of milk bottles with a pathogen, which in turn made many people very sick.

An example of chemical cross contamination would occur if the allergen control program was ineffective and products were cross contaminated with allergens. A chemical cross contamination hazard could occur if nonfood-grade chemicals are used where food-grade chemicals are required. Would the misuse of chemicals (wrong chemical, wrong amount) fit this category? Historically, there have been several recalls that were traced to the use of chemically treated pallets. Organizations must make the decision as to whether it will or will not (of course recommend keeping the hazard out of the operation) limit its pallets to nonchemically treated pallets. Once this is decided and implemented, then the organization must communicate the requirement to the supplier and ensure that the agreement is in place. The pallet supplier must ensure compliance and provide a record confirming that it provides only nonchemically treated pallets. Communication and approval of suppliers are addressed in more detail in Chapters 3 and 13, respectively. *Dr. Tatiana Lorca, manager, Food Safety Education and Training, Ecolab, Food & Beverage Division*, provides the following guideline not only for the use or nonuse of chemically treated pallets, but also for performing the risk analysis to determine how an organization should react to industry areas of concern that could relate to its products and processes.

None of the applicable standards (SQF 2000, BRC Food or [ISO 22002-1]) require that the pallets used to ship chemicals meet any specific certification or requirement. Regardless of the standard, it comes down to the potential risk that pallets supplied by [the] chemical provider (non-food products) pose to [the organization's] finished products, raw materials, packaging and ingredients. [Depending on the product, process, and results of the risk assessment], a requirement on the safety of pallets should be restricted to those used to ship finished goods, food products, and those used to stage raw materials/ingredients/packaging within the plant. As a best practice, [the organization] should not be using pallets which previously held non-food chemicals to store finished product, raw materials, ingredients or packaging. Remember to use common sense when dealing with pallets to avoid potentials for cross contamination.

Physical cross contamination may occur if the glass and brittle plastic program is not effective (see section on Glass and Brittle Plastic).

A company using treated wooden pallets could result in the contamination of a product. What if someone put his/her dirty gloves on a conveyor or on a piece of equipment, and then these somehow fell into the product? Obviously, this would be an example of physical contamination. Those gloves may have had multiple uses before being introduced into the product. Ever think about microbial pathogenic cross contamination? Some organizations have strict rules that require associates to change clothes (including their shoes) when entering process areas. For facilities that do not address this contamination, ever wonder where someone's shoes have been when worn outside the plant and then introduced into areas of product processing?

Basic GMP requirements do not allow wood in processing areas. This includes wooden handle brooms, wooden ladders, and wooden tools. Many processing facilities still use wooden pallets. Just be sure that these are in good repair. Wooden splinters scattered from pallets or wooden splinters left behind on a product when a top pallet is moved are very common. If operators aren't extremely careful, splinters can enter the ingredients when opened.

More specific information related to cross contamination opportunities and control are addressed in other PRPs.

KEY POINTS

a. Every effort must be made to prevent cross contamination.
b. Communication about the prevention of cross contamination opportunities must be relayed to all associates and must become everyone's responsibility.
c. To some degree most of the PRPs are related to cross contamination.
d. Records confirming that observations related to existing and potential cross contamination opportunities (i.e., GMP inspections, etc.) must be maintained in compliance with the record control program (Chapter 7).

COMMON FINDINGS/NONCONFORMANCES

a. The organization had not performed a complete analysis to identify and address any existing and/or potential sources of cross contamination.
b. A policy was not in effect for the use of loose fastenings (e.g., drawing pins and staples) in production and storage areas.
c. Wooden splinters were observed scattered on top of a semi-opened bulk tote of ingredients.

d. The food safety program stated that due to the nature of this product, "microbiological cross contamination" was not a concern; however, the justification for this decision and/or supporting objective evidence had not been recorded.

---

*Additional Food Safety Management System References (not all inclusive)

FSSC 22000:2010

     ISO 22002-1:2009 Section 10.0

SQF 7th Edition: Section 2.4.7

BRC Issue 6: Section 4.9

IFS Version 6: Section 4.12

*Note: Always review the standard or visit the website for your chosen standard to ensure that all related requirements are identified and addressed.

ALLERGEN CONTROL

An allergen is defined as a food protein substance that an individual may have a reaction to causing the body's immune system to react resulting in an allergic reaction. An allergen control program identifies and controls the presence of allergens in product either naturally or per a controlled recipe or specification. Depending on the process and the products manufactured, this can be a challenging program. Products that contain allergens must be accurately labeled identifying their presence. The process to manufacture products with allergens must be designed to absolutely prevent any chance of cross contamination with nonallergen products.

For example, an organization that produces many different dry product blends may have any number of combinations of allergens in its products. This organization, by design, must define its process to totally protect against cross contamination between products with allergens and those without allergens. Any cross contamination between products with different allergens must be prevented. For example, a product that might contain wheat and milk products (product one) cannot be cross contaminated with a product that also contains peanuts (product two). If the ingredient label for product one does not identify peanuts as an ingredient, and if cross contamination is not 100% prevented, an unsuspecting consumer with a peanut allergy very likely will end up with a serious trip to the hospital from consuming the undeclared peanut product. Typically, consumers with food allergies are very sensitive. Very small amounts of an allergen can and do cause a very serious, and often life-threatening, reaction. Some organizations actually include a warning statement stating that the products are produced in a facility that does manufacture products containing allergens; the products may be unintentionally exposed to an allergen. This type of statement may be similar to the following: "This product may contain peanuts" or "this product may contain egg products."

The best protection is to have dedicated manufacturing lines, but this is not always possible. Equipment such as weighing pans, blending buckets, scoops, and cleaning utensils should be color coded to ensure use only in allergen products or nonallergen products, but not both. Some organizations are designed so that the allergen processing and packaging areas are completely separate from the nonallergen areas. Associates from each specific area wear colored uniforms or bump hats identifying which area they are assigned to. Traffic between the areas is either not allowed or controlled so that it flows from nonallergen- to allergen-containing areas.

All allergen-containing materials that are handled or produced by the organization must be identified. This includes ingredients, finished products, and also products that may be stored and distributed by the organization, but not produced at the site. This is done through the development of the food safety/HACCP programs. Allergen ingredients or allergen-containing ingredients may be addressed through the hazard analysis for ingredients. The hazard analysis for the process steps identifies those steps that could introduce an allergen as a chemical hazard. Once this is identified, the food safety/HACCP team must evaluate the hazard, its risks, and significance and the PRPs in place to control or eliminate the hazard (Chapter 17).

Once an allergen is in the facility, the potential for cross contamination can happen anywhere in the process, from the receipt of the ingredient, through raw material storage, through blending, filling, packaging, finished product, storage, and distribution.

As mentioned previously, color coding of equipment and utensils is effective. Associates must be educated and trained to understand the requirements and the seriousness of what could happen if not controlled. Production runs on shared equipment should run all nonallergen products first, and then the allergen-containing products last, prior to cleanup. This way, if there is any cross contamination, it is nonallergen to allergen rather than vice versa. This can still be tricky if there are multiple products with different allergens. Before an organization makes a commitment to manufacture products with allergens, there must be careful and effective design review to absolutely ensure that this can be done without cross contamination between products. The only way that this can really be guaranteed is to have dedicated equipment and associates in areas that totally separate the nonallergen products from the allergen products.

Cleaning and product changeover procedures must be validated to confirm the effectiveness of totally removing any allergen residues from previous productions. This is critical to the effectiveness of the program. There are many different allergen test kits that can be purchased and used for this purpose. The organization must review the industry options and make the best choice possible for its products. A program for the application of the test kits must be defined and implemented with trained associates responsible for performing and recording the application. Records must be maintained that provide the objective evidence (proof) that test results are in compliance. Once, this auditor was doing an audit and the operator was using a well-known test kit to confirm that any remaining residue after cleaning was at an acceptable level. The operator stated that he wrote the test result on a post-it note and placed it on his supervisor's desk. The supervisor stated that he entered this information into the database. During this audit, several requested dates confirming action had been performed were not listed in the database. The supervisor's comment was that maybe he missed or lost the post-it note before he had a chance to enter the information. This was not difficult to believe considering the number of papers on his desk during the audit. There wasn't any record that the equipment was allergen free after the cleaning cycle. Therefore, no evidence was available to demonstrate the effectiveness of this process. Possibly, the operator did run the test faithfully, so chances were that the equipment was free of allergen residue. However, the supervisor couldn't prove it because the records were not there. They also could not prove the operator ran the test.

Keep in mind that any change in the allergen program, such as additional allergen products being added, would trigger action in compliance with the change management program (Chapter 15). Through the change management program, the food safety team would evaluate the new product, its ingredients, and its required processes through the hazard analysis, revising all aspects of the food safety/HACCP program as required, ensuring the continued production of safe products.

As of the printing date of this text, the FDA had identified eight allergens that are considered *key* allergens (account for about 90% of food allergies) that must be controlled. The eight allergens are peanuts, tree nuts, dairy, eggs, soy, wheat, fish, and crustacean shellfish.

The Canadian Food Inspection Agency (CFIA) has identified eggs, mustard, peanuts, seafood (fish, crustaceans, shellfish), sesame, soy, sulphites, tree nuts, and wheat.

In 2004, the U.S. Congress passed the Food Allergen Labeling and Consumer Protection Act (FALCPA), which mandated specific allergen labeling requirements along with specific GMP requirements, which primarily focused on preventing cross contamination between allergen-containing products and those not containing allergens. The FALCPA clearly defined that food products that did not meet its requirements would be considered adulterated and/or misbranded as defined in the U.S. Federal Food, Drug and Cosmetic Act (FDCA). FSMA (2011) adds "allergens" as a separate hazard group.

Allergen control begins with the receiving of the raw ingredients. An allergen control program must be defined as it relates to the organization's products and allergen ingredients. Typically, storage can be as simple as being sure that all allergens are stored separately. Raw ingredients that contain different allergens must also be stored separately in a manner that prevents cross contamination. This is a critical compliance area, especially when partial containers of raw ingredients are returned to storage. All containers must be properly labeled, but those that contain allergens must be stored separately in designated areas to prevent cross contamination. Designated areas must be clearly labeled. Ingredients or raw materials containing similar allergens must be stored together. Allergens containing raw ingredients must be stored in lower tiers so that if a container should leak, then it would not contaminate a non-allergen ingredient. It is critical that storage areas be clean and well kept, compliant with GMPs and GHPs.

All associates must be educated and trained in the allergen control program. Associates with specific responsibilities related to allergen control require additional training based on those areas of responsibility. This training must be included from orientation through the ongoing food safety and prerequisite awareness training. Frequency of training is determined by the food safety team. However, the training may vary depending on the subject matter information, audit findings, and changes in the processes. The training must be presented to all associates at least once a year. Further detail is defined specifically for the PRPs for training (Chapter 8) and for the food safety program—The Presteps (Chapter 16).

As stated previously, the existing and potential hazards as related to allergen control are addressed through hazard analysis (Chapter 17), which identifies the PRPs that would be in place to ensure the hazards are controlled. Typically, the organization's allergen control PRP defines requirements for scheduling production and changeover between products and cleaning. ISO 22002-1:2009 10.3 NOTE provides a summary of possible sources of cross contamination:

- "Traces of product from the previous production run which cannot be adequately cleaned from the product line due to technical limitations; or
- When contact is likely to occur, in the normal manufacturing process, with products or ingredients that are produced on separate lines or in the same or adjacent processing areas."

Finished products containing allergens must be stored in a manner that protects against cross contamination. This may be caused from product leaks or damage to a non-allergen finished product. Even if the product is sealed, contamination of the outside surfaces of a finished product container could result in either contaminating the hands of the person handling the product or spilling inside once the product is opened.

Recommend reviewing the SQFI guidance document "RE: 2.8.3 Allergen Cleaning and Sanitation Practices" for excellent guidelines on allergen control with emphasis on effective cleaning and sanitation practices related to allergen management.

If the organization uses rework, further precautions must be taken. ISO 22002-1:2009 10.3 clearly defines under what conditions allergen-containing rework can be used:

- "In products which contain the same allergen(s) by design; or
- Through a process which has been demonstrated to remove or destroy the allergenic material."

Rework of product that contains allergens must be clearly identified so that it is only used in product that has the same allergen. In addition to ensuring that cross contamination between allergens is nonexistent, reworked product must always be properly labeled with its source (lot number, etc.), ensuring that if there is a recall related to ingredients, then records will link to the products affected. There have been some very serious incidents over the years when the lot traceability of ingredients that were in rework were either not accounted for or just forgotten. A recall may have occurred relating to a product containing a specific ingredient lot number. Most of the product was accounted for and recalled; however, some of that tainted ingredient was in a batch of improperly labeled rework, which was not recalled but consumed, resulting in the potential to cause illness or harm (i.e., a food safety event) (Chapter 14).

Some organizations do not manufacture products with allergens, but do purchase and distribute allergen-containing products. This relates directly to storage practices for finished products. Allergen-containing products, whether manufactured on site or purchased to be sold, must be stored in a manner that does not result in a cross contamination event should that product leak, break, become damaged, or in some way spill. This includes the possibility of contamination from products that contain different allergens. An allergen-containing product that is damaged or that may be contaminated must be managed in compliance with the control of nonconforming materials and products program (Chapter 9). Storage practices related to identified or potential hazards are evaluated through the food safety/HACCP programs hazard analysis (Chapter 17). Allergen control records must be identified and maintained in compliance with the record control program (Chapter 7).

Allergens may also be present in vending machines or brought to work in the associates' lunches. Some organizations have removed all allergen-containing products from the vending machines and have prohibited associates from bringing lunches with allergen products. This can be a tough requirement to enforce, and many question the value of its enforcement. Most organizations have chosen to educate their associates about allergens, so they are aware of specific situations should they occur. They also limit food consumption to the lunch areas and the locations of the vending machines. Associates are not allowed to bring any food product into the processing areas. These guidelines have been very successful; however, it must be a management decision. Some organizations have been strictly certified as allergen free. If this is the case, then obviously no allergen-containing item could be allowed on the premises. The program must be defined, implemented, and associates trained with records maintained to confirm compliance (Chapter 7).

KEY POINTS

a. Organizations must have an allergen control program to protect against cross contamination.
b. Programs must be in place, even if the organization does not manufacture any allergens.
c. Existing and potential hazards related to allergens are evaluated through the food safety/HACCP program's hazard analysis.
d. Storage practices must define requirements for storing allergen containing raw ingredients and finished products in a manner that prevents cross contamination.
e. All associates must be trained in allergen management as part of the food safety training program. Initial training must be provided during orientation.

f. Rework that contains allergens must be identified to prevent cross contamination and also to ensure that effective traceability is not lost (Chapter 14).

g. Records must be identified and maintained in compliance with the record control program (Chapter 7).

h. Color coding to ensure that dedicated equipment and utensils are used for allergen products is an excellent means of control.

i. Allergens must be declared on the product's label according to federal law, which went into effect in 2004 with the passing of the FALCPA. The FALCPA clearly defines that food products that do not meet its requirements will be considered adulterated and/or misbranded as defined in the U.S. FDCA.

j. FDA has identified the big eight allergens (milk, soy, fish, shellfish, wheat, eggs, tree nuts, and peanuts). CFIA identifies these eight, plus sesame seeds as the ninth allergen. It is stated that these allergens account for 90% of all food allergies.

COMMON FINDINGS/NONCONFORMANCES

a. Organization did not have a defined plan for the control of allergens.

b. Although it was stated that allergens were uniquely identified with a green sticker, there were four cases of a product, which contained soy, eggs, and dairy products, without the green sticker which were stored above open drums of rework product.

c. Rework was not properly labeled identifying the allergen product stored in the rework bin.

d. Records were not available to confirm that rework product was being used in the same allergen mix.

e. Records that were maintained to confirm that the returned finished products containing allergen had been destroyed rather than returned to inventory were not available.

f. There was considerable spillage of dry product on the floor under partial boxes of dried egg powder. Defined storage requirements for allergen control required immediate cleanup of any spillage.

g. Records confirming that associates responsible for allergen ingredient storage had been trained in related requirements were not available.

h. Production records confirmed that a dairy-based product had been manufactured prior to nondairy products; however, cleaning records were not available to confirm cleaning of the equipment between these two products.

i. Although it was stated that confirmation via the approved test kit was being performed daily prior to running a non-allergen product, records for the previous 2 weeks to confirm that this procedure was in compliance were not available.

j. It was stated that associates were allowed to consume food products only in the break rooms or other designated areas; however, there was an empty bag of roasted peanuts on the work station in the pasteurization area.

---

*Additional Food Safety Management System References (not all inclusive)

FSSC 22000:2010

    ISO 22002-1:2009 Section 10.3

SQF 7th Edition: 2.8.2

BRC Issue 6: Section 5.2

IFS Version 6: Section 4.20

*Note: Always review the standard or visit the website for your chosen standard to ensure that all related requirements are identified and addressed.

GLASS AND BRITTLE PLASTIC

Foreign material hazards are the most frequent food safety occurrences. Because such a contamination affects only a limited number of individuals, these hazards do not get the same publicity that a pathogenic biological outbreak affecting thousands of people in multiple states receives. However, any food safety hazard is serious. Every effort must be made to be proactive, taking every possible precaution to avoid contamination by foreign materials. The glass and brittle plastic program is designed to prevent product contamination from glass and brittle plastic. Brittle plastic is defined as plastic that is made from acrylic resins that break into pieces when subjected to blows beyond its impact resistance. Depending on the organization's design, some also include control of ceramic breakage. If ceramic breakage is a potential hazard, it must be addressed. The extent of defined requirements depends on the organization's risk from possible contamination. If glass is used as a container or in the manufacturing process, then this would increase the likelihood of contamination and may require a documented procedure. However, for those operations that have a limited amount or no glass, a form or checklist may be sufficient.

Defined requirements must state that glass is not allowed in the facility including in processing, packaging, storage, shipping, and laboratory areas. Wherever possible, glass must be protected or removed from the areas. For example, mirrors used to guide forklifts are now generally constructed of a polished steel or plastic. Clock faces are available with a hard plastic cover rather than glass. Light bulbs must be shielded to prevent shattering on breakage. This may be accomplished by either purchasing plastic-coated bulbs or ensuring that all fixtures have a hard plastic shield that would protect against shattered glass should a bulb break.

Exposed glass from windows or equipment (i.e., sight glasses on bulk storage tanks) must be identified and then monitored to confirm that they are intact (not missing, cracked, or broken). This is best done by creating and maintaining a detailed inventory list of all locations where glass and brittle plastic is located in the facility. This list then becomes the basis for the glass and brittle plastic monitoring program, which is done by operators, supervisors, and maintenance.

Associates must be encouraged to get in the habit of confirming that any glass or brittle plastic in their area remains undamaged and in place. In addition, organizations must implement and schedule a glass and brittle plastic audit (recommend a minimum of monthly) to confirm all is intact. The glass and brittle plastic audit is in addition to evaluating these items during the GMP audit. This is not overkill; it is imperative that these potential hazards be monitored beginning with the daily routine of the responsible area associates and ending with the verification and audit programs.

Glass such as windows and tiles that may be part of the plant design must also be monitored for breakage, cracks, and missing pieces. This can be accomplished effectively through the preventive maintenance (PM) program. During the PM check, the technicians should also confirm that no potential issues are lurking. For example, if a glass tile is loose, it can be removed and replaced before it actually falls out of place or breaks. Requirements for maintenance activities such as the management of the PM program are discussed in the "Equipment Suitability, Cleaning, and Maintenance" section. The requirement for identifying food safety–related preventive and corrective maintenance activities as a top priority is critical. Assigning glass and brittle plastic repairs is often omitted and not a priority. Maintenance associates must be trained and educated to make these types of repairs in a manner that prevents contamination to production lines, equipment, and exposed products.

The procedure must also include requirements for handling any glass breakage. A breakage incident and subsequent cleanup activities required to ensure the safety of the product must be recorded. If a concern arises related to a possible product contamination, then that product must be identified and managed in compliance with the control of nonconforming materials and products (Chapter 9).

If the location of glass or the use of glass in the process (packaging in glass bottles) is present, and it presents a risk within a process step, then this is recorded along with identified control measures through the food safety/HACCP program's hazard analysis. The hazard from the glass bottle or container is initially addressed through the hazard analysis of materials, which identifies the potential hazards originating at the supplier's operation. Once the glass bottles or containers are received and accepted, then these and any related hazards become the responsibility of the organization to identify and control through its hazard analysis. For example, to protect against breakage, procedures must define requirements for the proper storage of empty glass containers. Procedures must also define what must be done should an incident occur that could contaminate or cross contaminate the products. Organizations have found it beneficial to create a separate form that is completed for any glass or brittle plastic event. The *glass and brittle plastic breakage log* is completed and maintained as a record identifying the incident, and subsequent actions taken to protect the safety of the product. This could be a major incident that requires product to be segregated, put on hold, and investigated, or it could just require a basic cleanup of the breakage in the area without any product being affected. Either way, this report provides the record of occurrences and confirmation that requirements are being followed (Chapters 16 and 17).

Organizations also define all required protocol to control and eliminate the potential for or an existing hazard should a glass container break during the filling operation. The PRP must be defined to ensure that controls and related issues identified through the food safety/HACCP hazard analysis are effectively addressed.

Associates must be trained to react quickly and effectively if a suspected issue related to glass or brittle plastic contamination occurs taking immediate steps to isolate possible contaminated product minimizing and/or preventing the possibility of contamination. As soon as the situation is contained or controlled against further contamination, then the occurrence must be brought to the attention of a previously identified designated person such as an area supervisor or a representative of the quality department. As stated previously, any contaminated or potentially contaminated product or material must be identified and addressed in compliance with the control of nonconforming materials and products program (Chapter 9).

KEY POINTS

a. Glass breakage is considered a food safety risk. The organization must take every precaution to prevent a glass or brittle plastic foreign hazard from coming into contact with a food product or a food contact surface.

b. The condition of glass lamps or bulbs can be monitored through the PM program. The UV lamps used in equipment lines and insect lamps must be protected to ensure that the breakage does not shatter.

c. Periodic inspections must be included as part of the GMP audits, daily area inspections, and also the PM program.

d. Records confirming that the monitoring program is in compliance with defined requirements and that the PRP for glass and brittle plastic is effective must be maintained in compliance with the record control program (Chapter 7).

e. The effectiveness of the glass and brittle plastic program, in addition to being evaluated through the prerequisite verification program (Chapter 18), can be monitored through reviewing glass breakage incident reports, customer/consumer complaints related to glass, and any recalls resulting from a failure of this program.

f. The organization must implement a strict policy prohibiting associates, visitors, and contractors from bringing glass or brittle plastic materials into the facility.

g. Any changes in the process, product, or system that may include the addition or use of glass or brittle plastic must be reviewed by the food safety team in compliance with the food safety/HACCP program (Chapters 16 and 17) and with the change management program (Chapter 15).

h. Any existing or potential nonconforming product issue related to the glass and brittle plastic program must be recorded and managed in compliance with the control of nonconforming materials and products program (Chapter 9).

i. Lights must be protected such that if the lamp breaks, the glass does not shatter. There are several choices to address this, which may include the following:
    i. Plastic globe shield
    ii. Bulbs with plastic coating
    iii. Plastic tube and end caps for fluorescent bulbs
j. Laboratory glass equipment must be contained in the laboratory. Any glass transferred from or to the laboratories must be contained in such a manner to prevent glass breakage and potential cross contamination.
k. Every effort must be made to prevent the use of glass in the process; however, should glass such as a laboratory beaker be required, then it must be recorded on the glass and brittle plastic inventory log.

COMMON FINDINGS/NONCONFORMANCES

a. The organization had a glass handling procedure; however, it did not address brittle plastic materials.
b. The organization did not address brittle plastic as a potential hazard.
c. Records were not available to confirm that the organization had performed an analysis of its use of glass and brittle plastic to identify potential hazards and ensure its control.
d. Bins used for rework were missing three large chunks of brittle plastic; however, records were not available to confirm that this had been identified and investigated internally.
e. An open box of overhead bulbs was stored in the ingredient storage area next to open drums of concentrate.
f. A pile of broken glass that appeared to have been swept in a heap was on the floor in the glass container receiving area. It also appeared that this glass was being scattered throughout the area by the forklifts.
g. It was stated that existing and potential hazards related to light fixtures had been addressed through the hazard analysis and also through the glass and brittle plastic program; however, records did not confirm that this program had been implemented.

---

*Additional Food Safety Management System References (not all inclusive)

FSSC 22000:2010

     ISO 22002-1:2009 Section 10.4

SQF 7th Edition: Section 11.7

BRC Issue 6: Section 4.9

IFS Version 6: Section 4.12

*Note: Always review the standard or visit the website for your chosen standard to ensure that all related requirements are identified and addressed.

SANITATION: CLEANING AND SANITIZING

An effective sanitation program eliminates or reduces many hazards while having a direct influence on identifying critical control points. Sanitation, without a doubt, is a significant PRP.

In today's automated world, we hear less about the basics of sanitation than we used to. *Jon Porter, J Porter & Associates Ltd., Mt. Pleasant, IA*, stressed these basics in four words: "Time, temperature, concentration, and mechanics." These variables affect cleaning regardless of what is to be cleaned. Jon emphasized that

> A specific length of time is required to clean a given surface and given soil also referred to as "out-of-place" matter. Grease on a gearbox is a lubricant, but on a food contact surface, grease is soil. So, soils vary, depending on the nature of the food and processing conditions.
>
> Time can be the most variable; the amount allowed for sanitation activities often is affected directly by production schedule. Can you compensate for the loss of necessary time? The variables of temperature, concentration, and mechanics can be altered, but changing variables can create problems. Increased temperatures of the solution or adding more chemicals are not the answer.

Basic cleaning and sanitizing requirements related to equipment design and good housekeeping requirements are defined in the organization's procedures for GMPs. It is recommended that more specific requirements such as those for CIP, sanitizing, Master Cleaning Schedule (MCS), and Sanitation Standard Operating Procedures (SSOPs) be defined in separate procedures. Wherever possible, requirements may be defined on the actual record being completed with the tasks. As with all procedures, the organization must take care to ensure that procedures are written where needed and that content is specific and to the point in order to provide the necessary guidance to ensure that tasks are performed consistently in a manner that is user friendly and provides effective and consistent results. Applying a process approach to the sanitation process is effective (Figure 10.1).

- Knowledge of sanitation requirements for the product being produced
- Effective communication with, and education of, those responsible for performing the activities
- Well-documented SSOPs that are maintained within a formal document control process
- Records defined and maintained to ensure that objective evidence confirming compliance to defined requirements are accomplished
- Audit program designed to monitor activities in a manner that enhances overall effectiveness
- An effective process that provides a means to address existing and potential issues effectively
- Top management and managers within all departments who communicate firm support of the importance of sanitation and the role that sanitation associates play in the success of the business

FIGURE 10.1 Requirements of a process approach to sanitation.

The science of cleaning and sanitizing and its application to a specific product or food contact service is residue or product specific *dirt* is very complex. Types of chemicals, times, temperature, and velocity depend on the soil left behind by the product type. Dairy products are very challenging, requiring a combination of both caustic and acid cleaners in a specific sequence to ensure that the equipment is clean or free of any product residue, whereas a company that makes artificial juice products could ensure equipment cleanliness (free of hazards) with a simpler process. The artificial juice product by nature of its ingredients requires a less complex method.

The requirements for the cleaning program must be defined with records maintained to confirm compliance and to confirm the effectiveness of the program. Whether or not a specific FSMS standard defines this as a requirement, most auditors will look for and confirm that the organization's cleaning and sanitizing program has been validated to ensure that the *correct* activities and chemicals are being used to eliminate the potential of food safety hazards in the products.

CLEANING AND SANITIZING CHEMICALS

Dr. Tatiana Lorca, Manager, Food Safety Education and Training, Ecolab, Food & Beverage Division, provides the following insight on the identification of and the use of cleaning and sanitizing chemicals that are best for a specific organization's products:

> [Although some may be concerned about the long term effectiveness of specific cleaning chemicals and sanitizers], there is no scientific evidence that microorganisms develop resistance to sanitizers over time. Although many organizations do seem to be concerned about this, there is, at this time, no scientific evidence that organisms can become resistant to a specific sanitizer after continual use to that chemical. However, within an operation, many factors can influence the efficacy of a sanitizing program including sanitizer coverage, exposure time and target microflora type. Some sanitizers are more effective against bacteria and others are more effective against yeasts and molds. Some sanitizers may be more effective under certain use conditions because they are more resistant to inactivation by soil load. Because some plants may have mixed microbial flora due to the variety of products manufactured, effective control over microbial populations as a whole can be achieved by using different sanitizers at different times and locations. For example, plant management could use one type of sanitizer within their Clean in Place (CIP) system and an alternative in their central sanitizer system, or use one type during the production week and a different sanitizer with residual activity while the plant is idle over the weekend.

THE SCIENCE OF CLEANING

The complete science of which chemical is best for which operation, cleaning frequency, specific targeted *soil* or *microflora*, and other characteristics of a specific process can be quite involved. Understanding the requirements is very important to the effectiveness of the program. It is encouraged that sanitation programs be created with the assistance of reputable chemical suppliers, and that where available, specialize in the organization's type of process and products. Many of these companies will also provide structured training programs on the use of their chemicals, and on the basic techniques required for an effective cleaning and sanitizing program. This type of support is invaluable.

A master cleaning schedule (MCS) is an excellent tool to identify and track basic areas that require cleaning (i.e., external surfaces, floors, drains, etc.) and should be created, implemented, and effectively maintained. Cleaning chemicals used for basic cleaning must be approved for food contact surfaces and be labeled as *food grade* or *approved for food manufacturing sites*. Material safety data sheet labels and other information must be maintained to confirm compliance.

Cleaning chemicals must be stored in designated areas, secured (i.e., locked), separate from nonfood-grade chemicals and listed on an approved chemicals list. Any change in a chemical must be approved through a structured process (change management) for the intended use by the food safety team. Many reputable cleaning companies assist with this information, as well as provide specific training for the company's associates relating to the proper use of their chemical products. Training must ensure that cleaning chemicals are used in accordance with label requirements, that records are in compliance with the record control program and confirm that cleaning requirements are met and effective.

Cleaning Utensils

Cleaning utensils must be

- Stored in designated areas
- Stored in a manner such that functional ends do not touch the floor
- Controlled by area of use to prevent cross contamination

Color coding for cleaning utensils is an excellent means to avoid cross contamination and is strongly recommended.

Critical Food Safety and Quality Suppliers

Cleaning and sanitation service suppliers must be identified and maintained in compliance with the critical food safety and quality supplier program (Chapter 13).

SSOPs

Related to basic cleaning and/or sanitizing procedures that may also be referred to as sanitation standard operating procedures (SSOPs), the following information should be defined:

- Specific areas, equipment, or utensils to be cleaned
- Specific chemical to be used for the specific cleaning and/or sanitizing activity

- Responsibility for the task
- Method and frequency of the task (reference cleaning schedule or checklist if applicable)
- Postcleaning inspection and responsibility
- Required records
- Required approval by supervisor or trained designee confirming task completed and acceptable
- Process performed to verify effectiveness of the activity, the responsibility, and the record for this task

Clean-in-Place (CIP)

Clean-in-place is a process line or equipment is designed to be able to have water and the cleaning chemical (detergent, caustic, acid, sanitizer, etc.) circulated through the equipment for a specific time, temperature, velocity, and chemical strength that is adequate for each application. Much thought, engineering, and testing goes into such a design to ensure its effectiveness. Cleaning chemicals and sanitizers are designed specifically for CIP cleaning under specific conditions. For other equipment, like tanks, a spray ball is engineered to provide effective coverage of the equipment through the cleaning process. It is critical that this type of system is confirmed effective, capable, and also engineered in a manner (i.e., separate lines) to protect against cross contamination. Unlike earlier CIP systems, systems should be designed such that the cleaning system can be separated from product lines to prevent cross contamination. As with all PRPs, a process that confirms effectiveness must be defined and supported with records that demonstrate compliance with specified requirements (Chapter 7).

Dr. Tatiana Lorca, Manager, Food Safety Education and Training, Ecolab, Food & Beverage Division, provides the following comment related to a revised BRC guideline requiring that the organization demonstrate and maintain additional records related to CIP detergent (caustic) reservoir tanks:

> Issue 6 of the BRC Global Standard for Food Safety has expanded the requirements for CIP systems 4.11.6.3. (CIP Performance) requires that "Detergent tanks shall be kept stocked up and a log maintained of when these are filled and emptied (effective January 2012)." You need to ensure and record that you are using the appropriate chemicals at the correct use concentration, and that the system is functioning as expected. What needs to be recorded depends whether you run a reuse CIP system. Records of chemical addition, when the detergent recovery tank is emptied, cleaned and recharged all need to be documented. If you run an automated system, the program will prompt the tank to be filled with the appropriate amount of chemical and the chemical concentration will be recorded by the system. There is no additional need to record filling or emptying the tank as the whole CIP process is recorded. Don't forget that you must verify you are using the appropriate concentration of chemical and this must be kept as a record. (ref. BRC Issue 6 4.11.1)

As mentioned previously, the four key factors of an effective CIP process are time, temperature, velocity, and chemical strength. These four parameters must be taken into account during the design of the process and the cleaning chemical of choice. Records confirming that the CIP function is compliant and effective must be maintained. This record should include the date, time, process cleaned, associate responsible for the function, approval from supervisor or other trained associate confirming all requirements have been met, and that the outcome is effective. This information can be listed on a recording chart that becomes the data record.

Water that is used for cleaning is considered a process aid thus must be evaluated for any potential hazards during the hazard analysis of the food safety programs. Requirements for handling and ensuring that the water supply is acceptable and free of any hazards must be managed through the organization's water treatment PRP. The requirement for water used as a processing aid depends on the water source, the nature of the product, process, and various other parameters that may or may not be unique to the organization's operation (See Utilities, Air, Water, Energy section).

As with all PRPs, being able to confirm effectiveness is key to not only an effective cleaning and sanitizing program, but also the overall value to the food safety program. A combination of effective measures such as adenosine triphosphate (ATP—a chemical compound and bioluminescence for hygiene monitoring) testing, postcleaning checks, and prestart-up inspections must be incorporated into the process with records maintained that provide the proof (objective evidence) (Chapter 7). Effectiveness should also be confirmed through GMP audits and internal audits. Results may be tracked/recorded on the PRP verification checklist (Chapter 18).

KEY POINTS

a. Organizations may want to evaluate the possibility of effectively managing and tracking performance of the cleaning and sanitizing program through a formalized software program.
b. An effective encompassing MCS must be created, implemented, and effectively maintained.
c. The cleaning program takes into account all areas requiring cleaning and sanitizing including the break rooms, restrooms, locker rooms, ceilings, external grounds, etc.
d. Basic cleaning requirements, including those for CIP systems may be defined in SSOPs.
e. Effectiveness and validation of cleaning and sanitizing programs must be performed and records maintained.

COMMON FINDINGS/NONCONFORMANCES

a. The master cleaning schedule form and the cleaning checklists form were not identified and maintained as controlled documents.
b. Completed MCS was not identified and maintained in compliance with the record control program.
c. The janitorial service company contracted to clean and maintain the offices, break room, and bathrooms was not identified and managed as a critical food safety and quality approved supplier.

d. Organization did not have a formalized color coding process for managing cleaning utensils as required by SSOP. This significantly increased the opportunity for cross contamination between those utensils used for raw product and those used for pasteurized product.

e. Through review of records and the interview process, it could not be confirmed that external service suppliers for cleaning functions were using chemicals listed on the approved chemical list.

f. Records were not available to confirm that the organizations defined requirements for food safety and reuse of food-grade chemicals had been communicated to the service suppliers.

g. Records were not available to confirm the verification of cleaning activities by a supervisor or trained team member as required by SSOP CL-2 Rev 1.

h. Records were not available to confirm that CIP systems had been validated for effectiveness.

i. There was no evidence of a defined process for monitoring cleaning and sanitation effectiveness.

j. It was stated that the effectiveness of the cleaning program was monitored through monthly GMP inspections; however, the inspection records did not provide evidence that the cleaning program was actually being evaluated during the GMP inspection.

---

*Additional Food Safety Management System References (not all inclusive)

FSSC 22000:2010

       ISO 22002-1:2009 Sections 8, 11

SQF 7th Edition: Section 11.2.13

BRC Issue 6: Section 4.11

IFS Version 6: Section 4.10

*Note: Always review the standard or visit the website for your chosen standard to ensure that all related requirements are identified and addressed.

EMPLOYEE PRACTICES, PERSONAL HYGIENE, AND EMPLOYEE FACILITIES

Requirements for employee practices, personal hygiene, and employee facilities directly relate to the defined requirements for FDA 21 CFR 110 for GMPs. Requirements of the food safety management standards are consistent between the standards. Specific personal hygiene requirements, whichever standard chosen, must be relative to the potential or existing process or product hazards.

ISO 22002-1:2009 13.1 states that "personal hygiene facilities [must] be available to ensure that the degree of personal hygiene required by the organization can be maintained. The facilities [must] be close to the points where hygiene requirements apply and [must] be clearly designated." BRC Global Standard for Food Safety Issue 6 states that "staff facilities [must] be sufficient to accommodate the required number of personnel, and [must] be designed and operated to minimize the risk of product contamination. The facilities [must] be maintained in good and clean condition."

It is critical that not only the restrooms and locker areas but also the designated eating areas are kept clean and in good repair. What message is sent to employees if the organization does not provide their associates a *nice* place to take breaks and enjoy their lunches? The lunchroom and break areas must be clean, well organized, in good repair, and GMP compliant. These areas must be on a cleaning program. Refrigerators, microwaves, and other storage areas must be well maintained. Many organizations have a scheduled clean-out every week ("all food is discarded every Friday at 3 p.m.") to ensure food doesn't collect.

Food and drink consumption in work areas must be eliminated wherever possible, with records of effective risk assessment and mitigation available for remaining areas. Spitting is, of course, prohibited within all internal areas of the facility.

IMPORTANCE OF *HANDWASHING*

It is said that frequent handwashing, done properly, is the single most activity that can control the spread of, not only food-borne illnesses, but also colds and flu sickness. Employee training must continually emphasize the importance of handwashing including the emphasis on proper techniques.

Some fun ways to educate on handwashing:

- Use of *glo-germ* and black light to monitor effectiveness of handwashing activity.
- Sample micro testing placing finger tips on a Petri dish using total plate count agar. Parameters may include comparing no wash, rinse only, soap only, sanitizer only, and combination of hot water, soap, and sanitizer.
- Present a *good* handwashing video; there are a variety of them available on the Internet. Recommend reviewing and presenting a sampling of various videos to reinforce.
- Share Dr Winter's food safety song set to the Beatles music now titled "you better wash your hands."
- Share a video clip from the TV series Seinfeld titled "the pie."

Clever means of communicating can be a very effective educational tool if presented in an effective and supportive manner.

Handwashing Equipment

Specific requirements for handwashing sinks, including their placement in the operation, depend on the FSMS standard of choice. Hands-free wash stations are recommended. At a minimum, the food safety team must perform a risk assessment related to placement of handwashing sinks in the facility and use this information to make a sound judgment on their locations. Be cautious about solving the handwashing sink access with the hanging of hand sanitizer units. The effectiveness of these products must be verified and validated. Most hand sanitizers are just that: sanitizers. Basic sanitation protocol is that a surface (including hands, gloves, etc.) must be clean before it can be sanitized. It is not possible to sanitize dirt; thus, effectiveness of sanitizer alone on dirty hands is questionable.

Dr. Tatiana Lorca, Manager, Food Safety Education and Training, Ecolab, Food & Beverage Division, provides the following guidelines related to handwashing requirements:

> Handwashing is vital to protecting the safety of the food product and minimizing the potential to transmit pathogenic microorganisms from the hands of food production workers (and anyone in the plant environment) to food, utensils, and packaging materials. None of the current Global Food Safety Initiative (GFSI) recognized programs identify a specific temperature for handwashing. Even though the regulations do not apply to a food processing environment, it is recommended you follow your local Food Code requirement for appropriate temperature (ref. U.S., Food & Drug Administration 2009 Food Code, 5-202.12): "A hand-washing sink shall be equipped to provide water at a temperature of at least 38°C (100°F) through a mixing valve or combination faucet. If the water is too hot or too cold, it is unlikely your associates will keep them under running water long enough to effectively wash their hands for 20 seconds or long enough to sing 'Happy Birthday' twice."

ISO 22002-1 13.7

Identifies the following criteria for when associates must wash their hands:

- Before starting any food-handling activities
- When returning to work after being out of the area for any reason
- Immediately after using the toilet or blowing your nose
- Immediately after handling any potentially contaminated material

---

There are many specifically defined employee practices, personal hygiene, and employee facilities requirements addressed in the FSMS standards. A summary of these requirements are listed in the following text (again, it is recommended that the standard of choice be reviewed in detail to ensure compliance):

- Clean work wear without buttons and outside pockets above the waist.
- Work wear that provides adequate coverage to prevent hair, perspiration, etc., from contaminating the product.

- Hair, beards, and mustaches worn in a manner that protects against causing possible product cross contamination.
- Gloves used for product contact surfaces must be clean, in good condition, and used in a manner that cannot contaminate or cause cross contamination.
- Enclosed shoes made of nonabsorbent materials.
- Personal protection equipment, where required, designed to prevent product contamination and maintained in clean condition.
- Where permitted by law, associates undergo a health exam prior to employment.
- Where permitted by law, associates inform management of any illness that could cross contaminate the product. Associates with wounds must cover these to prevent contamination with the product. BRC Global Standard for Food Safety Issue 6 7.2 requires that open wounds are protected by a colored Band-Aid and, if possible, the use of a metal detectable company-issued Band-Aid. Records must verify that each new batch of Band-Aids has been tested to confirm that the metal detector does detect and reject the Band-Aids when actually in the product.
- Educate associates to understand that they must make every effort to refrain from coughing and/or sneezing when in the process areas or any area where food contact or food contact items are present.
- Controlled use and storage of personal medicines to prevent the possibility of cross contamination.
- Proper handwashing, including frequency based on the potential risk of the areas of responsibility.
- False fingernails, nail polish, and false eyelashes not allowed.
- Smoking, eating, drinking, or chewing tobacco not allowed in process areas (limited to designated areas).
- Jewelry including watches, rings, earrings, and exposed body piercing (including the tongue and mouth areas) not allowed.
- Loose items above the waist such as pens clipped to uniform shirts or stuck behind the ears not allowed.
- Food, rubbish, dirty clothing, product contact tools, and equipment not allowed to be stored in personal lockers.

If the organization decides to require hair covering only in specific zoned areas, then these areas must be clearly identified and associates trained to comply with the requirements. Reasons/justifications for identifying these zoned areas must be defined in documented procedures or in the HACCP program. In most situations, it is recommended that hair covering be worn in all areas of the operation. Trying to manage different areas can become a huge challenge with nonconformances frequently identified just because associates and visitors find it difficult to understand the zoning requirements.

Records confirming compliance to defined requirements for employee practices, personal hygiene, and employee facilities must be identified and maintained as defined for the record control program (Chapter 7).

KEY POINTS

a. Food consumption must be limited to designated areas and not allowed in any area that could result in cross contamination. Organization must ensure that *zoned* areas are sufficiently identified and managed effectively.

b. Organization must evaluate and confirm process for food consumption in process areas, eliminating wherever possible, with records of effective risk assessment and mitigation available to justify decisions.

c. Employee areas such as rest rooms, lunch rooms, and locker rooms must be maintained in clean and good condition compliant with FSMS requirements.

d. If smoking is allowed, then it must be contained in a separate area that is not located such that it could cause cross contamination or create a health risk for nonsmokers (breathing secondhand smoke).

e. Management must do an evaluation of the handwashing facilities in the processing and packaging areas to ensure ready access for associates. Handwashing sinks must be used for handwashing only and not for equipment washing and storing.

f. Sinks must be labeled with posted signs identifying the purpose of the sink (i.e., "handwashing only" or "equipment cleaning only—not a handwashing sink").

g. Signs must be posted in all areas where handwashing is required (i.e., "Associates must wash hands before returning to work."). The food safety management standards recommend *hands-free wash stations*; however, at this time, these are not a requirement.

h. Requirements for employee hygiene must be compliant with 21 CFR 110.

COMMON FINDINGS/NONCONFORMANCES

a. Requirements for personal hygiene were defined in the GMP procedure; however, this procedure was maintained by the human resource department; and not identified and maintained as a system controlled document.

b. It was stated that requirements for employee practices and hygiene were defined in a corporate-level document; however, this location was not able to produce this document stating that it had not been incorporated into their system.

c. The management system referenced a controlled document for defined GMP requirements; however, upon review of this document, only *people* safety (OHSAS) requirements were defined.

d. Visitor log, which provided the objective evidence that visitors and contractors were trained in the organization's GMP and food safety requirements, was not identified and maintained in compliance with the record control program.

e. Evaluation of the associate restrooms indicated that although basically clean, they were poor repair (stained floors, broken units, ceiling tiles missing).

f. The restroom designated for transport drivers was very dirty, towels on floor, and toilet leaking. It was stated verbally that this had been identified on the previous three internal GMP inspection reports but had not been fixed.

g. Organization did not have a defined and implemented program regarding associates working with exposed product while sick or with open wounds.

h. Requirements for not wearing nail polish around exposed product were not defined in the GMP procedure. Associates with polished artificial nails were observed on the exposed product packaging line.

i. GMP storage procedures stated that no food, trash, medicine, or drinks were allowed in personal lockers; however, peanut butter (allergen) crackers, dirty dishes, candy, candy wrappers, and opened bottles of soda were found in both the women's and men's locker areas.

j. Records were not available to confirm that the uniform company was identified and maintained as a critical food safety and quality supplier (Chapter 13).

k. It was confirmed that associates working in production areas were wearing uniforms with snaps rather than buttons; however, maintenance associates, subcontracted maintenance associates, management, and supervisors were observed wearing dress and golf shirts with buttons. It was stated verbally that requirements for no buttons did not apply to these groups.

l. Posted procedures stated that associates must wear disposable gloves when handling any product surface including empty bottles and lids. This requirement also stated that new gloves must be put on prior to undertaking the activity; however, associates were observed wearing the same gloves continually while not only handling empty bottles, but also handling corrugated boxes of lids and picking up discarded bottles from the floor.

m. Defined GMP personal hygiene procedures required that hair nets and beard snoods be worn in specifically zoned high-risk or critical areas (i.e., mixing, filling, and packaging); however, there was no evidence of these zones being formally identified. Three associates and two subcontractors were observed in the packaging area not wearing hair nets and beard snoods (both contractors had beards).

n. Evidence was not available to confirm that management had implemented a *clothing* or *work wear* policy.

o. Restroom in the break room was dingy, dirty, and in poor repair (i.e., there was trash on the floor and a broken handle on the sink).

p. The approved smoking area was adjacent to the visitor entrance. Although there were three cigarette disposal units in the area, there were numerous discarded cigarette butts on the ground, which presented a negative impression to visitors, contractors, and associates entering through this area.

q. Evidence was not available to confirm that a defined program for cleaning and maintenance of personnel lockers had been defined and implemented.

r. Records were not available to confirm that air filters on recently installed hand dryers had been included in the PM program. Note that hand dryers had been added throughout the facility; however, there was no record confirming that this installation had been reviewed and accepted by the food safety team prior to the installation (Chapter 15). (Relates also to design and change management.)

s. Evidence was not available to confirm that requirements for personal hygiene had been formally defined.

---

*Additional Food Safety Management System References (not all inclusive)

FSSC 22000:2010

     ISO 22002-1:2009 Section 13

SQF 7th Edition: Section 11.3

BRC Issue 6: Section 7

IFS Version 6: Sections 3.2, 3.4

*Note: Always review the standard or visit the website for your chosen standard to ensure that all related requirements are identified and addressed.

REWORK

Rework may be defined as a product or an ingredient that originally did not meet specifications but was either brought back into conformance or used in the process at a specific percentage to ensure that the finished product remained in specification. Typically, product not meeting food safety–related specifications is destroyed; however, in some instances (i.e., repasteurize), food safety–related issues can be corrected. Conformance of rework is managed as defined in the control of nonconforming materials and products program (Chapter 9).

The organization must define its process for *rework*. The addition of rework and any potential or existing food safety hazards is evaluated by the food safety/HACCP team in the food safety/HACCP programs (Chapter 17). The addition of rework into the process must be identified on the food safety/HACCP flow diagram and confirmed as accurate by the food safety/HACCP team during the verification of the flow diagram (Chapter 16).

Rework may also be referred to as *reprocess* material. Some food industry sectors may refer to *rework* as *regrind* or *repackage*. Typically, repackaged material occurs when a case of sealed product containers becomes damaged; associates then sort these containers and *repackage* those containers that are not damaged. Once the *repackage* is complete, then the product can be released for sale.

Rework material must be handled, stored, and used in a manner that

- Product safety, quality, traceability, and regulatory compliance are maintained
- Protects it from biological, chemical, and physical contamination
- Protects against cross contamination especially as related to allergen cross contamination

In order to protect against cross contamination, rework must be properly identified to ensure that non-allergen containing rework is stored separately from allergen-containing rework. Also, rework containing different allergens must be stored in a manner to prevent cross contamination.

Rework related to the allergen control program is specifically addressed in "Allergen Control" section.

Traceability (Chapter 14) and lot number identification must be maintained so that accurate product and ingredient information is included with the finished product containing specific batches of rework. Product recall requirements including those requirements for performing mock recalls are defined in "Product Recall" (Chapter 14). Crisis management as related to a product recall is also defined in Chapter 14. Records of traceability must be identified and maintained in compliance with the record control program (Chapter 7).

The rework classification or the reason for rework designations must be recorded and at a minimum, include product name, unique identification (i.e., production date or lot number) shift, line or origin, reason for rework, operator's initials, and supervision's initial of approval.

When rework is used as an in-process step, the acceptance criteria for the rework must be defined. Additionally, the process step for addition, the method, and any preparatory stages must be defined. Any rework requiring the removal of packaging includes control(s) to ensure that the removed packaging does not become a contaminant.

ISO 22002-1:2009 14.3 states that

- "Where rework is incorporated into a product as an 'in-process' step, the acceptable quantity, types, and conditions of rework must be specified. The process step and method of addition, including any necessary preprocessing stages, must be defined.
- Where rework activities involved moving a product from filled or wrapped packages; controls must be put in place to ensure the removal and segregation of packaging materials and to avoid contamination of the product with extraneous matter."

IFF Version six states the following related to rework:

- 2.2.11 states that "specifications [must] be available and in place for all raw materials (raw materials/ingredients, additives, packaging, materials, rework).
  - Specifications [must] be up to date, unambiguous and be in compliance with legal requirements and, if existing, with customer requirements."
- 4.3.10 states that "the [organization must] ensure that in the event of changes to product formulation, including rework and packaging material, process characteristics are reviewed in order to assure that product requirements are complied with."
- 4.18 states that "traceability [must] be ensured at all stages, including work in progress, post treatment and rework."
- 5.3.3 states that "all rework operations [must] be validated, monitored and documented. These [must] not affect the product."
- 5.10.1 states that "a procedure [must] exist for the management of all nonconforming raw materials, semi-finished and finished products, processing equipment and packaging materials. This [must] at a minimum include a decision about the further use (e.g., release, rework/post treatment, blocking, quarantine, and rejection/disposal)."

KEY POINTS

a. Organization must define its process for rework. Depending on the size and the various processes, it may be most advantageous to identify and maintain by each area or specific process.

b. Ensure that rework is properly labeled and stored to prevent cross contamination.

c. There may be several terms used to identify rework such as repackaged and reprocess.

d. When rework is used as an in-process step, the acceptance criteria for the rework must be defined.

e. Rework material must be stored, handled, and used in a manner that product safety, quality, traceability, and regulatory compliance are maintained.

COMMON FINDINGS/NONCONFORMANCES

a. Unmarked bins of rework were stored in several areas of packaging.

b. Records were not available to confirm that the organization was recording lot numbers of ingredients used in product identified as rework.

c. Procedures stated that organization did not use rework; however, there were two bins stored in the ingredient weighing room labeled as *rework*. Operator stated that this was added back to each batch at approximately 10% of the total batch; however, records were not available that identified the use of this *rework* nor were there any lot number or other identification recorded that would be used for traceability.

d. Organization did not separate rework containing allergens and rework not containing allergens. The operator stated that since it was added back at a low percentage, it did not have to be identified.

---

*Additional Food Safety Management System References (not all inclusive)

FSSC 22000:2010

ISO 22002-1:2009 Section 14.0

SQF 7th Edition: Section 2.4.7

BRC Issue 6: Sections 3.9.3, 5.25

IFS Version 6: Sections 2.2.11, 4.3.10, 4.18, 5.3.3, 5.10.1

*Note: Always review the standard or visit the website for your chosen standard to ensure that all related requirements are identified and addressed.

WAREHOUSE, STORAGE, AND SHIPPING

Requirements for warehouse, storage, and shipping activities are very straightforward, addressing the protection of the product from damage, deterioration, and loss throughout the entire process. Requirements may be defined in area procedures or work instructions. They may also address and include the process from receiving ingredients through shipping and distribution until the responsibility is transferred to the customer.

Having control of stored ingredients, packaging, and finished product is obviously critical to all aspects of the organization and especially relating to food safety. This is accomplished through compliance with many different PRPs. Examples of those that related to storage, warehouse, and shipping include, but are not limited to, GMP, GHPs, protection against cross contamination, calibration, pest control, sanitation, traffic control, allergen management, crisis management, regulatory, shelf life control, temperature control, inventory control, delivery temperature, lot control, traceability, critical food safety, and quality supplier management. That said, ensuring that customers receive the correct product, in specification and on time, is also critical to the business.

Edward Link (2008) (Quality Pursuit) in An ISO 9000 Pocket Guide for Every Employee supplies the following thoughts on management's role related to the control of finished product through storage and distribution:

> The treatments provided to the storage and movement of product should be carefully considered. You cannot recover the cost of customer dissatisfaction from an insurance company. Late deliveries and the arrival of damaged goods are of no use to the customers. Assuring [that the defined] requirements of the standard are met [will function as] the assurance against customer dissatisfaction in this category. (p. 81)

A program must be defined, implemented, and maintained that addresses the following:

- Method for handling the product to protect against cross contamination and any existing or potential food hazard
- Product protection related to food safety and quality after final inspection and test depending on the customer agreement
- Method for handling product (rotation) to protect against damage, deterioration, and ensure first in first out (FIFO) rotation
- Authorization for transfer of product and shipment
- Control of packing, packaging, and marking processes (including materials used) that ensure activities conform to specified requirements
- Evaluation of the condition of the product in storage at defined intervals
- Inspection requirements related to transports before receiving ingredients and before loading
- Inspection of storage areas
- Finished product and raw ingredient storage parameters including potential risk through storage such as cross contamination

Suppliers who provide shipping services (i.e., transports, etc.) must be identified and maintained in compliance with requirements for the critical food safety and quality supplier program. These requirements may be defined, either by the responsible department or communicated from the department responsible for the process.

Those responsible for assigning transport or shipping companies must have ready access to the critical food safety and quality approved shipping suppliers, and be trained in the related responsibility.

External warehouses that are subcontracted for storage and distribution must also be identified and managed through the critical food safety and quality supplier program. Requirements must be clearly communicated to these suppliers such as those for shipping temperature last loads transported and/or GMP compliance. Records demonstrating compliance must be identified and maintained as defined in the record control program (Chapter 7).

Designated storage areas must be used in such a manner to prevent cross contamination, damage, and deterioration of the products while awaiting shipment.

STORAGE OF HAZARDOUS MATERIALS

Storage requirements must also address regulatory requirements for handling hazardous materials including cleaning chemicals and any other chemicals that either might be used in the process or that may come in contact with the shipping containers. Storage of incoming materials such as glass bottles must also be handled in a manner to ensure that the final product is not jeopardized.

EQUIPMENT CALIBRATION

Equipment used to measure temperature and humidity requiring calibration must be maintained in compliance with the calibration program (Chapter 11). Periodic *verification* such as monitoring recording charts or thermometers at a defined frequency should be specified. Frequencies and methods depend on the importance (potential risk) of the requirement or possibly a storage requirement of the customer. It is critical not to lose sight of the fact that it is everyone's responsibility to ensure the safety and the quality of the product. Storage and shipping conditions must ensure that product remains safe and in specification.

STORAGE REQUIREMENTS

Warehouse storage areas must ensure that any product or material that may have been identified as potentially unsafe, not meeting specifications, or, for whatever reason (damage, leaking, code date, etc.), not acceptable for use or shipment to the customer in compliance with the control of nonconforming materials and products program (Chapter 9). Those responsible for ensuring compliance must be qualified in a manner that ensures that any material and/or product identified as nonconforming is identified and segregated in a manner that protects the product from inadvertent use or shipment. Storage of nonconforming material or product must also be such that it protects approved finished product from being cross contaminated.

Many times warehouses are used to store nonfood items such as chemicals and packaging. Those responsible for storage practices must ensure that ingredients, materials, and finished product are safe from potential hazards and protected against becoming nonconforming. Like items must be stored with like items (i.e., all packaging stored in same areas, etc.). Storage of incoming materials such as glass bottles must be in a manner that protects against cross contamination. Items including finished product, packaging, or ingredients cannot be put in jeopardy. Chemicals must be stored in a separate area and secured (see Sanitation: Cleaning and Sanitizing section). Storage must be in compliance with regulatory requirements for handling hazardous materials such as food-grade cleaning chemicals, maintenance chemicals, discarded chemicals, pesticides, etc.

CONTROL OF THE FINISHED PRODUCT

Some organizations define that the product may be shipped prior to final approval as long as the product remains within its control. In other words, it is not actually released to the customer even though it may be in transit. If this is the method of practice, then procedures must clearly define the control the organization has on this product until its final approval. It is recommended that the operation, at a defined frequency, actually test the process to ensure and confirm its control. Remember, an auditor may ask the shipping operator how he/she knows which products to ship for a specific order and how he/she knows the product has been approved for shipment. It is the organization's responsibility to make every effort to ensure that its end products are safe for consumption and meet customer requirements. Potential and existing food safety hazards are reviewed by the food safety/HACCP Team and addressed in the food safety/HACCP programs (Chapters 16 and 17).

KEY POINTS

   a. Typically, food safety requirements related to storage, warehouse, and shipping practices are defined in and monitored through the organization's PRPs such as GMP, GHPs, pest control, storage practices, protection against cross contamination, chemical storage and handling, allergen control, traffic control, and many others.
   b. Operator competency training must include an understanding of food safety/PRP compliance.
   c. The process for identifying product to be shipped and confirming its approval to ensure that the correct product reaches the right customer should be defined in area procedures.
   d. Operator competency must include knowledge of the requirements for communicating to the customer (may be advising supervisor) should a problem surface that would affect the delivery of the order on time and in specification.
   e. Transports must be inspected prior to loading to ensure cleanliness and identifying any issue that could cause harm to the product. Results must be recorded (i.e., bill of lading, shipping document, etc.) with records maintained in compliance with the record control program.
   f. It is critical to ensure that the requirement for confirming and documenting seal numbers is defined, implemented, and effective.

COMMON FINDINGS/NONCONFORMANCES

a. Although procedures defined requirements for storing finished products, records were not available to demonstrate compliance to these requirements.

b. There was no defined method, nor was there evidence that product was being evaluated at defined intervals to ensure that it was not damaged, out of date, or otherwise unacceptable.

c. Suppliers of shipping services or distribution warehouses were not being identified and maintained in compliance with the critical food safety and quality supplier program.

d. Storage work instruction WARE-02 Rev 02 defined that product was stored and used according to the FIFO method. However, several bags of raw ingredients with a March 2013 production code date were noted in the production area while pallets of the same product dated with a November 2012 production code were stored in the raw storage area.

e. Although it was stated verbally that trailers were inspected for acceptability criteria prior to loading finished product, records did not confirm that this was being done.

f. It was observed during the evaluation that several pallets of packaging were damaged and/or shrink-wrap protection torn. There was no means evident protecting this material from cross contamination.

g. The glass container storage area was located next to raw ingredients storage. There was evidence of several broken containers and loose glass in the area. There was no provision defined for addressing these situations to protect against the possibility of cross contamination with a food safety hazard.

h. It could not be demonstrated through the interview process or review of training records that associates responsible for shipping had been trained on the process for trailer inspection prior to loading.

i. Approved suppliers for shipping services were being maintained on an approved list in compliance with purchasing requirements; however, this information was not readily available to those personnel arranging the transports.

j. Although it was stated that the warehouse and other storage areas were being maintained in GMP compliance, there were several violations noted during this evaluation. Also, training records were not available to confirm that associates responsible for maintaining compliance status had been trained in food safety related PRPs.

k. Segregated hold areas contained product identified with both hold tags and release-for-shipment tags.

l. The purchase order for warehouse services referenced that requirements were defined in work instruction WARE-01 Rev. 2; however, the warehousing service company could not produce this document upon request.

m. Pallets of finished product were stored next to the maintenance used equipment storage area creating the opportunities for cross contamination of the finished product.

n. It could not be confirmed that nonconforming product and materials were identified and stored in a separate area in compliance with the nonconforming materials and products procedure.

o. It was stated verbally that the shipping operator was responsible for the application of seals once a transport was loaded to protect and control the finished product; however, the operator stated that it was the responsibility of the guard. There was no evidence that the guard service had been trained in the process nor was there evidence that the guard service was identified and maintained in compliance with the critical food safety and quality supplier program.

p. This operation accepted returned goods from customers; however, there was not a separate area for storage of these goods presenting the possibility of cross contamination.

q. Forklifts were very dirty, especially those forklifts that were traveling both internally and externally to the building. Tires were wet and dirty tracking debris through both storage and packaging areas.

r. The *Rejected Trailer Log* was not identified and maintained in compliance with the record control program.

---

*Additional Food Safety Management System References (not all inclusive)

FSSC 22000:2010

    ISO 22002-1:2009 Section 16.0

SQF 7th Edition: Sections 2.4, 11.6

BRC Issue 6: Section 4.14

IFS Version 6: Sections 4.9, 4.14, 4.15

*Note: Always review the standard or visit the website for your chosen standard to ensure that all related requirements are identified and addressed.

---

11 Calibration

Calibration is a term applied to reflect equipment that is used to control, measure, test, and confirm the accuracy of the results. ISO 22000:2005 8.3 states that "The organization must provide evidence that the specified monitoring and measuring methods and equipment are adequate to ensure the performance of the monitoring and measuring procedures. Where necessary to ensure valid results, the measuring equipment and methods used

   a. Must be calibrated or verified at specified intervals, or prior to use, against measurement standards traceable to international or national measurement standards; where no such standards exist, the basis used for calibration or verification must be recorded
   b. Must be adjusted or readjusted as necessary
   c. Must be identified to enable the calibration status to be determined
   d. Must be safeguarded from adjustments that would invalidate the measurement results
   e. Must be protected from damage and deterioration"

Results of the calibration and verification activities must be identified and maintained in compliance with the record control program (Chapter 7).

DEVELOPING THE CALIBRATION PROGRAM

Over the years developing a compliant and effective calibration program has proven to be one of the toughest hurdles. Compliance has also proven to be an efficient and structured prerequisite program providing a strong foundation for an effective and compliant food safety program. There are many parts to this program that must be identified, implemented, and maintained consistently. It is also an excellent example of a program that should not only be applied to food safety related equipment, but also include monitoring and measuring equipment that is used to "demonstrate conformance of product to specified requirements" (ISO 9001:2008 7.6). Equipment referred to as *critical* includes equipment that is used to verify food safety limits and also equipment used to confirm compliance to quality, regulatory, and customer requirements. It is essential to the overall process that this equipment performs to the degree of accuracy necessary to ensure that the product is within specification and meets the customer's requirements.

Russ Marchiando, Packaging Plant Manager—Wixon, Inc, stated that when first developing their calibration program to be in compliance with their quality management system (ISO 9001)

We never had a regular calibration schedule for most of our measuring and testing equipment. Often, the calibration that was done was *hit or miss* on an infrequent basis. I can just imagine the amount of product that was given away for free before this system was put into place. We had a great deal of resistance from lab managers when they were asked to establish regular calibration programs for their equipment. Many could not understand the need to calibrate equipment to the extent of traceability to a national standard. However, once established, positive control became obvious. Now our regular calibration schedule includes cleaning and preventive maintenance for the equipment, ensuring it's properly functioning.

Calibration of equipment that is used to check product conformity to specifications can often be overlooked in non-structured management systems, which was somewhat true in our case before we implemented our ISO 9001 system. While we had programs in place for calibrating some of our measuring and test equipment like balances and scales, pH meters and other analytical equipment, we had over-looked the calibration of some production and lab equipment such as thermometers, humidity monitoring equipment, and product screening equipment. Also, we did not have required actions and tolerances formally defined. The requirements of the standard not only helped us define what equipment should be calibrated, but it also provided a structured approach to assuring applicable equipment is in calibration by defining a schedule of when specific equipment is to be calibrated. The ISO standard dictated the necessity of calibrating inspection, measuring, and test equipment in a way that is traceable to a national standard (where applicable) or known standard which built the foundation enhancing our food safety program by providing us with the confidence of having an effective HACCP calibration prerequisite program. Our current calibration program has evolved over the past ten years providing us with good, documented requirements along with records that provide evidence of an effective program. The calibration status of our *critical* equipment is much more visible than it was before our implementation effort. (*Tim Sonntag, VP Quality and Technical Services, Wixon, Inc.*)

IDENTIFYING CRITICAL EQUIPMENT

The first step is to identify the *critical equipment*. Critical equipment is that equipment that is used to verify measurements and testing related to food safety, quality, regulatory requirements, and additional customer requirements. This may not include all equipment used to measure and test. There is a difference! Each process is different, and it is up to the management of the system to identify specifically the equipment that is used to demonstrate that the product meets specified requirements. It is very important to understand that the equipment that is not identified as critical still must be accurate. It is separating the critical equipment into a specific category that requires particular actions to maintain compliance. Most organizations use the same system for both categories (critical and noncritical) with less stringent actions, tolerances, and the like, defined for noncritical. An example of this would be a hand-held refractometer used in blending a batch of fruit punch. The hand-held refractometer is used by the operator to monitor the blend and confirm that the mix is close to being in specification. However, the final go-no-go measurement is performed in the laboratory on a *calibrated* refractometer. Of course, the hand-held refractometer must be as accurate as possible, or the efficiency and possible waste could affect the process. The hand-held refractometers are verified and adjusted routinely as defined in the area procedures or work instructions. In doing so, they do not require the degree of scrutiny, traceability, and record keeping that is required for the critical equipment.

Guidelines that can be used in the identification of the critical equipment in addition to any references documented in the food safety program (hazard analysis and prerequisite programs) may include

- An analysis of what the measurements are used for
- The required tolerance and seriousness of the action plan should a piece of equipment be found to be out of tolerance (i.e., recall, product reworked, etc.)
- The placement of the equipment and its design
- Frequency of use
- Historical data on the suitability of the equipment in the current environment (may include historical data on the results of previous calibrations)
- Results confirming compliance to specified requirements
- Other data that would be used to support the findings
- Specific measurements required by the customer
- Equipment used to test or monitor an activity that is later confirmed by equipment testing at a later step in the process

Although not a requirement, it is recommended that noncritical equipment be identified in a manner that distinguishes it from critical equipment. This alleviates confusion. Identification can be done effectively by applying a different color tag or sticker on the equipment. Procedures or work instructions must define the differences and make certain that associates are trained to understand this difference. The schedule for calibration can be maintained through the preventive maintenance program that is designed to notify associates when calibration is due. This schedule is an excellent tool to trigger requests for calibration ensuring that it is done before the defined time frame expires.

The following must be defined for critical equipment requiring calibration either in a formalized controlled procedure, work instruction, or on a calibration matrix:

- Method for unique identification (equipment numbers, serial numbers, etc.).
- Method for which the calibration status is to be identified such that it is readily evident to the associates using the equipment (labels, plaques, logs, etc.).
- Frequency of tests required to ensure accuracy and suitability.
- Methods for testing (reference appropriate work instructions).
- Identification of the standard (i.e., national, international, or known) to be used.
- Responsibilities and required training for associates performing the calibration activities.
- Tolerances related to the required degree of accuracy for the tests being performed.
- Identification of external service suppliers, if appropriate.

- Activities to be performed should the equipment be found reading out of tolerance.
- Requirements for protecting the equipment.
- Records to be maintained demonstrating compliance to defined requirements (Chapter 7).

CALIBRATION MATRIX

Although not specifically required, it is recommended that the management team responsible for calibration maintain a matrix that not only identifies the equipment, but also includes all the important aspects of the program (i.e., the unique identification number, frequency, tolerances, date done, date due, etc.) mentioned in previous section. This matrix becomes an effective tool for maintaining an efficient process. Figure 11.1 is a sample of a calibration matrix. The organization must determine if this or a variation of this fits their purpose.

SCHEDULING CALIBRATION

It is recommended that the *next due date* be clearly defined as the last day in the month that the calibration check is due. For example, if scales are calibrated every 6 months and the last calibration date was January 15, 2013, and the next due date is labeled as July 15, 2013, on the calibration sticker. There's no doubt that a July 20, 2013, date would be unacceptable. To save numerous headaches, it is recommended that the label either reflect the month and the year only (i.e., July 2013) or the last day of the month due (July 30, 2013). Identifying the date in this manner makes accountability much clearer and easier to maintain. Ensure that related procedures and work instructions clearly define this requirement and be absolutely certain that the calibration service provider understands these requirements. Many nonconformances have been written because of this very issue. If the sticker states overdue on July 15, then the operator must be trained not to use this equipment on July 16. Remember, the system must clearly define what

Equipment	ID No.	Frequency	Method	Tolerance	Location

CAL-02-01

FIGURE 11.1 Calibration matrix.

is required, and then the operator must perform the task as required and have records to prove that it is done (Newslow 2001, p. 142).

PROTECTING THE EQUIPMENT

Equipment must be used in the environment for which it is designed. Efforts must ensure that the environment is equivalent with the capability of the equipment, consistent with the required measurements, and that it does not affect the accuracy. A laboratory balance designed for use in a controlled atmosphere situation would most likely be affected in a wet and/or dusty processing area. The equipment must also be protected from tampering with calibration adjustment or any other situation that may invalidate the calibration status.

CONFIRMING EQUIPMENT IS IN TOLERANCE

Activities and responsibilities must be defined for evaluating product that has been tested on equipment that is found to have been measuring out of its defined tolerance range. Product dating back to the previous acceptable calibration test must be evaluated to ensure its conformance status. Any product found to be nonconforming as a result of these activities must be addressed as defined in the system's procedures for the control of nonconforming materials and products (Chapter 9).

This particular requirement can produce many headaches and subsequent system nonconformances if not carefully defined. When evaluating test equipment and setting requirements, be sure to consider the effect that an out-of-tolerance situation would have on the product. The defined tolerances must be realistic. The process must distinguish between tolerances that require equipment repair from those that directly affect the safety of the product. Some organizations require additional backup checks that can be used to verify that the product is actually acceptable. This can save considerable time and effort from having to retrieve and evaluate months of product data. Be careful when defining requirements. Define what must be done, then do it and support the action with records. Records must be maintained in compliance with the record control program (Chapter 7). Set realistic and useful parameters.

TRACEABILITY TO NIST OR A KNOWN STANDARD

Records must be maintained for critical equipment that demonstrates that it was calibrated using equipment that is certified and traceable either to a recognized standard (i.e., National Institute of Standard and Technology [NIST]) or a known standard. The use of equipment traceable to a NIST standard is the most common and most efficient method. This group was established by Congress to provide official measuring criteria and standards to ensure product reliability. NIST standards are considered the official measuring certification for inspection, measuring, and test equipment. If a NIST standard is unavailable, then records must link to a known standard that ensures accuracy and repeatability of the equipment reading. Examples of a known standard may include a gas liquid chromatography, which may be calibrated by a curve created according to the manufacturer's instructions. This curve is used to calibrate to a *known* standard rather than to a NIST standard. Another example would be the standardized *color tiles* used to confirm the accuracy of a colormeter provided from the manufacturer. Requirements for creating, identifying, and maintaining the known standards must be defined (Newslow 2001, p. 144).

REQUIREMENTS AS EXTERNAL DOCUMENTS

Manufacturing manuals and other similar type documents that are identified as the source of required information must be established and maintained as external documents. Requirements for external documents can be complex and must be in compliance with the document control program (Chapter 6).

EXTERNAL CALIBRATION SERVICE SUPPLIER

It is very common to contract an external calibration service company to perform calibration activities. The external supplier of calibration services must be identified, approved, evaluated and maintained in compliance with the critical food safety and quality supplier program. The purchasing document, which may be a purchase order or a contract, must clearly identify the product ordered. The product is the service of performing *calibration.*

It is absolutely essential that records provided by the external service supplier define the necessary proof or compliance. Review the criteria and records required for compliance with the supplier. An individual should be assigned the responsibility of working with this supplier and ensuring that activities and records meet the organization's defined requirements. Records received from external service suppliers must be carefully reviewed to ensure that all out-of-tolerance situations are addressed internally. Many times, that is overlooked. Frequently, records are filed without confirmation that the equipment tested was within tolerance limits. Requirements for reviewing and signing these records before filing must be defined.

CALIBRATION RECORDS

Records must be maintained to demonstrate compliance to the record control program (Chapter 7). Records required for the calibration program are comprehensive and must include the following information:

a. The unique number (identification) of the equipment.
b. Date calibrated and date the next calibration is due (confirms defined frequency).
c. Required tolerances (how accurate must the reading be).
d. Actual reading *as found* (confirms instrument tolerance status).
e. Reading *as left* (confirms tolerance status at completion of the calibration exercise).
f. Evidence that the validity of results of product tested, on equipment that is found reading out-of-tolerance during its calibration, has been evaluated with appropriate actions taken as a result of this evaluation.

g. The unique identification of the test equipment used to perform the calibration (records must be traceable to national, international, or a known standard).

h. Identification of the person(s) performing the test.

i. If performed internally, this provides the trail to confirm that this individual has met the training criteria (competency) for this activity.

j. If an external service company is being used, then trail would lead to confirmation of compliance with the critical food safety and quality supplier program.

MANAGEMENT COMMITMENT AND EFFECTIVENESS

Management must provide the necessary resources to ensure a compliant and effective calibration program. Without this accuracy, the safety of the product can be in jeopardy, along with noncompliance situations related to quality, regulatory, and customer requirements. As a result of this, the product and the brand name of the organization can be put in jeopardy. There is also an indirect threat when calibration of instruments used for process control measurements is not appropriate. Management must apply appropriate expertise and resources are available to understand and ensure that an effective calibration prerequisite program is defined and maintained.

Every associate (employee) is responsible for food safety. *Edward Link (2008) in an ISO 22000:2005 Pocket Guide for Every Employee (Quality Pursuit)* describes the employee's role when using critical equipment:

> When an employee is about to use an instrument to make a measurement that reflects or affects the safety of the product, it is his/her responsibility to know that the label or record indicates that the calibration is still valid. Also, should any event (like dropping a measurement device) occur that might cause the device to go out of calibration, it is the responsibility of anyone having knowledge of this event to take the steps necessary to assure that the device is calibrated again. (p. 97)

KEY POINTS

a. It is important to apply stringent requirements to the *critical* equipment that affects food safety, quality, regulatory, and customers' final product requirements.

b. Equipment that is used for process monitoring is still important, and the means for maintaining it must be defined appropriately with records available to demonstrate compliance. Requirements for process monitoring equipment may be maintained in the preventive maintenance program rather than through the formalized calibration program.

c. Over time, it may be determined that the defined calibration requirements are either too stringent or not stringent enough. It is important that the justification and program revision dates are noted especially when revising such aspects as frequencies and tolerances.

d. Recommend the use of a calibration matrix for identifying requirements and tracking status/compliance.

e. A calibration matrix can provide a structured document for identifying requirements.

f. Calibration program may include equipment used by multiple departments (i.e., maintenance, quality). It is recommended that the responsible departments define and manage the *critical* equipment within their control.

g. A preventive maintenance program can be used effectively to schedule calibration activities.

h. Equipment that is identified as *critical* is dependent and unique to the specific operation, its food industry segment, and the food safety management standard of choice. Each operation must evaluate its process, make its decisions (justify them based on facts), and support compliance with objective evidence (records).

i. The external service supplier must be identified, approved, and maintained as a critical food safety and quality supplier.

j. Calibration records must be identified and maintained in compliance with the record control program (Chapter 7).

COMMON FINDINGS/NONCONFORMANCES

a. Not all critical equipment related to food safety had been identified.

b. Review of records from the external service supplier indicated that equipment had been found out-of-the tolerance range in several instances; however, there was no evidence that the validity of product tested with this equipment since the last acceptable calibration had been investigated.

c. Several pieces of *critical* equipment were being used within the process with past due dates on the calibration stickers. No records were available to confirm whether the required calibration had actually been performed on time.

d. Two examples of *critical* equipment used in the process had past due dates on the calibration stickers. However, review of records indicated that the equipment had recently been calibrated and was not overdue.

e. Status of calibration for critical equipment was not readily available to associates using this equipment.

f. Records provided by the external calibration service company did not provide identification traceable to a national standard (NIST) or a known standard.

g. Serial numbers listed on the master calibration list did not match the serial numbers on the equipment actually being calibrated.

h. Some scales and thermometers used to demonstrate compliance to critical test parameters were not identified in the calibration program as critical equipment.

i. Scales used in the process blend area had two different conflicting calibration status labels.

j. Lot numbers that provide NIST traceability for pH buffers were not recorded on the calibration log although defined as a requirement in the calibration work instruction for pH meters.

k. In some instances, pH buffer was being transferred into dispensing bottles without including the lot number of the buffer that provided the link back to the pH buffer label for NIST traceability.

l. Pantone color standards were being used to confirm color; however, evidence could not confirm that these standards were being controlled.

m. Quality department had defined the calibration process; however, it could not be confirmed through record review and interviews that competency training to calibration requirements had been completed and was effective.

n. Records did not confirm that the lot numbers for the HPLC calibration standards provided the link to the *known standard*.

o. Site did not have a formally defined process to ensure compliance with the food safety standard's requirements for calibration.

p. It could not be confirmed that the use of and maintenance of calibration software was being controlled in compliance with the document control program.

q. Required acceptable tolerances for equipment requiring calibration were not defined.

r. The process for communicating tolerances to external service providers performing calibration, and the actions to be taken when equipment was found to be out-of-tolerance, was not formally defined.

REFERENCE

Newslow, D.L. 2001. *The ISO 9000 Quality System: Applications in Food and Technology*, Wiley-Interscience, New York.

---

*Additional Food Safety Management System References (not all inclusive)

FSSC 22000:2010

ISO 22000:2005 Section 8.3

SQF 7th Edition: Section 11.2.10

BRC Issue 6: Section 6.3

IFS Version 6: Section 5.4

*Note: Always review the standard or visit the website for your chosen standard to ensure that all related requirements are identified and addressed.

12 Pest Control

Having an effective pest control program is critical no matter what food safety standard is chosen. Food companies have had or attempted to have good pest control programs for many years. Many times what some feel are *good* programs in reality are not. Companies that are thought to be reputable are hired and sent into the operation to take care of the pests with little or no supervision or accountability to the organization. The pest control program, if outsourced to a professional company, must be managed through the approved program for critical food safety and quality suppliers.

It is important that management identify an organization's pest control subject matter expert. It is recommended that there be a primary and a backup person identified. These positions should have training or familiarity with pest control protocol. This can be learned through assistance with an independent pest control expert consultant or through a reputable pest control service company. When initially choosing a pest control service supplier, ensure that the requirements are well defined and communicated. Their performance must be monitored continuously with any issues addressed effectively and in a timely manner. Review the process and products to ensure that the organization is receiving the best possible protection. The frequency of visits may depend on the products, the time of year, building design, building age, amount of internal control required, and the organization's processes.

Most pest control operators (PCOs) are on either a weekly or semi-monthly visit frequency. Some stretch it to monthly, but again this depends on the organization's individual situation. In large facilities, the PCO may visit weekly doing a portion each week so that over a specific time period, everything is checked according to the contract. This also allows the PCO more intense exposure to the organization's operation should something surface between visits. It is important that the operation also maintains a *Siting Log* for anything observed between visits along with a contact number should the PCO need to be called in prior to the next scheduled visit.

More and more pest control companies are using an electronic software system to record their findings. Specific issues may be written on a trip report, but the overall process is checked electronically with either a CD left on site or providing website access to the visit data. This software is also used to develop quarterly reports that provide trended data or any other issue of concern for compliance. These quarterly reports have become an excellent tool to manage and monitor the effectiveness of a company's pest control program. Ensure that these, along with all other pest control records, are identified and maintained in compliance with the record control program (Chapter 7).

Most pest control service suppliers, even though they may have an effective software program, also maintain a hard copy book that contains most, if not all, of the required records. It is critical that this binder is kept up-to-date, identified, and maintained in compliance with the record control program (Chapter 7). Pest control records at a minimum, should contain the following:

1. Contract (also a controlled document) between the organization and the pest control company, which may include commitment for service frequency and activities.
2. Operator licenses and training materials.
3. Certificate of insurance.
4. Map of bait stations, insect light traps, and other apparatus (recommend that this is updated at least annually unless an issue requires it more frequently), dated and initialed (also a controlled document).
5. Hard copy visit reports (may also have CD access or online access to reports), which includes any identified issues that require correction.
6. Trend reports and recommendations.
7. Approved pesticide list and labels for the approved pesticides.
8. Records of any pesticide used, including pesticide usage logs.
9. Material Safety and Data Sheets (MSDSs) and other chemical information documentation. The MSDS document is designed for all chemicals to provide critical information on the safety, approval, and application of pest control chemicals.

When designing the organization's pest control program, ensure that effective pest control measures are in place to prevent contamination of the products from the receipt of incoming items through the release of finished products. Effective application must be applied at all stages of receipt, manufacture, and distribution; this includes all buildings, warehouses, and external grounds on the premises. Some define requirements for only those areas used for storing or processing food products, ingredients, and/or packaging material; however, be careful not to omit adjoining buildings that could harbor pests or insects if they aren't treated. These critters do travel. Depending on whether the organization has any co-packers, distribution warehouses, etc., requirements should also address those locations. It is best to have everything defined in one document, whenever possible.

When designing or evaluating an existing program, it is important to do the research. Information abounds on what makes the best and most effective pest control program. Most reputable pest control companies have well-trained entomologists who can assist the organization in evaluating the program and in making the best decisions. Many documents, texts, and articles are available for reference; however, it is recommended that an independent expert pest control consultant be hired to assist the organization in working with reputable pest control companies (may want to interview two to three). The best possible economical, efficient, and specific-to-the-organization program should be designed. The effectiveness of the pest control program, as with all of the prerequisite programs, must be monitored through a good manufacturing practice (GMP) inspection program and an internal audit program. A simple test to check the thoroughness of inspections is to place a business card or rubber mouse into a rodent trap with a note asking that the item be returned to the organization's representative. The electronic ID or the inspection tag should be located inside the units requiring the PCO to open each cover to scan as proof of checking the device. Of course, nothing works better than having a company representative or guide accompany the PCO during visits, but in our lean manufacturing world, this is not always practical or possible.

The complete pest control program is, of course, important to the overall effectiveness of an organization's food safety program; with that said, this program must be approached as a *preventive* program. Pest issues can and must be controlled through effective preventive measures that center on good design, cleanliness, and good housekeeping practices that all relate to continual compliance with GMPs.

Important attributes to focus on and use to build an effective program that is best for the organization include ensuring that

- Associates are trained in and consistently apply GMPs and other sanitary practices
- Identified issues or potential issues are corrected in a timely manner
- A program exists that provides associates with the opportunity to report and record any pest activity
- The PCO understands and abides by the organization's procedures including the approval of chemicals and pesticides used in the facility

The external pest control service company must be managed as a critical food safety and quality supplier. The PCOs must be trained in compliance with the organization's requirements for site visitors and contractors.

KEY POINTS

a. Records must confirm that the PCO is trained in all specific food safety system requirements. This includes being trained to understand the organization's controlled document for pest control and maintaining records continuing this training. If a hard copy controlled document is provided, ensure that this is identified through the document control distribution process.

b. External service supplier for pest control must be managed as a critical food safety and quality supplier.

c. Many companies have design and cleaning challenges with older buildings, but these issues must be managed effectively to ensure cleanliness and to protect against pest and rodent infestation.

d. Many organizations require their PCO to replace the light bulbs in the insect light traps at least annually. Ensure that records of this are maintained. A sticker can be placed on each unit identifying the last time the bulb was changed. The UV lights must be shatterproof, identified, and managed through the glass and brittle plastic program.

e. Pest control records must be complete, and maintained in compliance with the record control program.

f. PCO findings must be addressed in a timely manner based on the potential risk of the findings, with records maintained to confirm this.

g. It is recommended that management identify an individual, along with their backup, from within the organization that is ultimately responsible for the pest management program. The individual(s) trained in pest control should be responsible for managing and overseeing the pest control program.

h. It is recommended that notes be made on the PCO report confirming any issue that has been reported to the company representative, what was done (issued work order) and track its status. All entries must be initialed and dated.

i. If birds are an issue or a potential issue, effective bird control to protect against a bird sourced hazard must be included in the pest control program.

j. Effectiveness of the pest program should be monitored and addressed through GMP inspections and internal audits.

k. Pesticides must be applied in accordance to regulations by a licensed PCO. It's not recommended that an organization store its own pesticide on site, but if it is done, then it must be stored, in a secured manner separate from any product, ingredient, and packaging of other chemicals.
   • Any pesticide brought on site by the PCO must be within their control (attended) at all times.

l. Only mechanical traps, extended trigger traps, or glue boards should be used inside interior areas. Effective location and spacing may vary depending on the process; however, it is generally recommended that interior devices are spaced every 20–40 ft along exterior perimeter walls and at both sides of each exterior doorway.

m. Exterior bait stations must be labeled, secured, and, if possible, raised off the ground using only approved bait designed for its purpose.

n. Generally, recommended placement of exterior bait stations is at 50–100 ft intervals along the perimeter of the buildings, but actual spacing depends on each individual organization's situation.

o. Insect light traps should be placed at the height and frequency recommended by the PCO or the expert pest control consultant for the targeted insect. Insect light traps must not be placed too close to product or packaging. It is recommended that placement be at least 10 ft from product and packaging.

COMMON FINDINGS/NONCONFORMANCES

a. The identified pest control company was managed through the corporate division; however, there was no means defined and implemented for the individual sites to communicate performance to the responsible corporate representative.

b. The pest control reports completed by the PCO were in several instances either hard to read or difficult to interpret.

c. Records were not legible and thus it could not be confirmed that the organization's representative had reviewed prior to PCO completing the visit's tasks as required by the defined procedure.

d. In some instances, it was stated verbally that issues were discussed between the organization's representative and the PCO; however, the PCO records did not reflect the discussion, next steps, or status.

e. Pest control documentation did not define or link to a document that defined the process for reviewing and approving pest control chemicals prior to use.

f. Records were not available to confirm that PCO had been trained in the current version of the corporate pest control document as required by that document.

g. The organization's procedure identifying *approved chemicals* did not address pest control chemicals.

h. The PCO's license had expired.

i. Pest control records did not include a current location map of all the devices (i.e., not all external bait stations were identified).

j. Device map did not identify location of the insect light traps.

k. There was no record of when the UV lights in the insect light traps had been replaced.

---

*Additional Food Safety Management System References (not all inclusive)

FSSC 22000:2010

ISO 22002-1:2009 Section 12

SQF 7th Edition: Sections 11.2.7, 11.2.11

BRC Issue 6: Section 4.13

IFS Version 6: Section 4.13

*Note: Always review the standard or visit the website for your chosen standard to ensure that all related requirements are identified and addressed.

13 Purchasing, Outsourcing, and Supplier Management

Purchasing, outsourcing, and supplier management relate not only to the actual purchasing activity, but also to the approval and monitoring of suppliers. Depending on the standard of choice, there can be different interpretations on exactly how encompassing this program has to be. It is recommended that no matter which standard is chosen the organization makes the decision to apply compliance requirements to critical food safety and quality suppliers. A critical food safety supplier provides ingredients, packaging material, product contact/processing aids, and/or services that are required for the production of safe products and also for meeting regulatory requirements related to food safety. A critical quality supplier is a supplier who provides ingredients, packaging material, product contact/processing aids, and/or services that are required for the production of the product that meets quality, customer, and any nonfood safety and regulatory requirements.

Compliance with this element has proven over the years to be a tough challenge. Organizations have struggled with understanding and implementing the requirements effectively. Although a challenge, the application becomes successful when the organization uses experience and common sense based on the business logic and needs of the organization. Purchasing and supplier compliance-related requirements are more frequently taken for granted or *assumed* in compliance than are the other food safety management system elements. Most organizations have a purchasing department with defined requirements, relating to cost and availability of the product. These are, of course, very important to the business, but the food safety management standards focus strongly on the organization's responsibility to ensure that suppliers provide food safe, hazard-free supplies that meet defined requirements. Organizations and individual departments must take ownership for supplier management.

The organization accepting the materials and services from the supplier is not absolved of the responsibility to ensure that it is safe and does not introduce any unknown, existing, or potential hazards into the organization's product. The organization must do everything it can to ensure that a hazard is not received from a supplier. There are many organizations (some very popular and well known) that are now out of business, bankrupt, and with top management in jail because they caused consumer illness and/or death due to a food safety event whose root cause was traced to a supplier. The only fault of the organization was that it was too trusting. They *assumed* that the product was hazard free. Having defined, implemented, and maintained effective critical food safety and quality supplier prerequisite programs are essential to the food safety program and the business overall. The production of safe food is, of course, the number one priority. It is difficult for an organization to stay in business if its suppliers are not able to consistently provide safe ingredients, materials, and services that meet the defined specifications. Frequently, in an established organization, these requirements are only informally understood and defined. "We have been doing business for years together. These guys are great. We do not have anything to worry about." A quality manager in a Midwest operation shares the following description of what his company went through when it first started the certification journey:

> Purchasing was by far the hardest part of implementing the system. Prior to implementation, the purchasing function was a jungle. There were poor controls and documentation requirements. Interpreting the requirements of the standard for purchasing was initially very difficult. The first procedures were very restrictive and as a result, nearly every audit showed massive nonconformances. After nearly 5 or 6 rewrites, it finally clicked (for me anyway) what the standard required. After that and very strict adherence to purchasing procedures by the purchasing coordinator, we were able to turn it around, developing a process that is both compliant and useful to our company activities. (*Anonymous*) (Newslow 2001, p. 112)

COMMUNICATION

The supplier management program must define the means for communicating food safety information, requirements, and issues with suppliers. If a supplier is required to provide an item based on the organization's current specification, then the program must include the process for providing the most current specification or procedure to the supplier. This relates to document control (Chapter 6) and external communication (Chapter 3). The supplier's location and contact person are responsible for maintaining the current specification or procedure, so that it is readily available to those preparing the product or performing the service and must be in compliance with related document control requirements. Very often during an audit, the auditor asks the auditee to contact the supplier to have them provide a copy of the current specification. This has improved over the years, but at best, the average compliance to this exercise is approximately 75%. It is difficult to consistently provide the organization with products in specification without having the current version of the specification. Many mistakes happen every day because of this, and many have the potential to affect the safety of the product. Ideally, receipt of out-of-specification ingredients and materials should be caught during incoming inspection (Chapter 15). Depending on the requirements and the incoming inspection program, some issues are definitely caught on receipt. It can be common for an organization to check only a few or sometimes none of the parameters on site, accepting (assuming) that the supplier has checked and confirmed that the product meets all the requirements. This lack of inspection may be effective 99% of the time, but who gets the 1% that is not correct? Programs must be defined, implemented, and maintained in a manner that is confirmed effective.

OUTSOURCED AND SERVICE SUPPLIES

The critical food safety and quality supplier program must also be applied to suppliers of product contact surfaces (equipment, etc.) and processing aids (cleaning chemicals, lubricants, food grade oil, etc.). This category must be evaluated, recorded, and continually monitored through the food safety program is hazard analysis (Chapter 17).

Effective communication and supplier management includes those critical food safety and quality suppliers who are providing outsourced services. Outsourcing is defined as a company that is external to the organization but under contract to perform a service critical to food safety and/or quality. An example of this may be a company contracted to manage the organization's preventive maintenance program. Examples of critical food safety and quality service and/or outsourced suppliers are identified in Figure 13.1.

Requirements related to the specific activity and parameters must be clearly communicated to service suppliers. For example, requirements for suppliers of transport or shipping services may include shipping temperatures, trailer conditions, content restrictions, and delivery timeliness. Storage temperatures, pest control, good manufacturing practices (GMP) compliance, and third-party inspections may apply to external warehouse storage facilities. The organization must decide the most efficient and effective means to communicate the food safety and quality requirements. For some, a purchase order could be an effective record; others could use a contract. A contract (agreement) that defines the current requirements, dated and signed, must be identified and maintained as a controlled document (Chapter 6). Because the contract is providing proof that the organization is meeting its defined requirements, it also becomes a record that must be maintained in compliance with the record control program (Chapter 7). Few documents are both a record providing proof and a controlled

Pest control	Calibration
Cleaning services	Consulting and training
External warehouses	Transport or shipping
Water treatment	Co-manufacturing companies
Waste disposal	Preventive maintenance
Recycling services	Laboratory testing services
Temporary employment agencies	Uniform/apparel suppliers
Security guard service	
Shredding and product destruction companies	
Maintenance contractors (i.e., manage filter changes, etc.)	
Equipment manufacturers performing preventive maintenance (i.e., lab equipment)	

FIGURE 13.1   Examples of food safety and quality critical service/outsourced suppliers.

document defining requirements that must be performed to be compliant with the requirements of the chosen standard of the organization. A completed purchase order or bill of lading would be considered a record. The blank versions would be a controlled document.

THOUGHTS ON GETTING STARTED

An effective means to define the critical food safety and quality supplier management program is to brainstorm with department management, key associates, and the food safety team to identify which suppliers must be included in this program. This session is normally included in the hazard analysis of ingredients, material processing aids, and product contact surfaces. Hazard analysis is performed in compliance with the requirements for the food safety/HACCP programs (Chapter 17). The information from this exercise is invaluable both in supporting and defining an accurate food safety program and also for those responsible to ensure that the suppliers are educated and understand the requirements and criticalness of this process. In most instances, service suppliers are not part of the actual hazard analysis, but managed through prerequisite programs. This information should also be identified and assigned during the brainstorming/hazard analysis sessions.

PROVISIONAL SUPPLIERS

In identifying the criteria for approval and evaluation of a new supplier, some organizations define a provisional time frame during which the supplier must perform and meet specific requirements in order to be approved. These suppliers would be identified as a *provisional supplier* until they either pass the test or fail. Failure would deny them the opportunity to be an approved supplier. The use of this type of program is the decision of the organization, and its value depends on the supplier's processes and products.

Of course, any new or potential critical food safety supplier is considered a change, and all changes related to the food safety program must be evaluated. When identified as necessary, the food safety/HACCP program, related documents, and all other requirements must be revised and implemented prior to the change being put into effect. The change management program is discussed in more detail in "Change Management" (Chapter 15).

The food safety/HACCP programs may also identify other required information such as a letter of guaranty, material safety data sheet, allergen declaration, Kosher certification. The letter of guaranty is a document that the supplier often presents stating that they have programs in place to manufacture products free of food safety hazards and that meet current specification, regulatory, and internal requirements for food safety and quality. An additional requirement to be performed by the food safety/HACCP team is to evaluate the potential risk of each ingredient, material, or service. Depending on the risk and the organization, some suppliers may require an audit prior to being approved. This does vary because not all suppliers are at a level of supply criticalness to require a physical inspection, but there are definitely many that do. Smaller companies may not be able to afford to send someone to the supplier or pay a consultant to do it. It is the hope of many that as organizations become more confident in the food safety management standards, that compliance with a third-party FSMS audit will reduce the number of other audits. There are many audit horror stories that an organization did very well on an external audit when, in fact, the organization had some very deplorable conditions. Performing an audit confirming GMP compliance is like cleaning house before company arrives. System compliance requires a different focus of interviewing, review of historical data, record review, evaluations, and top management commitment that has been recorded making it difficult to clean up just before an audit. Evidence (records, interviews, etc.) of system compliance is tough to fabricate completely, and a good, experienced auditor is able to assess the true status of the system. Also, the system requirements that management review meetings, internal audits, corrective and preventive action programs provide the compliance foundation for assessing the true status and effectiveness of a management system.

GRANDFATHER CLAUSE

Considering that an organization may have many existing suppliers who have been performing well, it may be hard to justify the value of starting over to evaluate and approve these suppliers. Related critical quality suppliers, existing suppliers with a good performance history can be *grandfathered* into the program. The criteria or qualifications for *grandfathering* must be defined, and a record linking this criteria to each supplier confirming compliance must be maintained. A sample of a *grandfather statement* follows:

> Individual suppliers having provided <u>organization name</u> with acceptable product in specification according to <u>organization name's</u> defined requirements for six (6) months prior to September 1, 2013, have been "grandfathered" as an approved supplier. A record confirming this is included in the supplier's file. All suppliers are evaluated at a minimum of annually from the date of approval. This record is also maintained in the supplier's file. The supplier record identifies the specific product(s) or service(s) for which the supplier is approved. For any additional product(s) or service(s), the supplier must complete the approval process as defined in the critical food safety and supplier program. Note that, related to the approval of food safety critical suppliers, the "grandfather" clause does not apply.

Some organizations create a master list of suppliers that includes the criteria statement, the individual supplier's names, date service began (if known), responsible associate(s), and a signature by the associate confirming compliance. Other organizations may create a form that is completed for each individual supplier, signed, and kept in the supplier's file. Existing suppliers *cannot* be grandfathered for food safety–related criteria. Grandfathered suppliers must still be evaluated for performance at a determined frequency as defined in the approved supplier program. Of course, the inclusion of critical quality suppliers is the decision of the organization and also depends on the food safety standard of choice. Related to food safety, suppliers must be monitored and evaluated to ensure continued compliance. Information related to supplier performance must be presented at the management review meetings, with any resource issues identified and addressed. An organization, its management, and associates must have strong confidence supported by objective evidence that the supplier is providing its products within specification and free of food safety hazards. Hazards associated with a product such as raw milk must be addressed through the hazard analysis, prerequisite programs, and where identified as significant, through a critical control point or an operational prerequisite program (ISO 22000, FSSC 22000—Only) (Chapter 17).

IDENTIFYING OWNERSHIP

Other than related to ingredients and packaging, it has proven effective for each department to take ownership of its suppliers. For example, if the quality department is responsible for the pest control service, then the quality manager or a designee should be responsible for evaluating the performance of that service. A service supplier that performs preventive maintenance work reporting to the maintenance manager should be reviewed by the maintenance manager or a designee. Each department should first identify the critical food safety and quality supplies and services used and then formally define the required criteria for effective performance. This may be defined in many ways such as a procedure or on an evaluation form. The suppliers must be evaluated, recommended at a minimum of annually, with results reported in compliance with the approved supplier program. In some locations, this information is sent to a central person or department responsible for ensuring compliance and data review, which is then reported at the management review meetings. In other locations, the responsible department manager or designee creates their own data and reports the status directly at the management review meeting.

MONITORING SUPPLIERS

The standards do not specifically require a list of suppliers, but a well-designed list can be very useful in providing and tracking required information. The list should include the name of the supplier, the products for which the supplier is approved, the date of the original approval, the date of the most recent evaluation, and the department responsible for this supplier. Some suppliers may, in fact, be approved for all items or services, but some larger companies, such as a chemical supplier, may only be approved for specific products. Ensure the relative information on each supplier is included in the file or on the list. *Dorothy Gittings, an independent consultant and third-party food safety auditor, D.L. Gittings Consulting Services, Inc.* provides a sample supplier spreadsheet (Figure 13.2), which has proven to be an excellent tool for identifying and tracking the required information.

Supplier issues are documented and addressed through the corrections, corrective action, preventive action, and/or continuous improvement programs (Chapter 4). Nonconformances related to materials and products are documented and addressed

	A	B	C	D	E	F	G	H	I
1									
2									
3	Supplier	Material/product	*Brief description of risk assessment conclusion (including potential risk to final product and ability to meet quality and food safety expectations, requirements, and specifications)*	Description of how supplier is assessed (example; onsite audit, third party audit)	Monitoring requirements	Authorization status	Date authorized	Next evaluation due	Person/s responsible
4									
5									
6									
7									
8									
9									
10									
11									
12									
13									
14									
15									
16									
17									
18									
19									
20									
21									
22									

FIGURE 13.2 Approved critical supplier list (sample).

through the control of nonconforming materials and products program (Chapter 9). Existing or potential food safety–related issues must be brought to the attention of the food safety team leader, backup food safety team leader, or a designated member of the food safety team immediately.

INELIGIBLE SUPPLIERS

The organization must also define the criteria for any supplier who has been removed as an approved supplier due to food safety issues, ineffective response to reported issues, and/or overall poor performance. The process for such a supplier to be approved for future work must be defined.

EMERGENCY INGREDIENT AND PACKAGING INFORMATION

If in an emergency situation, the operation must purchase supplies from a supplier who has not been approved, then management must immediately involve the food safety team to evaluate and provide input in this decision based on the risk(s) to the safety of the specific product in question. Product made from the emergency supply must remain within the control of the operation until it can be confirmed safe and it meets all quality, regulatory, and customer requirements. This clause is frequently written into an organization's approved supplier program; however, its use must only be in very extreme situations. The product must be controlled, and records must be maintained demonstrating compliance.

RESPONSIBILITIES FOR THOSE SUPPLIERS NOT APPROVED BY MANUFACTURING SITE

It is very common for larger companies with a corporate purchasing team to identify and approve its ingredient, packaging, and some service suppliers through the corporate function. The individual manufacturing sites then order supplies through a central system, only able to obtain supplies from those suppliers approved by Corporate. The fact that a group outside of the control of the manufacturing site (i.e., corporate, central purchasing, other sites, or customers) approves and controls a supplier does not absolve the site from responsibility for ensuring that food safety and quality requirements are met. In addition, *all* ingredients, packaging materials, processing aids, and product contact surfaces, no matter what group is responsible for approval, must be identified and evaluated for existing and potential hazards through the food safety/HACCP program's hazard analysis (Chapter 17). Many times, the approving group can assist with this information, but whether provided or not, the hazard analysis must be performed by the food safety/HACCP team in compliance with the standard of choice. There must also be a defined program for the site to report back to the approval group (i.e., corporate purchasing) on the performance of the suppliers especially in the case of having issues or a nonconformance.

MORE THOUGHTS

As a brief review, an effective critical food safety and quality supplier program either defines the following requirements or makes a reference to where this information is defined:

- Criteria required to identify which suppliers are critical food safety and quality.
- Required information for the food safety/HACCP program (i.e., identification, information for hazard analysis, required control, ingredient characteristics, country of origin, and specific process requirements); always reference the standard of choice for this information.
- Criteria not only for the approval process, but also for monitoring and control through evaluation of ongoing supplier performance.
  - Ongoing evaluations may be based on the potential risk of the item or service being supplied.
  - The supplier's continued ability to meet defined requirements.
- Type and extent of control that must be applied to a supplier depending upon the existing or potential food safety risk of the product or source.
- Process for addressing a supplier who has a nonconforming situation, whether food safety or quality related.
- Process for reporting supplier issues for those suppliers identified and not controlled by the manufacturing site (i.e., corporate, customer, etc.).
- Criteria and/or limits of noncompliance that would result in removing a supplier from approved status.
- Required objective evidence (i.e., records) that is maintained to demonstrate compliance.
- An *inactive* time period that would require the reevaluation of suppliers prior to use, in other words, the period of time elapsed that would require a supplier to go through the approval process again (i.e., inactive for 2 years) before its products or services can be sourced.
- Responsibilities for managing and ensuring the effectiveness of the critical food safety and quality supplier program.
- Records demonstrating compliance to the defined requirements for this program must be identified and maintained (Chapter 7).

KEY POINTS

a. The food safety/HACCP team identifies, evaluates, and records existing and potential hazards related to ingredients, product contact surfaces, processing aids, and packaging materials in the hazard analysis, which is part of the food safety/HACCP program (Chapter 17).
b. *Grandfathering* existing suppliers can be done relating to critical quality suppliers; however, it cannot be applied to critical food safety suppliers.

c. Results of supplier evaluations, related to food safety issues (may depend on the standard of choice) must be reported at the management review meetings.

d. Purchasing documents (i.e., purchase orders, contracts, bills of lading, and procedures) must define the product ordered.

e. Service suppliers of critical food safety and quality supplies must be included in the approved supplier program.

f. Records related to critical food safety and quality supplier activities are identified and maintained in compliance with the record control program (Chapter 7).

g. Supplier issues should be recorded and addressed through the corrections, corrective action, preventive action, and/or continuous improvement programs (Chapter 4).

h. Nonconformances related to products and materials are documented and addressed through the control of nonconforming materials and products program (Chapter 9).

i. Existing or potential food safety–related issues must be immediately brought to the attention of the food safety team leader, backup food safety team leader, or a designated member of the food safety team.

j. Supplier approval evaluations must be current based on defined requirements. Food safety expectations, requirements, specifications, and performance for critical food safety and quality suppliers must be defined.

k. Critical food safety and quality suppliers must be evaluated at a defined frequency, which, at a minimum, is recommended to be done annually. The food safety management standard of choice may define the minimum accepted frequency for evaluation.

l. The organization may want to design a form to record supplier evaluations. For example, a company that has outsourced for waste handling may have a specific form to evaluate cleanliness, safety, environmental compliance, work quality, timeliness, coordination, and response to the organization's needs.

m. It is recommended that depending on the size of the organization, its operation, variety of products, materials, and services managed, that a process flow diagram be created that tracks the different scenarios and required outputs of specific service suppliers (i.e., pest control, etc.).

COMMON FINDINGS/NONCONFORMANCES

a. Organization had not identified and approved its service suppliers critical to food safety and quality.

b. Organization had defined that the supplier approval program did not apply to its location because suppliers of ingredients and packaging materials were approved and managed through the corporate office.

c. The organization stated that there had been issues with the pest control company, however, the management team indicated that because it was a corporate-approved supplier, they had no recourse for correction.

d. The organization did not have a defined program for reporting back to the corporate buyer on any issues related to food safety or quality for the suppliers controlled by that position.

e. The organization's approved supplier program stated that any nonconformance issues would be recorded on the corrective action form and submitted to the supplier for attention; however, records did not confirm that this had been done for the last three shipments of damaged and leaking product. The supervisor stated that the supplier was called, and the product shipped back. The supervisor thought as long as they received credit that he would not have to issue a corrective action.

f. Documentation defined that approved suppliers must be evaluated at a defined frequency; however, records were not available to confirm that this had been completed as required by the controlled document.

g. Records did not confirm that a sampling of suppliers listed on different purchase orders were in compliance with the approved supplier program.

h. The listing for approved suppliers identified the company but not the items for which it was approved. XX Chemical Supply and YY Flavor House manufacture a wide variety of items. It was stated that those companies were approved only for specific products, but documentation did not identify these products.

i. Requirements defined that the supplier must provide products per the current specification. However, only the specification number was identified on the purchase order. Upon request, three of five suppliers contacted were unable to provide the most current specification version as identified on the master list of documents.

j. Suppliers of services such as pest control, calibration, and product testing were not identified and maintained as critical food safety suppliers.

k. A specific supplier of raw materials had not been used for over 3 years; however, it was still listed as an approved supplier. Referenced approved supplier procedure stated that any supplier inactive for greater than 2 years would be removed from the list and required to go through the approval process in the future.

l. It could not be demonstrated that all departments responsible for performing purchasing activities (i.e., evaluation of suppliers) had been trained and were performing their required responsibilities as defined in the approved supplier procedure.

m. Records did not confirm that the critical supplier (food safety and quality) issues (i.e., trends) were being addressed through the formalized corrective and preventive action program as required by the approved supplier procedure.

n. Records were not available to confirm that the critical food safety and quality suppliers were being evaluated annually as required by the approved supplier procedure.

REFERENCE

Newslow, D.L. 2001. *The ISO 9000 Quality System: Applications in Food and Technology,* Wiley-Interscience, New York.

*Additional Food Safety Management System References (not all inclusive)

FSSC 22000:2010 Additional requirements

      ISO 22000:2005 Section 5.6.1

      ISO 22002-1:2009 Section 9

SQF 7th Edition: Section 2.3

BRC Issue 6: Sections 3.5, 3.6

IFS Version 6: Section 4.4

*Note: Always review the standard or visit the website for your chosen standard to ensure that all related requirements are identified and addressed.

14 Crisis Management/ Emergency Preparedness

Having a *crisis management* or *crisis response* program is not new to most organizations. These types of programs emerged 20 plus years ago when organizations wanted to have structured programs to deal with crisis situations. A crisis can be anything from an adverse weather situation (i.e., tornado, hurricane, earthquake, etc.) to a fire, bomb scare, flood, or product recall. Associates were trained to know what to do immediately if a crisis surfaced. Operators, guards, and others that were responsible for monitoring incoming calls were trained to know exactly who to call and what to do if a crisis call came in over the phone.

Every organization must have a structured program that is clearly defined and communicated throughout the organization as to what to do if a crisis arises. It is important to have a current list of key contacts to call for various conditions. A crisis is not the time to figure out what to do or who to call. There should be a key person identified as the crisis manager leader or coordinator with an alternate or backup person also identified. Criteria required for training and competency for these positions must be defined, all responsible associates trained to meet this criteria, and records maintained demonstrating compliance. Crisis and potential crisis situations must be reviewed during the management review meetings.

As indicated earlier, it is important to practice responses to different crisis situations (e.g., a fire drill). Fighting a real fire is not the time to find out that the defined plan for reaction does not work. Another example of a practice situation would be a *mock recall*. A *mock recall* is literally practicing a recall. Recalls and mock recalls are discussed further in the "Product Recall" section. Again, do not wait until there is a real situation where a potentially unsafe product has been shipped to the consumer to find out that the organization's recall program does not work effectively.

ISO 22000:2005 5.7, entitled "Emergency Preparedness and Response", is addressed in the management responsibility section of this standard. As stated previously, most organizations have a defined *crisis management* program that identifies what to do during a crisis. What is often not included is the requirement to perform a thorough evaluation after a crisis to confirm the safety of the product. Was the product affected during the crisis? If so, what needs to be done to ensure that no product that has or could have an issue is released? Common sense tells us that this is being done, but it is an activity that many times is not defined as a requirement, nor is a record of the evaluation identified and maintained. Top management must ensure that requirements are clearly defined, in compliance and effective. This must include defining responsibilities for and criteria to be met for mock recalls, mock drills, contact lists, and the records maintained that provide the evidence of compliance and effectiveness. The individual food safety standards do have a varied list of requirements related to crisis management and the protection of the product during a crisis. The food safety team should review the specific requirements of the chosen standard and ensure compliance. Controlled security and building access is critical and must be part of the organization's security program ("Security" section in this chapter).

Although most companies do their mock recalls on a defined basis, many do not challenge their crisis program. It is recommended that this be done through a mock fire drill or, possibly, a mock tornado drill. Of course, effectiveness is important for the primary safety of the associate. This exercise must also ensure that in a real situation, the product either is confirmed safe or potentially unsafe, segregated, and managed in compliance with the procedure for the control of nonconforming materials and products (Chapter 9) prior to returning to production. A record must be maintained as proof of the activity and status of compliance. If compliance status is not achieved, additional practice sessions should be performed. Do not wait for a true emergency to find out that there are issues with the program. Further discussion related to protecting the safety of the product during a crisis may be discussed in other prerequisite programs.

Some organizations may define specific levels of incidents based on their potential risk to either individuals, the product, or brand name. Reactions based on these parameters are used to determine the seriousness or potential seriousness of the incident and required actions to be taken. The extent and application of this program is determined by the organization. The Food and Drug Administration (FDA) does define its recall criteria into three levels with mandatory compliance of recalls in the United States. More detailed information related to recalls is addressed in the "Security, Food Defense, and Bioterrorism" section of this chapter.

A crisis such as a recall may result in the news media or regulatory agency becoming involved either by calling or showing up in person. Ensure that all associates understand the requirements for communicating with these groups. Communication skills are a must. No matter what the true situation is, the initial communication comments can make or break the organization.

An injury may be considered a crisis. Associates must be trained to know exactly what to do should an associate be injured or become critically ill. The structure for this type of incident may be different from a tornado drill; however, the important factor is to have associates trained and empowered in a manner that promotes the effectiveness of the program.

Remember, test the process. One organization had a program that had someone call the facility reporting an illness from their product. Although there was a defined program in place, it took many tries before the person on the receiving end of the call finally handled such a call correctly. Practice, practice, and then practice again to ensure that should a real situation ever occur, the team is ready and confident in the required protocol for success.

Records related to crisis management must be maintained in compliance with the record control program (Chapter 7).

KEY POINTS

a. The organization must identify a crisis management leader and backup leader who are trained in the predefined requirements with records maintained to confirm compliance.

b. Both the crisis management leader and the backup leader must perform *mock* sessions to not only confirm the effectiveness of the program but also to build confidence in managing a crisis.

c. The organization must ensure that all required activities are met in the performance of mock drills.

d. The defined procedure must include a current crisis management contact list that associates have access to at any time.

e. If product during or after a crisis is identified as potentially unsafe or for any other reason not fit for use then this product must be managed in compliance with the procedure for the control of nonconforming materials and products procedure.

f. Associates representing all levels of the organization must be trained to know what to do (who to contact) should they identify an existing or potential crisis situation. The food safety management standards focus on the protection and the requirement to evaluate the status of the product from a food safety viewpoint with proof of required actions to be defined and records maintained demonstrating compliance.

g. Depending on the crisis, it might be necessary to contact regulatory officials with specific information. Ensure that this information is identified and defined in the program so that in a crisis, it is accessible and accurate.

h. ISO 22000:2005 5.6 has specific requirements for both external communication and internal communication as related to food safety–related concerns. Depending on the risk or potential risk of a situation or event, communication may link to the defined requirements for crisis management.

i. The organization must perform as many mock recalls and mock drills as necessary until management is confident that the programs have been effectively defined and effective.

COMMON FINDINGS/NONCONFORMANCES

a. Records were not available confirming that a crisis leader and backup leader had been trained in the crisis management program as required by the procedure for crisis management.

b. Although records confirmed the training of a backup crisis management coordinator, it could not be confirmed that the person in this position had actually performed a mock crisis event to confirm the effectiveness of the training.

c. Records confirmed that a successful mock fire drill had been performed; however, records did not confirm that the safety of the product had been evaluated prior to returning the system to production.

d. Records confirming that requirements as defined in the crisis management procedure were being performed as scheduled were not available.

e. Lunch time relief operator was unable to demonstrate familiarity with required actions should an incoming phone call communicate a crisis situation.

f. The guard was able to state the primary organization's contact should a crisis become evident at the guard shack; however, he did not know what to do if that person was not available. Note that the list of contacts for the organization did have information as to the contact protocol; however, this information had not been communicated to the guard company.

g. Organization stated in its procedure for crisis management that a broken light bulb in the process area became a potential for glass contamination; however, records were not available confirming that such a recent incident had been addressed in compliance with the referenced document.

h. Crisis management procedure stated that all recorded crisis situations would be recorded as a corrective action. Records were not available for five of seven recorded crisis situations confirming that a corrective action/preventive action (CAPA) had been initiated.

i. Records stated that evaluation of the safety of the product after a fire drill performed on January 1, 2013, had failed to meet the defined requirements; however, there was no evidence of an action plan including raising a CAPA as required by the referenced procedure.

j. Records were not available to confirm that associates including the crisis team leader had been trained to know how to direct the news media should a representative arrive asking for specific information about the organization or an incident that may have occurred.

k. Records were not available to confirm that the guard service was being managed in compliance with the critical food safety and quality supplier program.

---

*Additional Food Safety Management System References (not all inclusive)

FSSC 22000:2010

       ISO 22000:2005 Sections 5.6, 5.7

       ISO 22002-1:2009 Section 18.0

SQF 7th Edition: Section 2.1.6

BRC Issue 6: Section 3.11.1

IFS Version 6: Section 5.9

*Note: Always review the standard or visit the website for your chosen standard to ensure that all related requirements are identified and addressed.

PRODUCT IDENTIFICATION AND TRACEABILITY

Product identification and traceability requirements are basically straightforward. Ingredients packaging and finished products must be identified from receipt through all stages of production and delivery. This must be done so that all products can be retrieved should a problem occur. Responsibilities for performing, monitoring, and ensuring effectiveness must be defined. Requirements for this prerequisite program also relate to those defined for crisis management (previous section) and for the control of nonconforming materials and products (Chapter 9):

> *Identification* is defined as "any suitable means that allows an organization to trace a service or [product]."
>
> *Traceability* is the "ability to trace the history and delivery of a service or [product] at any point while the organization has responsibility." (Newslow 2001, p. 123)

Typically, for many years, food manufacturers have had well-defined programs that identify their products from raw materials through the finished product. The use of technology has provided many effective programs that can track the whereabouts of 100% of the product, many times within 2 hours or less.

Requirements for mock recall are defined in more detail in product recall (see "Product Recall" section). Product identification and packaging traceability begin with recording the lot number and supplier of the raw ingredients and packaging, then assigning an internal lot or batch number to the in-process product, which may be a new lot number or linked to the date code of the finished product. Finished products are usually identified and tracked via the date code. Records for the production of a specific finished product date code must trace to the lot or batch numbers of the ingredients used in the product. Requirements for traceability relate to what most in the food industry refer to as the *recall* process.

Of course, depending on the product being manufactured, identification and traceability activities may vary. The organization must define the method it feels is the best for its process. It is up to the processor to know the operation, know what is required, and then define these requirements appropriately.

The system does not demand the impossible, but it does require that, as appropriate to the operation requirements be defined. An example of one situation may be in receiving bulk material. Raw milk is tested, approved, and received into a bulk silo. The loads going into the silo are identified, but once blended, unique identification is lost. The system in this case would define a method that would provide an estimate as to what loads were received and when the silo was actually used. In some processes, retrieving could involve a lot of *extra* product; however, there are often situations where it is the only way to ensure that all tainted product is retrieved.

Edward Link (2008) in An ISO 22000:2005 Pocket Guide for Every Employee (Quality Pursuit) provides a good description regarding employee responsibilities and traceability records:

> The requirement for traceability has some very serious roots. The ability of your company to identify what is in your products could be required to protect

humans from injury, illness, or even death. It could also be required just to limit the level of customer (the next non-consumer link in the food chain) dissatisfaction on problems that somehow evade detection of your FSMS. Regulatory agencies have assured that management is well aware of the importance of traceability. All employees responsible for creating records that include information about traceability must do so with the upmost care. Our hope is that these records never have to be used. In the event that they are required, they need to be accurate. (p. 35)

KEY POINTS

a. A well-defined and tested program for ensuring the effectiveness of the organization's product identification and traceability must be in place.
b. Records confirming the effectiveness related to the product identification and traceability program must be maintained in compliance with the record control program (Chapter 7).
c. The organization must ensure that effective procedures are in place such that ingredients and packaging materials can be traced via their lot numbers back to the supplier's production information.
d. Records must provide tracking of raw ingredients and packaging from the time of receipt to the finished product.
e. Requirements must also link to those for related prerequisite programs such as crisis management (See "Crisis Management" section), product recall (see "Product Recall" section), and The Control of Nonconforming Materials and Products (Chapter 9).

COMMON FINDINGS/NONCONFORMANCES

a. A process had not been defined and implemented for identifying and tracing product from raw ingredients through processing and distribution.
b. It could not be clearly demonstrated through review of training records that associates responsible for recording incoming lot numbers of raw ingredients and packaging materials were competent to perform the defined requirements (Chapter 8).
c. Review of a sampling of records for batch preparation did not provide evidence that lot numbers were being recorded as required by operational procedure OP-2 Rev 1.
d. Incoming procedure Rec-2/Rev 1 required that the manufacturer's lot numbers be recorded for all ingredients upon receipt; however, records were not available to confirm that this activity was in compliance.

---

*Additional Food Safety Management System References (not all inclusive)

FSSC 22000:2010

    ISO 22000:2005 Section 7.9

    ISO 22002-1:2009 Section 15.0

SQF 7th Edition: Section 2.4.7

BRC Issue 6: Section 3.9

IFS Version 6: Section 4.18

*Note: Always review the standard or visit the website for your chosen standard to ensure that all related requirements are identified and addressed.

PRODUCT RECALL

ISO 22002-1:2009 15.1 provides the following statement: "Systems must be in place to ensure that products failing to meet required food safety standards can be identified, located and removed from all necessary points of the supply chain." The program for product recall links directly with most organization's program for crisis management and also with product identification and traceability. An effective product recall program is critical to have in place and known to be effective. We hope to never have to apply it to a real recall.

What is really required to have an effective program? First review the requirements of the FDA, the U.S. Department of Agriculture (USDA), or the Canadian Inspection Agency as listed on their individual websites (www.fda.gov, www.usda.gov, and http://www.inspection.gc.ca) for current government information and requirements for performing a recall. This information has proven to be very helpful to organizations defining their own internal programs plus it is critical that the organization at a minimum is familiar with the government requirements and how to access them if needed. An emergency is no time to learn how to drive through the government websites.

The crisis management program requirements are discussed in detail in this chapter. A recall is a crisis by almost every definition. It is recommended that the crisis management program have a crisis management leader and a backup leader trained to know exactly what is required in any crisis. As part of this, the program must have a list of key individuals to be contacted during a crisis. These individuals make up the crisis team and, depending on the crisis, are contacted to assist with required actions.

A product recall is defined as an action performed by the manufacturer and/or the distributor to remove product from the market that may be unsafe having the potential to cause health problems or, more seriously, death. FDA has three classes of recalls, which are identified later in this text. A product recall may be put in place for many different reasons. Most recalls are initiated by the manufacturing company. The recall may be the result of a product incident that either contains or has the possibility of containing a food safety hazard that may be harmful to the consumer. A recall may also be due to an incident related to quality such that it could harm the company brand or reputation and/or in conflict with regulatory requirements. The actual process for doing a recall may vary by organization, but a program must be defined, implemented, and confirmed effective.

An organization does not want to find out during a real recall that their program is not 100% accurate and effective. It is recommended that an organization perform a practice or mock recall at least quarterly. Most companies do this annually, but many things can change in a year including supplier management and associates in specific positions of responsibilities. The frequency of the mock recalls must be dependent on what is felt to be adequate to monitor and test the process. Performing the mock recall quarterly keeps associates practiced and should a glitch in the system be festering, it can be identified and fixed before it becomes a much larger problem. Also different situations result in different outcomes; doing this exercise quarterly increases the likelihood of finding any glitches. In today's world of computers and databases, asking to account for 100% of the product in 4 hours is reasonable. What used to take 48 hours, thanks to technology, may now be completed in 48 minutes.

It is recommended that at least quarterly, the assigned responsible manager or designee either physically purchase a product or review production and shipping logs to identify a product and its code date such that enough time has occurred to ensure that it has reached the market place. It would lose the effectiveness of the exercise if the product had been recently produced and still in the warehouse. Once the target product is identified, then issue the mock recall request to the responsible associates. Records must include the total number of products produced and the total number accounted for through the distribution chain.

It is hoped that the operation is never put in a position where this process has to be used for a real recall situation. The practice recall exercises are an invaluable tool to test the system. They provide a good foundation and learning curve. Results of these exercises must be maintained in compliance with the record control program demonstrating compliance and proof of the status of the process. Any problems identified must be recorded and addressed through the CAPA program. It is very likely that during the certification assessment, the auditor will request that a mock recall be performed. Lot numbers at random will be chosen and the request made.

The mock recall exercise must also be applied to raw ingredients and materials identifying the lot numbers and amounts of all of the ingredients for the identified product.

The organization must be prepared should it be contacted by an ingredient supplier stating that a specific lot shipped had been found to contain a food safety hazard. The organization must know what products that lot number went into and where the product is so that it can be retrieved and dealt with appropriately. In late 2008 and early 2009, a very intense recall relating to peanut paste and peanut butter produced at Peanut Corporation of America's (PCA) Blakely, Georgia, and

Plainview, Texas, facilities began. Within a very short time, this recall extended to over 3600 products dating back to January 1, 2007. The point of this example is that organizations must identify every ingredient in some manner that will allow it to be traced to the final product should a recall ever happen. Most likely if an organization cannot trace the lot number of the ingredient, then it will be required to pull all products made with that ingredient off the market. This is almost a sure recipe for disaster that could and has resulted in bankruptcy and organizations going out of business. The consumer must be protected from food safety hazards. If ingredient lot numbers are not identified so that the organization knows which products to recall, then the only way to be sure is to pull all products with that ingredient. One client that owns a small bakery told us that during this period when FDA was looking for every operation that could possibly have used the peanut products, an FDA officer visited the bakery and asked for the source of their peanut butter. The manager had confirmed with the broker that it was not peanut butter sourced from PCA; however, there were no records to prove this. The FDA officer made the bakery manager contact every customer that was sold a product with peanut butter as an ingredient. Each customer was asked if any product was left and were any illnesses reported. Because there were no records stating the tainted peanut butter had not been used or identifying what peanut butter was used, the FDA officer had to error on the side of caution and assume that the tainted peanut butter was in the products. Being a bakery product with short shelf life, all the products had already been consumed.

By the time the PCA recall and fiasco was over, at least 714 people (9 known deaths) in 46 states were confirmed ill with *Salmonella typhimurium* infection after consuming peanut and peanut butter products produced by PCA. Again, it was traced to over 3600 products. What happened with this company is a discussion for supporting the criticality of effective prerequisite programs. Frankly, PCA is not the only example. There are many more examples that are just as ridiculously alarming. The focus of this chapter is for the organization to understand the necessity of knowing its products and ensuring that an effective and compliant product recall program is in place, hopefully to never be needed, but to be there and ready.

The USFDA website provides the following information relating to recalls (http:/www.fda.gov/safety/recalls/ucm165546.htm):

Recalls are actions taken by a firm to remove a product from the market. Recalls may be conducted on a firm's own initiative, by FDA request, or by FDA order under statutory authority.

- *Class I recall*: a situation in which there is a reasonable probability that the use of or exposure to a violative product will cause serious adverse health consequences or death.
- *Class II recall*: a situation in which use of or exposure to a violative product may cause temporary or medically reversible adverse health consequences or where the probability of serious adverse health consequences is remote.
- *Class III recall*: a situation in which use of or exposure to a violative product is not likely to cause adverse health consequences.

- *Market withdrawal*: occurs when a product has a minor violation that would not be subject to FDA legal action. The firm removes the product from the market or corrects the violation. For example, a product removed from the market due to tampering, without evidence of manufacturing or distribution problems would be a market withdrawal.

In addition to the government websites, there are many information sources (i.e., internet, text books, and universities) available for defining such a program.

If an issue should occur with the food safety/HACCP program such as a deviation from a CCP critical limit, as defined in the food safety program (Chapter 17), the process must be stopped and product protected from consumption in compliance with the procedure for control of nonconforming materials and products (Chapter 9). If the organization determines that unsafe or potentially unsafe product has been shipped, then the product recall program must be initiated. It is up to the organization to ensure that an effective program is in place and that there is objective evidence (records) confirming its effectiveness (Chapter 7).

ISO 22000:2005 7.10.4 defines the following requirements for *withdrawal*:

"To enable and facilitate the complete and timely withdrawal of lots of end products which have been identified as unsafe. Top management must appoint personnel having the authority to initiate a withdrawal and personnel responsible for executing the withdrawal, and the organization must establish and maintain a documented procedure for:

   a. Notification to relevant interested parties (e.g. statutory and regulatory authorities, customers and/or consumers);
   b. Handling of withdrawn products as well as affected lots of the products still in stock;
   c. The sequence of actions to be taken.

Withdrawn products must be secured or held under supervision until they are destroyed, used for purposes other than originally intended, determined to be safe for the same (or other) intended use, or reprocessed in a manner to ensure they become safe. The cause, extent, and result of a withdrawal must be recorded and reported to top management as input to the management review. The organization must verify and record the effectiveness of the withdrawal program through the use of appropriate techniques (e.g. mock withdrawal or practice withdrawal)."

ISO 22002-1:2009 15.2 states that "Where products are withdrawn due to immediate health hazards, the safety of other products produced under the same conditions must be evaluated. The need for public warnings must be considered."

The BRC Global Standard for Food Safety, Issue 6 (July 2011) 3.11.1 requires the organization to have a documented program to "effectively manage incidents and potential emergency situations that impact food safety, legality, or quality." It also requires the "consideration of contingency plans to maintain business continuity." The contingency plans must include addressing the "disruption to key services such as water, energy, transport, refrigeration processes, staff availability, and communications."

The BRC Global Standard for Food Safety, Issue 6 (July 2011) 3.11.3, requires that "the product recall and withdrawal procedures be tested, at least annually, in a way that ensures their effective operation." Also 3.11.4 states that "in the event of a product recall, the certification body issuing the current certificate for the site against this standard [BRC] shall be informed within three working days of the decision to issue a recall."

Records related to defined product recall requirements must be identified and maintained in compliance with the record control program (Chapter 7).

KEY POINTS

a. Most often, withdrawals are managed through the organization's crisis management program. An organization's procedure for product recall must also link or reference to the crisis management program for related detail on that aspect of the process.

b. Requirements for performing a mock recall must be defined and usually included within the product recall procedure.

c. It is recommended that mock recalls be performed quarterly. Many may feel this is too frequent, but it is best, especially for complex operations, to perform these more frequently, to have the opportunity to test different scenarios, and to also ensure effectiveness of the program.

d. Records confirming that mock recalls are successfully performed in compliance with defined requirements must be maintained as defined in the procedure for the control of records (Chapter 7).

e. Criteria must be defined for the performance of the mock recall including links to CAPA programs and increased frequencies if the *mock* trial does not meet the criteria.

f. When performing the mock recall, it is recommended that the investigation goes back to a lot number of the ingredients and packaging. Ensure that the program would be effective should a supplier notify the organization of a possible food safety hazard in an ingredient.

g. Records related to defined product recall requirements must be identified and maintained in compliance with the record control program.

h. Each organization must be familiar with regulatory requirements for recall. Government websites are an excellent source for this information.

i. Traceability and lot identification must include raw materials. Bulk raw materials are hard to identify for tracing purposes because loads get comingled upon receipt, but it is up to each organization to evaluate the situation and define a process for which incoming materials can be identified and traced if necessary. For bulk materials, this will most likely require pulling extra data to ensure that required lots are assured to be included. Whatever the process, it must be defined and tested with records maintained to demonstrate compliance.

j. Associates must be trained and confirmed competent in related product recall responsibilities.

k. Management must ensure that there is a contact list that applies to product recall. The crisis management contact list would be acceptable as long as the product recall procedure references or links to the list for the required information.

l. Withdrawals or any related incident must be recorded and reported at the management review meetings. Management must initiate a corrective action to address any issues that must be addressed to avoid future reoccurrences.

COMMON FINDINGS/NONCONFORMANCES

a. A mock recall had not been performed in compliance with defined requirements.
b. Referenced procedure required that mock recalls be performed quarterly; however, records confirmed that the quality manager was performing recalls once per year stating verbally that the frequency was approved by management. Records confirming the approval were not available.
c. The relationship between the crisis management contact list and the product recall program did not appear to be understood by those responsible.
d. Records confirming that associates had been trained in related responsibilities were not identified and maintained in compliance with the record control program.
e. The mock recall did not include traceability back to the raw ingredients and packaging materials. It was stated verbally that this could be done successfully, but this could not be substantiated through record review.
f. Although mock recalls were being performed at the defined frequencies, records did not confirm that noncompliance results were being addressed and corrected. Review of related records did not provide evidence of any actions taken when defined criteria was not achieved.
g. Organization implemented a new warehouse management software system that was designed to enhance the traceability program resulting in more effective product tracking. The new system had been active for 6 months; however, there was no evidence that this change had been reviewed and food safety/HACCP programs revised to reflect this change. Also, there was no evidence that a mock recall had been performed to confirm the effectiveness of this new process.
h. Records confirmed that recently recalled product had been segregated and then sent for destruction by a third-party supplier. Records were not available to confirm that this destruction had in fact been completed. Also records were not available to confirm that this service supplier had been identified and being maintained in compliance with the critical food safety and quality supplier program.
i. A mock or practice recall was performed using finished product code dated March 1, 2013. Only 80% of the total products produced were accounted for and located. Records did not confirm that a CAPA had been implemented, as required when less than 100% of product was accounted for. This was a requirement of an internal procedure OP-3 Rev 1.

---

*Additional Food Safety Management System References (not all inclusive)

FSSC 22000:2010

     ISO 22000:2005 Section 7.10.4

     ISO 22002-1:2009 Section 15

SQF 7th Edition: Section 2.6

BRC Issue 6: Section 3.11

IFS Version 6: Section 5.9

*Note: Always review the standard or visit the website for your chosen standard to ensure that all related requirements are identified and addressed.

SECURITY, FOOD DEFENSE, AND BIOTERRORISM

The topic of this section includes not only the security of the facility, but also the organization's processes for food defense and bioterrorism.

    ISO 22002-1:2009 18. 1 and 18.2 state that

- "Each establishment must assess the hazard to products posed by potential acts of sabotage, vandalism or terrorism and must put in place proportional protective measures." (18.1)
- "Potentially sensitive areas within the establishment must be identified, mapped and subjected to access control." (18.2)

BRC Global Standard for Food Safety Version 6 (4.2) states that

- "Security systems shall ensure that products are protected from theft or malicious contamination while under the control of the site."
- "The company shall undertake a documented assessment of the security arrangements and potential risks to the products from any deliberate attempt to inflict contamination or damage. Areas shall be assessed according to risk; sensitive or restricted areas shall be defined, clearly marked, monitored and controlled."

SQF Code Version 7 (2.7.1) states that "The methods, responsibility and criteria for preventing food adulteration caused by a deliberate act of sabotage or terrorist-like incident shall be documented, implemented and maintained." This standard (2.7.1) goes on to require defined responsibilities, methods, and a food defense protocol be developed to protect the food from deliberate adulteration caused by a "deliberate act of sabotage or terrorist-like-incident and shall be documented, implemented and maintained"

Most organizations already have a crisis management program to deal with any such situation, but the standards require a proactive focus on food defense to prevent an incidence. Bioterrorism threats that are related to the food supply are a realistic concern; prevention is the key.

SECURITY

The design of the facility relating to secure access is directly related to an organization's program for security. For example, a facility completely enclosed with a fence with access controlled through a guard entrance may have a different program compared to a company without the guard entrance, fence enclosure, and located on a main street. It is up to the organization to define its program for securing the operation and ensuring that no unauthorized person has access to the facility. Most organizations with a situation similar to the latter description secure the facility through card access or limited access such as through reception and shipping areas only. Depending on the size of the organization, it may have a separate security program or combine it with other programs such as food defense, biovigilance, and bioterrorism. Requirements must be defined and associates trained to ensure that everyone has a strong and effective understanding of the requirements and what to do if a situation or event happens. Requirements may also be defined as part of other programs such as the visitor and contractor communication (Chapter 3) and the crisis management program.

CONFIRM THAT THE FACILITY IS SECURE

Management must perform security assessments to ensure compliance with defined requirements. This assessment must include testing the *secure* access of the facility at various times of the day. In addition to testing the entrances, it is recommended that management assign an individual(s) unknown to the associates to enter and wander the facility measuring the time it takes before being stopped and questioned. Ensure that records of this test are maintained to demonstrate status (Chapter 7). Any issues should be recorded and addressed through the CAPA programs (Chapter 4).

FOOD DEFENSE AND BIOTERRORISM

Food defense is the collective term used by the FDA, USDA, DHS, etc. "to encompass activities associated with protecting the nation's food supply from deliberate or intentional acts of contamination or tampering. This term encompasses other similar verbiage (i.e., bioterrorism (BT), counter-terrorism (CT), etc.)" (www.fda.gov/Food/FoodDefense).

Bioterrorism is defined as "terrorism involving the intentional release or dissemination of biological agents. These agents are bacteria, viruses, or toxins, and may be in a naturally occurring or a human-modified form" (http://en.wikipedia.org/wiki/Bioterrorism).

As a result of the events of September 11, 2001, Congress created the Public Health Security and Bioterrorism Preparedness and Response Act of 2002 also known as the Bioterrorism Act, which was signed into law on June 12, 2002, by President Bush. Among many aspects of this act, it assigned FDA with the responsibility for the protection of the food supply (Title III, Subtitle A) (www.FDA.gov).

CARVER + shock is a type of risk assessment that aids an organization in being proactive in protecting its processes from intentional contamination and sabotage. Information related to the CARVER + shock risk assessment can be found in Chapter 15: "Risk Management" section and on the FDA web page for Food Defense and Terrorism (www.fda.com).

PAS 96:2010

For additional reference and assistance in addressing the requirements of this element, it is recommended that the Publicly Available Specification (PAS) 96 titled "Defending food and drink" be reviewed and used as a guide in defining the program. PAS 96:2010 provides guidance on how to protect a food company from the attack and what to do to mitigate the effect should an attack occur.

FOOD DEFENSE PLAN

FDA states that a "Food Defense Plan identifies steps to minimize the risk that food products in your establishment will be tampered with or intentionally contaminated." FDA states that "a documented set of procedures improves your ability to respond effectively. A Food Defense Plan helps maintain a safe working environment for employees, provides a safe product to customers, and protects the bottom line" (www.fda.gov/fooddefense). This makes good sense from all perspectives.

ALERT

The USDA has a program identified as ALERT that provides guidance on what to do to be alert and proactive in preventing a deliberate contamination of the product or products. ALERT "identifies five key points that industry and businesses can use to decrease the risk of intentional food contamination" (www.cfsan.fda.gov/alert). Figure 14.1 explains the ALERT acronym.

A: ASSURE—assure that supplies and ingredients are from safe and secure sources
L: LOOK—watch or *look* after the security of the products and ingredients
E: EMPLOYEES—know (background) of employees and visitors
R: REPORTS—evaluate the effectiveness of the security programs protecting the products
T: THREAT—preplan who to contact and what to do if a threat or suspicious activity occurs
ALERT

FIGURE 14.1 FDA *ALERT.* (From http://www.fda.gov/Food/FoodDefense/Tools EducationalMaterials/ucm295898.htm)

ALERT: Programs to *assure* that supplies and ingredients are from safe and secure sources may include

- Approved supplier program (critical food safety and quality suppliers) (Chapter 13)
- Secure shipping and receiving requirements (i.e., sealed containers, transports, etc.) (Chapter 10, "Warehouse, Storage, and Shipping")
- Incoming inspection programs that require specific security evaluations before acceptance (Chapter 15, "Incoming Inspection")
- Open lines of communication (external communication) programs with suppliers and customers (Chapter 3)
- Training, awareness, and competency programs to ensure associates are educated in the requirements for security and food defense (Chapter 8)

ALERT: Programs that watch or *look* after the security of the products and ingredients may include the following:

- Secure shipping and receiving requirements (i.e., sealed containers, transports, etc.) (Chapter 10, "Warehouse, Storage, and Shipping")
- Incoming inspection programs that require specific security evaluations before acceptance (Chapter 15, "Incoming Inspection")
- Warehousing and storage procedures that ensure that materials and ingredients are accessible to only authorized associates (Chapter 10, "Warehouse, Storage, and Shipping")
- Lot identification and traceability programs ("Product Identification and Traceability")
- Secure access to the facility (Chapter 10, "Construction and Layout of Buildings")
- Training, awareness, and competency programs to ensure associates are educated in the requirements for security and food defense (Chapter 8)

ALERT: The following programs may relate to knowing your associates or *employees*:

- As much as legally allowed perform background checks or at a minimum check references (Chapter 10, "Employee Practices").
- Define and manage an effective program to control visitors and contractors. Important also to check references for contractors that may work unescorted. May take extra time but could be well worth it (Chapter 3).
- Implement controlled access even internally to sensitive areas allowing entry for only authorized associates (Chapter 10, "Layout of Premises and Workspace").
- Establish an identification process such that anyone can confirm if someone is in an unauthorized area for that position. An example may be the color

of a bump cap or an identification on the uniform (Chapter 10, "Employee Practices").

- Training, awareness, and competency programs to ensure associates are educated in the requirements for security and food defense (Chapter 8).

ALERT: Examples of programs that may be used to create *reports* that provide the objective evidence of the effectiveness of the security programs that protect the products may include the following:

- Evaluating the effectiveness of the security program through the internal audit program and the PRP verification program (Chapters 5, 10 and 18).
- Monitoring the industry and using this information to educate associates. Training, awareness, and competency programs to ensure associates are educated in the requirements for security and food defense (Chapter 8).
- Ensure that records are maintained as justification for any actions or confirmation of effectiveness and compliance (Chapter 7).

ALERT: Establish a plan to ensure that associates know who to contact and what to do if a *threat* or suspicious activity occurs:

- Most companies have this addressed in their crisis management program, but ensure that requirements for a potential or real threat including both internal (company management) and external (police) contacts are included in the program.
- Manage any potential product or ingredient through the control of nonconforming materials and products program ensuring that the item is not inadvertently shipped (Chapter 9).

Remember ALERT and take advantage of the programs and suggestions accessible at the following websites: www.cfsan.fda.gov/alert and www.cfsan.fda.gov/fooddefense.

Guidelines for enhancing food safety are provided in Figure 14.2 (www.cfsan.fda.gov/alert).

- Don't buy food products that have been damaged, dented, or opened.
- Be alert to abnormal odor, taste, and appearance of a food item.
- Don't eat a food product if you have any doubt about its safety.
- Report the food product if you suspect tampering.
- Call the USDA's meat and poultry hotline (800-535-4555) if the suspected food product contains meat or poultry.

FIGURE 14.2   How you can enhance food security. (From http://www.fda.gov/ForConsumers/ConsumerUpdates/ucm094560.htm)

FDA TRAINING VIDEOS

An effective training tool is the FDA DVD titled "Employee First," which is available on the FDA website in English and other languages. The same website has videos, posters, miscellaneous handout material that can also be used as training tools. It is recommended that at a minimum, security training that would include viewing the referenced FDA DVD be provided during new employee orientation and at least annually. Training must be defined and administered in compliance with the organization's training, awareness, and competency (Chapter 8). Training records must be identified and maintained in compliance with the record control program (Chapter 7).

Basically, requirements are similar no matter which standard is chosen emphasizing, that the organization define a program that identifies and implements an effective proactive method to protect the organization's products from potential acts of sabotage, vandalism, or terrorism. Also the organization must identify and ensure that *sensitive* areas are protected by controlling access to these areas. Although requirements for compliance with this element may be defined in relatively short paragraphs compared to other elements, this does not make this less serious. An effective, proactive program must be defined, implemented, and confirmed effective. Additional requirements and actions required may also relate to the crisis management and emergency response programs (also addressed in this chapter). Records must be maintained demonstrating that requirements are compliant (Chapter 7).

KEY POINTS

a. An effective training tool is the FDA DVD titled "Employee First," which is available on the FDA website.
b. A Food Defense Plan or related program must be defined, implemented, and confirmed effective related to being proactive in protecting against deliberate sabotage or adulteration of the organization's food product.
c. Management must perform security assessments at predetermined frequencies to ensure compliance with defined requirements. Records confirming test and outcome must be maintained in compliance with the record control program (Chapter 7).
d. The FDA website along with additional referenced documents (i.e., PAS 96) is available to provide guidance in developing and implementing the program.

COMMON FINDINGS/NONCONFORMANCES

a. Records were not available to confirm that associates had been trained in the organization's security requirements.
b. The security procedure stated that every associate must watch the FDA DVD "Employee First" at a minimum during new employee orientation and also annually; however, records were not available to confirm that this was being done.

c. The organization did not have a proactive program defined related to preventing and monitoring for intentional sabotage or adulteration.

d. Although a program for food defense was defined and a documented procedure available, responsibilities to ensure compliance and to provide guidance should an event occur had not been identified and training completed.

e. Records were not available to confirm that the Food Defense Plan had been evaluated and tested to confirm compliance and effectiveness.

---

*Additional Food Safety Management System References (not all inclusive)

FSSC 22000:2010

    ISO 22000:2005 Sections 5.6, 5.7

    ISO 22002-1:2009 Section 18.0

SQF 7th Edition: Section 2.7.1

BRC Issue 6: Section 4.2

IFS Version 6: Sections 5.9, 6

*Note: Always review the standard or visit the website for your chosen standard to ensure that all related requirements are identified and addressed.

REFERENCE

Newslow, D.L. 2001. *The ISO 9000 Quality System: Applications in Food and Technology*, Wiley-Interscience, New York.

15 Prerequisite Programs (PRP)
Part Two

When discussing food safety, food safety programs, and food safety standards, the subject of prerequisite programs (PRPs) continues. This chapter is a continuation of the discussions on PRP programs and the requirements as defined by specific food safety standards along with best practices and recommendations for the development, implementation, and maintenance of effective programs. It is again recommended that the organization ensures a strong familiarity with the standard of choice as it relates to the organization.

CHANGE MANAGEMENT

The management of change program has and will continue to be referenced through many different elements and programs discussed in this text. Dealing with changes and ensuring that changes are communicated to essential associates is not a new concept. It has, however, been a tough process for organizations to consistently comply with over the years. It used to be very common during an audit to ask about change management and be told that it is tracked through communication with the related departments, but when requesting records to confirm this, they were either sparse or nonexistent. A director of quality for a very large corporate client told me once that he had a great process for recording and tracking change, but it was very difficult to get everyone to use it. He went on to say that out of 25 incidents in the previous 3 years, 22 of them occurred because their defined procedure was not followed. Fortunately, these *incidents* were not food safety or otherwise critical issues, but at a minimum, they were quality issues. It is hard in today's world to think about such a serious requirement not being addressed. Just because an avoidable incident was not a food safety concern, it still costs money to evaluate, rework, or destroy the product. Companies are also concerned about the *brand* name, regulatory, and customer requirements. It does not matter which food safety management system (FSMS) is chosen; changes must be identified and evaluated prior to implementation. This is not an option. It is not something that is done when there is time or a responsible person gets around to it. Changes must be evaluated based on their impact on food safety and then on the operation overall. Records must be maintained confirming the review, the required actions, and actions that are completed and effective.

Auditors seek out changes through their physical evaluation (plant tours, area audits, etc.), interviews, and record review (i.e., management review meeting minutes). It is not up to the auditor to decide whether a change is critical or significant. It is the auditor's responsibility to confirm that the organization has a defined procedure, that the procedure is followed, effective, and that records are maintained demonstrating compliance. Records related to the management of change must be identified and maintained in compliance with the record control program (Chapter 7).

The *management of change, change management, commercialization,* or whichever name the organization decides to call this program can be accomplished through many different levels of complexity. The process does not necessarily translate to a better or more efficient process just because it is more complex. I have seen some complex processes that were extremely confusing, not user friendly, and very difficult to maintain compliance. However, there have also been some complex programs that were user friendly providing excellent information that added much value to the organization and its overall system. The theory then is that it is up to the organization to define, implement, and maintain a program that effectively evaluates, communicates, and records change, so that no change is implemented until it is evaluated, and any issues, including those related to food safety, are identified and addressed.

There are many different types of programs that can be developed. Internet searches, workshops, and textbooks can provide some excellent guidance in developing and maintaining an effective program. Jon Porter always said that "without the Internet, HACCP would not be possible." Some companies may use a *stagegate* approach similar to what a research and development (R&D) department may use when developing a new product. Typically, the *stagegate* program refers to specific phases, levels, or milestones. The project or process must satisfy the requirements of each stage or gate prior to moving forward. Other organizations may use a decision tree–type flow diagram that divides the changes into processes (i.e., packaging, maintenance, etc.). Various steps of the change and related decisions are tracked as defined in the flow diagrams. Some organizations keep it simple, tracking a change through the completion of a form or checklist. The checklist describes the change, the reason for the change, and any initial concerns or issues. This checklist then travels to each department manager. Each responsible department manager or his/her qualified designee reviews the potential change based on his/her department's responsibility and role in the process for that change. Review authorities most likely would include the food safety team (FST) leader, FST, and the quality manager, who could evaluate the change and related data to determine if it could affect food safety. If the answer is "yes," that it is, or could be related to food safety, then the FST must meet to evaluate actions the facts related to the change, make any decisions, and implement required actions related to the change. This may include process changes, modifications, food safety/HACCP program revisions, training requirements, document revisions, or any other related system decisions that would be required prior to the change being implemented.

In defining a program for change management, it is recommended that the implementation team use the following information as the foundation for the program review: History of changes, their impacts, and their positive and negative outcomes. The process flow decision tree is a valuable tool, but it is recommended that the process also identify and define particular groups or types of changes.

For example, there could be two basic types of changes:

1. Group 1 is a simple change not requiring review by the FST. An example would be replacing a pump with the exact same model.
2. Group 2 is a more complex change defined as a critical change. An example would be adding new equipment or changing an ingredient that would definitely require a review of the current food safety/HACCP program, and changes made as determined necessary by the FST. Another example would be replacing an existing metal detector with a new one (different manufacturer, tolerances, operating requirements, etc.).

Depending on the change and related processes, additional groups or levels may be identified with specific descriptions and criteria defined to provide effective guidance through the decision-making process. Also, software programs may be available to aid in identifying, evaluating, approving, tracking, and recording the process. What works best for the organization is the most important. Just because a program works for one organization does not make it the best choice for another.

It is easy to ignore or consider a change unimportant when, in fact, if the due diligence had been completed effectively, a change to the food safety/HACCP program may have been required. For example, maintenance announces that they have purchased and are going to install a new and improved metal detector to replace the current equipment on line 2. This particular piece of equipment is very different from the previous one, thus the FST must evaluate the equipment requirements, the testing requirements, verification standards, and the related procedures. If this equipment is linked to a critical control point (CCP), then it must be validated before being put into use (Chapter 17). What may be required cannot be determined until the evaluation is actually performed. This evaluation must be thorough and effective with records demonstrating compliance, as defined in the record control program (Chapter 7).

Responsibilities for the change management process must be defined. This should be done by each department taking ownership for its role in the process. Engineering may have a specific set of responsibilities relating to the equipment; choosing the equipment, installation, preventive maintenance, equipment capabilities, etc; whereas, the human resource department may need to work with other departments to ensure that training and competency requirements, if different, have been defined and that training resources are allocated to ensure the competency of the operators. Of course, those responsible for quality (quality manager) and food safety (FST) must evaluate the change based on how it may impact food safety and any other aspect of the process related to their levels of responsibility.

The program must be clearly defined. It must be tested through either actual use or a mock change drill to ensure that the process works for the organization. Some organizations refer to this change as commercialization because it may involve a new product or a new process introduced from corporate. A change is a change no matter what it is called or how minor it is. There must be a process to evaluate the change, at a minimum, for its impact, or potential impact, on food safety in compliance with the standard of choice. It only makes good business sense to also ensure its success related to quality, process parameters, customers, regulatory, and any internal organizational requirements.

ISO 22000:2005 is very specific in its requirements related to *change* and the communication of change. The program must be clearly defined to ensure the communication with associates of any issues having an impact on change, and also ensuring that the FST is notified in a timely manner of any changes. The FST ensures that information is included in updating the FSMS. Top management ensures that information on changes is included as an input to the management review.

ISO 22000:2005 defines the following requirements related to managing and communicating *change*.

- ISO 22000:2005 4.2.2 for document control states that "controls must ensure that all proposed changes are reviewed prior to implementation to determine their effects on food safety and their impact on the food safety management system."
- ISO 22000:2005 5.3 b defines that FSMS planning implies that the integrity of the FSMS is maintained when changes to the FSMS are planned and implemented.
- ISO 22000:2005 5.6.2 section for internal communication defines that "in order to maintain the effectiveness of the food safety management system, the organization must ensure that the food safety team is informed of changes in a timely manner, including but not limited to the following:
  a. Products or new products;
  b. Raw materials, ingredients and services;
  c. Production systems and equipment;
  d. Production premises, location of equipment, surrounding environment;
  e. Cleaning and sanitation programs;
  f. Packaging, storage, and distribution systems;
  g. Personnel qualification levels and/or allocation of responsibilities and authorizations;
  h. Statutory and regulatory requirements;
  i. Knowledge regarding food safety hazards and control measures;
  j. Customer, sector, and other requirements that the organization observes;
  k. Relevant enquiries from external interested parties;
  l. Complaints indicating food safety hazards associated with the product;
  m. Other conditions that have an impact on food safety."

An ISO 22000:2005 8.5.2 defines that for updating the FSMS the following which indirectly relates to the management of change applies: "Top management [must] ensure that the FSMS is continually updated. In order to achieve this, the FST must evaluate the FSMS at planned intervals. The team must then consider whether it is necessary to review the hazard analysis, the established operational PRP(s), and the HACCP plan (CCPs)." The FST may monitor improvements and changes to the system through management review meetings, the food safety meetings, continuous improvement programs, staff meetings, employee one-on-one discussions, and HACCP team meetings (if different from the FST meetings). As always, records must be identified and maintained in compliance with the record control program.

KEY POINTS

a. The FSMS takes the identification, communication, and evaluation of changes including those to the product and/or process, seriously to protect the safety of the product.
b. The organization must have an effective means to manage change, whether simple or complex.
c. Records related to the change management program must be identified and maintained in compliance with the record control program (Chapter 7).
d. Responsibilities for this program must be assigned to qualified associates who have the authority to ensure that changes are put in place and effective, prior to implementation of the change.
e. Changes must be evaluated and addressed as related to food safety prior to implementation.

COMMON FINDINGS/NONCONFORMANCES

a. It could not be confirmed through the review of records and interviews that management of change was being monitored as it relates to food safety.
b. There was no evidence that the implementation of an x-ray monitoring device as a CCP (line 6) had been confirmed effective and validated prior to the start-up of that line.
c. Records were not available to confirm that management associates performing responsibilities related to the change management program had been trained in the process.
d. Records related to recent equipment changes were either not available or incomplete.
e. During this evaluation, maintenance associates were not able to access the source (i.e., electronic) of the most current version of the change management procedure (Note: A hard copy of the document was presented, but it was revision 2, and the most current version at the time of the audit was revision 5).
f. The organization had recently expanded to a second shift; however, there was no evidence that this change had been evaluated and addressed related to the food safety program and the change management program.
g. The organization had recently outsourced an additional warehouse storage building for finished product, but there was no evidence that this change had been evaluated related to the requirements for the management of change program.

---

*Additional Food Safety Management System References (not all inclusive)

FSSC 22000:2010

　　ISO 22000:2005 Sections 4.2.2, 5.6.2, 5.2, 8.5,2, 7, 5.8

SQF 7th Edition: Sections 2.1, 2.3

BRC Issue 6: Sections 1, 2, 5

IFS Version 6: Sections 4.3, 2.1, 2.2

*Note: Always review the standard or visit the website for your chosen standard to ensure that all related requirements are identified and addressed.

DESIGN CONTROL

Requirements for design control are addressed specifically in ISO 9001:2008 7.3 for quality management systems. The relationship of *design control* to food safety is also important. Typically, design control (also referred to as design development and control) links to the engineering and R&D departments. Until the ISO 9001:2000 revision, manufacturing organizations actually had a choice as to whether or not to include design control (R&D, engineering departments, etc.) in the scope of approval. With the 2000 revision, the option was removed. Any manufacturing facility that included a design development department within the scope of its operation had to include this department in its certification process. The FSMS addresses design control, as its responsibilities affect the production of a safe product. In the development process, this links directly to the food safety program is hazard analysis related to ingredients, in process, and finished products. This also relates to the evaluation and approval of critical food safety and quality suppliers. In the past, products were developed and introduced to the manufacturing sites, possibly without directly evaluating food safety parameters. Ideally, this evaluation must be done through the development process and then communicated to the manufacturing site so that the FST, through the change management program, is able to evaluate the change prior to production. The FSMS makes it very clear that changes must be evaluated, and any issues related to food safety, food safety hazards, and food safety program overall (hazard analysis, procedures, records, etc.) must be addressed prior to actual implementation of the change. The management of change is addressed in more detail in the previous section "Change Management".

Whether the change is a food safety–related revision, quality revision, or new product implementation, let's review design control. Design control must be a logical approach to product revisions and new products. Compare design control to building a new house. The first step is deciding exactly what type of house is needed, the location, and then the architectural drawings. House building is a step-by-step process requiring numerous interactions between many subcontractors with diverse backgrounds. A cement slab must be correctly poured before the walls are erected. Timeliness and precision during the building of the house require teamwork that ultimately affects the entire project. It would be useless for the roofer to show up ready to add the roof on the first day of groundbreaking. The process must be coordinated in a simple, but effective manner.

Design control of a new consumer product also requires a distinct plan. Anyone who has ever worked in an R&D department can share many stories of marketing and sales folks showing up asking for a new or different product but not really knowing or being clear as to what they want. For example, a new affordable fruit punch for kids is developed that the developer hopes they'll love, but doesn't know for sure. It takes time, experience, and expense to develop a product, but without careful planning and determining the market for the fruit punch, it won't sell. The product most likely will fail. This may seem like an overexaggeration, but it is meant to communicate the necessity of having a distinct plan. Define the parameters. What is the desired outcome? Before the project begins, determine whether the desired output is possible. A plan must be developed, as it would with house building, which identifies the various stages of the project, what milestones must be achieved at each stage, and who would be involved and responsible at each stage. Compared to building the house, stages must be identified (i.e., the electrical inspector who must approve the wiring before the dry wall goes up.) Requirements must address the control and verification of product design at each stage of the process to ensure that it meets defined requirements. Depending on the extent of the project, the requirements vary. Inspections and specific stages for building an additional room in an existing house are much less involved than building an entire house from the ground up. This is the same with product development. A new special formula vitamin drink would require many more stages and interactions than a project to evaluate a product package for cost savings. It is common for many organizations to define these requirements in a documented procedure or work instruction. It is recommended, however, that the initial step be a flow diagram of the process identifying specific requirements and decision points including responsibilities at each stage based on the complexity of the project. Due to the potential complexity of, but also in some instances, the simplicity of each project, the flow chart with decision points can be efficient. Records confirming that required activities have been performed throughout the various stages must be identified and maintained in compliance with the record control program (Chapter 7).

The requirements for recording the design control program including design changes performed must include the identification of any food safety hazards (existing and potential) or concerns. Activities and outputs from this process must be evaluated for any existing or potential risks with records reviewed and approved by a responsible and competent (qualified) associate prior to its implementation. The effectiveness of the design control process must be confirmed.

Through experience, R&D teams are typically cautious when contemplating compliance with a structured management system standard; however, once the requirements are reviewed and the logic applied, it is frequently discovered that the process is already in place. Logic and common sense must be applied to the program. ISO 9001:2008 7.3 clearly identifies the required design development protocol as follows:

"Step 1: Plan and control the design and development of the product.
  a. Identify the development stages.
  b. Review, verify, and validate each stage.
  c. Define the responsibilities and authorities.
  d. Identify and manage the interfaces between different groups involved to ensure effective communication and clear assignment of responsibility.

Step 2: Inputs relating to product requirements must be determined and records maintained in compliance with the record control program. Inputs must be reviewed for adequacy with requirements that are "complete, unambiguous, and not in conflict with each other." Inputs must include
  a. Functional and performance requirements
  b. Applicable statutory and regulatory requirements
  c. Information derived from previous similar designs (if applicable)
  d. All other requirements that may be appropriate to the specific project

Step 3: Outputs must be in a form suitable for verification against the inputs and approved prior to release of the product or project. Outputs, at a minimum, must include
  a. Records confirming input requirements are met
  b. Food safety related requirements identified, recorded, and where possible, control requirements communicated
  c. Product acceptance criteria defined, recorded, and communicated
  d. Specific product characteristics essential for its safe and proper use defined and communicated

Step 4: Design control process must be reviewed per defined requirements at predetermined frequencies and at suitable stages. Review, at a minimum, must include the following:
  a. Evaluation confirming design requirements are met.
  b. Identification of any problems or potential problems or concerns.
  c. Proposed necessary actions to address problems or concerns.
  d. Verification must be performed per defined requirements to ensure that outputs have met input requirements.

Step 5: Validation must be performed per defined requirements to ensure that the resulting products are capable of meeting the requirements for the specified application or intended use (where known). The following relates to the validation requirements for this step.
  a. Validation must be completed prior to the delivery or implementation of the product (or project).
  b. Design and development changes must be identified, reviewed, validated, and approved before implementation.

    c. This review must include evaluation of the effect of the changes on constituent parts and product already delivered. It must also focus on any existing or potential food safety hazards or other issues."

Records confirming compliance with each stage of the design program must be identified and maintained. Records, at a minimum, must include verification, validation, and confirmation of changes as related to food safety, quality, regulatory, and customer requirements (Chapter 7).

Keep in mind that not all development and design changes follow each stage. It is up to the responsible developer, product manager, or team member(s) to evaluate, plan, and record the design control process plan for each project. Whether the design team is part of organization's manufacturing site, a corporate department, or a customer team, the FST at the manufacturing site has the ultimate responsibility to evaluate the change or new product for existing or potential food safety–related hazards. As previously stated, the process at the manufacturing site is commonly managed through a formalized program for the management of change, which is addressed in more detail in the "Change Management" of this chapter.

KEY POINTS

    a. It is important to apply common sense and logic to the design development and design change programs also referred to as design control.

    b. Typically, a formalized design control program is managed through an R&D department and/or an engineering department.

    c. Manufacturing locations most frequently manage the evaluation of a change or potential change through its management of change program, which is discussed in more detail in the "Change Management" section.

    d. Design requirements must be clearly defined and actions performed as defined with records maintained in compliance with the record control program (Chapter 7).

COMMON FINDINGS/NONCONFORMANCES*

    a. Requirements for design development were not formally defined.

    b. Records were not available to demonstrate that the design process was being performed in compliance with defined procedures.

    c. Evidence was not available to confirm that all interface activities had been performed as identified in the design plan for the new *product*.

    d. Records were not available to confirm that changes were being evaluated for potential and existing food safety hazards by the FST at the manufacturing site prior to initial product run.

    e. Records did not confirm that the *new* suppliers for "X" ingredient(s) were identified and managed as a critical food safety and quality supplier.

---

* Note in a food safety system these types of findings would most likely be recorded against the change control element.

> ***Additional Food Safety Management System References (not all inclusive)**
>
> FSSC 22000:2010
>
> > ISO 22000:2005 Sections 5.6, 5.8, 7, 8.5
>
> SQF 7th Edition: Section 2.3
>
> BRC Issue 6: Section 5.1
>
> IFS Version 6: Sections 2.2, 4.3
>
> **Note: Always review the standard or visit the website for your chosen standard to ensure that all related requirements are identified and addressed.*

CUSTOMER AND CONSUMER COMMUNICATION (CUSTOMER SATISFACTION)

The definition of and the distinction between the customer and the consumer are dependent on the design of the organization. Many organizations have direct contact with the consumer or the end user. Others may deal directly with brokers or large grocery chains that, in turn, have direct consumer contact. Some organizations may do a combination of both. This requires the organization to evaluate its processes to ensure that there is an effective and well-defined line of communication with the customer and also with the consumer as this relates to the organization's process. For example, if the organization produces products used for ingredients, then it is unlikely that there would be a link with the consumer. Organizations that sell directly to the consumer usually include a toll free number on the label, which provides the communication link. The process for managing customer complaints vs. consumer complaints could be the same process, or it could be two entirely independent departments and communication links. However an organization is designed, it must clearly define the processes for communicating with those receiving, manufacturing (if an ingredient), storing, and consuming the product. The defined process must address both handling individual concerns and also trending feedback data for existing, potential, and continuous improvement opportunities.

Customer communication is done in many different ways depending on the organization and its product. In the world of quality management system standards, it is about making sure that, when an order is taken, the requirements related to the order are understood, and the organization is able to meet them. A food safety and quality based management system focuses on not only the production and delivery of a safe product, but also a product that meets all of the customer's needs and expectations. Customer requirements must be understood and confirmed doable prior to making a commitment to provide the order. This includes not only meeting defined specifications, but also being able to provide the quantity per customer-requested schedule. A process must be clearly defined for addressing and resolving any *differences* between the requested order and the organization's ability to provide it. Good examples of where differences may arise would be quantity, cost, or ship date. For example, if a customer requests 5000 cases to be delivered Friday, and it is known at the time of receiving the order that only 3000 cases will be available, this must be resolved prior to accepting the order. Resolution may be that the customer chooses to postpone the delivery until the requested quantity is available or amend the order to the available amount. This type of up-front resolution is important. If it is not possible to meet the requested shipment quantity even though agreed upon, then the organization has failed to meet the customer's needs and expectations. If this is known at the time of receiving the order, then it must be resolved at that time. It is recommended that orders link directly to the customer specification. Customer specifications, if maintained by the organization, must be managed in compliance with the document control program to ensure that the organization always has the most current version. This also relates to any other controlled documents provided by the customer. Documents required by the organization but provided by the customer in order to produce the customer's products must be identified and maintained as a controlled external document (Chapter 6).

The FSMS requires that a food safety program for communicating to the customer related to food safety is defined and that evidence of effectiveness is recorded. If there is a hazard common to a required ingredient, then the receiving organization must ensure that it has a control in place to eliminate or reduce this hazard to an acceptable limit. For example, raw milk is known to have pathogens. The facility receiving this product must ensure that it has a control in place, which happens to be pasteurization, and also that programs are in place ensuring that there are not any situations where the raw milk could contaminate pasteurized milk. The organization must ensure that its process is effective (and have records to prove it) in controlling the hazard or potential hazard, thus producing a safe product. It would definitely be an issue if customer service accepted an order for a product that required a special type of equipment for pasteurization, and the organization was not equipped to perform that function. These issues must be identified and addressed prior to accepting the order. Fortunately, in today's world of food safety, there is a visible increase in organizations that are proactive, ensuring through change management and R&D programs, that any concerns are resolved far in advance of the actual production.

Further discussions related to customer/consumer communication related to food safety have been included in the chapters for management responsibility (Chapter 3), various PRPs (Chapters 10 and 15), crisis management (Chapter 14), and the control of nonconforming materials and products (Chapter 9).

As the management system standards have evolved, a strong emphasis has been placed on measuring customer satisfaction. This may be considered more of a quality management system focus than food safety; however, this concept is important to discuss. An organization must have a process in place that provides it with a proactive measurement of customer satisfaction. The answer to being satisfied is not that there are no customer complaints or the "we do not hear anything so they must be happy" approach. This can be a tough one. If the sales force is part of the organization's team, these representatives can informally ask their customers questions and submit the feedback to management. Organizations usually create a combination of several processes. It is recommended that the top management team approach this using their business planning process for answering the following questions:

- "How do we know our customers are satisfied?"
- "How can we get value-added feedback for continuous improvement?"

Consumer satisfaction measurements can be tough to capture. Typically, if something is really wrong, a consumer issues a complaint. If no complaint is raised, it is assumed that everything is okay. This is not always true. Years ago, I heard an industry specialist make a statement related to the consumer stating that: for every person that takes the time to issue a complaint, there are 99 others who are not happy but did not take the time to complain just switched to another manufacturer. Most organizations have tried surveys, but the response rate is typically not very good. Time is at a premium and no matter how good the intentions, most companies just don't take the time to complete them. Even as a consumer, it is very difficult to find time to fill out all those surveys. A personal issue comes to mind relating to making a complaint. After repeated issues with a particular prescription pet food and unsuccessful attempts trying to contact the manufacturer, changing to another brand seemed just easier. The new brand is an excellent product with excellent customer service so it was an easy choice for our family. However, it was very disappointing from a professional viewpoint to use a product for 14 years and not receive a courtesy call back.

Be prepared to identify, test, and try various methods and use the results from the combination of these measurements to make process adjustments accordingly. Records related to customer communication must be identified and maintained in compliance with the record control program (Chapter 7).

Charlie Stecher, an independent consultant and third-party auditor, Anchar, Inc. shares his observations that

> many organizations struggle with how to meet the requirement for measuring customer satisfaction. Most seem to pick one of their measurements as customer complaints which is not a true measure of satisfaction. I ask them what does your customer think is important? That is when I get the surprised look or one of the light bulbs just comes on.

Some may say that they don't need to worry about measuring customer satisfaction because that is only a requirement of the quality management system. No matter which standard is chosen, it makes good business sense to ensure the satisfaction of the customer. Related to food safety, the organization must do whatever it takes to build the confidence and trust of the customer in the organization's ability to produce a safe food of the highest possible quality in compliance with the customer's requirements.

A plant manager of a large ingredient company, who must remain anonymous because of company policies, described the benefits of this process:

> The sales department would promise the world, for obvious competitive reasons, but we didn't have a prayer of routinely meeting the promise, so we would get close and then someone (frequently me) would wave a wand and presto it was shipped. The requirement for having a defined process ensuring that we can meet customer requirements prior to accepting the orders, forced us to get together with the customers to discuss

and agree upon product requirements, our capabilities, and the parameters that really mattered. We are better off as a result. The benefits of this communication and awareness were immeasurable. (*Anonymous*)

KEY POINTS

a. It is critical that the organization focus on the needs of their customers; by doing so, they can ensure the accurate delivery of a safe product, in specification, and on time.

b. Organizations have found that by having a program that focuses heavily on meeting customer's needs and expectations, they positively affect the level of compliance attained during customer audits.

c. Organizations must define a method to measure consumer satisfaction.

COMMON FINDINGS/NONCONFORMANCES

a. Through the interview process, it could not be clearly demonstrated that some departments, such as shipping, understood the process for communicating with the customer (either directly or through customer services) should a problem arise in meeting the customer orders.

b. Customer orders referenced specific specifications that were not readily available.

c. Customer orders referenced Chocolate Powder, item 44352013; however, records of production indicated that the product had been produced to customer specification 44342011. Records did not identify the issue or provide any evidence that it had been identified, discussed, and accepted by the customer.

d. It could not be confirmed, through record review, that the organization was reviewing customer and consumer complaints for food safety–related issues. Records confirmed that all complaints were handled equally as time allowed.

e. It could not be confirmed that processes for managing customer and consumer complaints were being evaluated for trends based on their potential risk.

f. Although it was stated that customer/consumer communication was being evaluated at the management review meetings, records were not available to confirm that the organization had a defined program for measuring customer satisfaction.

---

*Additional Food Safety Management System References (not all inclusive)

FSSC 22000:2010

    ISO 22000:2005 Sections 5.6, 5.8

SQF 7th Edition: Section 2.1

BRC Issue 6: Section 3.10

IFS Version 6: Section 4.2

*Note: Always review the standard or visit the website for your chosen standard to ensure that all related requirements are identified and addressed.

INCOMING INSPECTION

The organization must define requirements for inspection and approval of incoming ingredients and materials based on existing or potential hazards (risks). The BRC Global Standard for Food Safety, Issue 6 (July 2011) 3.5.2 clearly defines that the "controls on the acceptance of raw materials shall ensure that raw materials do not compromise the safety, legality, or quality of products." This standard clause 3.5.2.1 goes on to state that "the company shall have a documented procedure for the acceptance of raw materials and packaging on receipt based upon the risk assessment." The acceptance testing must include "one or more of the following parameters:

- Visual inspection on receipt
- Certificates of conformance—specific to each product
- Certificate of analysis
- Product sampling and testing

A list of raw materials and the requirements to be met for acceptance shall be available. The parameters for acceptance and frequency of testing shall be clearly defined. The procedures shall be fully implemented and records maintained to demonstrate the basis for acceptance of each batch of raw materials."

The FSMS requires a defined procedure for incoming inspection based on the potential risk of the item. Inspection must be defined to include those parameters critical to food safety that must be performed at a frequency that optimizes the safety of the product being produced. This relates to the approval of, communication with, and ongoing evaluation of critical food safety and quality suppliers (Chapter 13).

Seal numbers must be verified with the identified numbers recorded on the bill of lading with any issues immediately addressed prior to accessing any item on the transport. Most organizations immediately reject the load if either the seals do not match or are broken, thus compromising the integrity of the load. Trailers must be inspected for cleanliness, holes, spillage, evidence of rodent or insect activity, and any other contaminants. Results of the trailer inspection must be recorded either on a specific form or, possibly, on the bill of lading. Records must include, at a minimum, notations related to seal numbers, holes, evidence of insects, rodents, unacceptable odors, and trash or debris (Chapter 7).

If receipt of a certificate of analysis (COA) or a certificate of conformance (COC) is a requirement, then this requirement must be clearly defined to ensure that no shipment is received without the proper paperwork. All requirements for COA and/or COC content prior to accepting the item (verification of data and test results) must be defined. Records must be maintained to confirm compliance (Chapter 7).

Ingredients and packaging material that is found to be either contaminated or in any way not in compliance must be either rejected prior to receipt or, if already received, then segregated and protected from unintended use. Specific requirements for the control of nonconforming materials and products are addressed in Chapter 9.

The organization must implement and manage requirements for approval, monitoring, and reporting issues related to supplier performance. Supplier issues must be recorded according to defined requirements. Issues related to meeting food safety, quality, customer, and nonfood safety regulatory requirements must be recorded and communicated to the associate responsible for communicating with the supplier. Requirements for the identification and maintenance of an effective critical food safety and quality supplier program are discussed in Chapter 13.

KEY POINTS

a. FSMS specified requirements for incoming inspection based on the potential risk of the item must be defined.
b. Incoming inspection parameters critical to food safety must be identified, and inspection performed at a frequency that optimizes the safety of the product being produced.
c. Incoming trailers must be confirmed as secure upon receipt with seals that match those listed on the supplier's shipping records (i.e., bill of lading). Results of seal confirmation must be recorded.
d. Incoming loads with either broken or seals that do not match should be rejected. Partial loads not sealed must be inspected to ensure that no food safety hazard exists.
e. Incoming material, depending on the item being received, may be accepted based on receiving COCs and/or COAs; however, the requirements for these documents including the specifications stated must be agreed upon prior to receipt of the product or material.
f. Formalized requirements related to the receipt of a COC or a COA for specific materials, ingredients, or products must be provided for each specific load. Suppliers must not send one document for multiple loads. Each load must have its own documentation upon receipt.
g. Records confirming that all requirements are met must be maintained in compliance with the record control program (Chapter 7).
h. Responsible associates must be trained and confirmed competent with records maintained confirming that requirements have been met. More information on training and competency is provided in Chapter 8.

COMMON FINDINGS/NONCONFORMANCES

a. Organization did not have a formalized procedure for evaluating and recording incoming materials to ensure that these were purchased from an approved supplier and that all requirements had been met (i.e., COA, inspections, etc.).
b. Records did not confirm that seals were being verified and recorded.
c. Organization did not have a consistent process in place for inspection or confirmation that incoming materials met defined requirements.
d. It could not be confirmed that bulk access points for ingredient storage were consistently secured (capped and locked) when not in use.

---

*Additional Food Safety Management System References (not all inclusive)

FSSC 22000:2010

>ISO 22000:2005 Sections 7.8, 5.6, 8

>ISO 22002-1:2009 Section 9.3

SQF 7th Edition Section 2.4.5

BRC Issue 6: Section 3.5.2

IFS Version 6: Sections 4.13, 4.14

*Note: Always review the standard or visit the website for your chosen standard to ensure that all related requirements are identified and addressed.

RISK MANAGEMENT

The information provided in this text has focused on compliance and applying a proactive approach to the production of a safe product. Decisions and applications must be applied, not only based on the organization's processes and segment of the food industry, but also on the effective application of experience and common sense. No matter what the focus, it all boils down to the *risk*. *Risk* has become the catchphrase of today's FSMS.

"What is the risk?"
"How likely is it to occur?"
"How significant is the outcome?"
"Can the risk be mitigated?"
"If the hazard cannot be mitigated, then how will the hazard be controlled?"
"What is the justification for the decisions?"

Risk assessments are commonplace in our everyday lives, both personally and professionally. The link of risk assessment to the food safety/HACCP program is discussed later in this text. Examples of life's activities that relate to risk assessments are listed in the following:

> a. Why do we wear seat belts? What is the hazard? What is the risk? What is the likelihood of needing a seatbelt? If we do need it, how significant would the outcome be if we did not wear a seatbelt? Are we going to measure the likelihood on the number of accidents based on the number of cars on the road, or the number of accidents that we could be involved in as individuals?
> b. Why should a motorcycle rider wear a helmet? What is the hazard? What is the risk? What is the likelihood of needing a helmet? There are statistics available for the percentage (likelihood) of accidents and the significance of the outcome.
> c. Why do we have to cook our hamburgers on the grill until they are medium well to well done? What is the hazard? What is the risk? What is the likelihood of getting sick if the hamburger is not cooked properly? Some people weigh the risk and significance based on the statistics and feel that it is

not significant enough to give up their love of meat cooked medium-rare. Where is the data? What is the basis of the decision? Can the decision be justified effectively? Are there mitigation opportunities?

d. Why must we wash our hands after using the restroom or prior to eating a meal? Again, what is the hazard? What is the risk? What is the likelihood of getting sick if we don't wash our hands? There are statistics available. Surveys claim that the occurrence of the flu would be reduced 30% if people were more diligent in washing their hands.

The list could go on and on, but the important factor is to understand that the application of the logic of *risk assessment* applies to everything we do. In applying the analysis to the organization, one must clearly define the focus, the justification, likelihood and the significance of the outcome.

Tools must be put in place to effectively identify and control the risk. Mitigation activities must be clearly documented and evidence of effectiveness recorded. Records related to all of the aforementioned must be identified and maintained in compliance with the record control program (Chapter 7). It all begins with the *risk assessment*.

What Is a Risk Assessment?

What is a risk assessment, and what role does it play in the food safety program? Each standard addresses risk in its own manner. The most important fact is that if a risk assessment is performed, then the risk is effectively and accurately identified and addressed. Although performing a risk assessment, specifically, may not be a requirement and possibly not considered a PRP, most, if not all, of the decisions to be made or already made are really about *risk*.

"Risk" is defined as the amount of harm that can be expected to occur during a given time period due to a specific harmful event (i.e., an accident). Statistically, the level of risk can be calculated as the product of the probability that harm occurs (i.e., that an accident happens), multiplied by the severity of that harm (i.e., the average amount of harm, or more conservatively, the maximum credible amount of harm). In practice, the amount of risk is usually categorized into a small number of levels, because neither the probability nor degree of harm can typically be estimated with accuracy and precision. (http://en.wikipedia.org/wiki/Risk_Matrix)

The risk assessment concept has been around for many, many years. The reader is challenged to perform an Internet search for *risk assessment* and very likely will be amazed at the results. The following definition was found (http://en.wikipedia.org/wiki/Risk_Assessment), which explains the vast use of *risk assessments* through many different industries and processes:

Risk assessment is the determination of quantitative or qualitative value of risk related to a concrete situation and a recognized threat (also called hazard). *Quantitative risk assessment* requires the calculation of two components of risk *(R)*, the magnitude of the potential loss *(L)*, and the probability *(p)* that the loss will occur. In all types of engineering of complex systems, sophisticated risk assessments are often made within safety engineering and reliability engineering when it concerns threats to life, environment, or machine functioning. The nuclear, aerospace, oil, rail, and military industries have a long history of dealing with risk assessment. Also, medical, hospital, and food industries control risks and perform risk assessments on a continual basis. Methods for assessment of risk may differ between industries and whether it pertains to general financial decisions or environmental, ecological, or public health risk assessment.

Frequently, the risk assessment is a difficult task because likelihood of occurrence and significance of outcome both require assumptions. When developing a food safety program, we stress that the system be based on facts; however, the facts are just not available in many instances. Jon Porter always said that "without the Internet, HACCP would not be possible." The Internet can be extremely helpful in searching for *facts* to use as the basis of the food safety/HACCP program decisions, but again, *facts* are not always available. Also, it is very difficult to assess what has not happened other than stating that, because there is no evidence of this ever occurring, the likelihood is low or remote.

An example of this would be that the likelihood of getting very sick from consuming raw milk is high. That statement is made based on the facts of why milk pasteurization was implemented approximately 80 plus years ago and continues today. The historical data (sourced from the Internet), would confirm/justify the decision for the high severity and high likelihood of death if the pasteurization step fails.

Risk assessment consists of an objective evaluation of showing how likely it is that the hazard will occur. Assumptions and uncertainties must be considered and presented in this process. Once the likelihood of occurrence is determined, then the significance of the outcome must be defined.

As stated previously, the difficulty in risk management is that the measures used (potential loss and probability of occurrence) can be very difficult to assess. The chance of error in measuring these two concepts is high. Risk with a large potential loss and a low probability of occurring is often treated differently from one with a low potential loss and a high likelihood of occurring. In theory, both are of nearly equal priority, but in practice, it can be very difficult to manage when faced with the scarcity of resources, especially time.

Risk assessments can be performed for any activity. Risk assessment related to food safety has been used to assess the *risk* and the *significance* of an identified hazard or potential hazard to determine the degree of control required to ensure the production of a safe product (Chapter 17).

Mitigation

Mitigation is the activity for evaluating a situation in order to reduce or minimize the severity of the impact or likelihood of a risk or an event. Mitigation is basically a result of the risk assessment. The mitigation process is used either when deviations from the defined FSMS are identified or if there is a situation that the organization needs to evaluate and justify not meeting the exact requirement of the standard. The approved GFSI schemes accept the mitigation process differently. It is very important to know the requirements of the standard of choice. Keep in mind that every effort must be made to be compliant. Mitigation is not an escape clause, but a realistic option to evaluate a situation to determine the degree of risk and what must be applied to ensure its control.

Mitigation may be defined as "the elimination or reduction of the frequency, magnitude, or severity of exposure to risks, or minimization of the potential impact of a threat or warning" (http://www.businessdictionary.com/definition/mitigation. html). Mitigation may also be thought of as an activity put into place to reduce the severity and/or seriousness of a hazard. Before delving into the *mitigation* activity, it is important to discuss the use of a risk assessment matrix. It is recommended that the FST create a matrix or form that is used to identify the hazard (or potential hazard), the risk, the likelihood of occurrence, and the significance of the outcome.

LIKELIHOOD AND SIGNIFICANCE

The development of a HACCP program, from the beginning, has focused on risk. Once a hazard has been identified, the likelihood of it occurring and the significance of the outcome must be evaluated. The *likelihood* of the occurrence and the *significance* of the outcome are the basic concepts that risk assessments are based on. The identification of risk (likelihood and significance) to the food safety/HACCP hazard analysis is discussed in detail in Chapter 17.

The focus of this chapter is on the application of the risk assessment concept and the process for the mitigation of the risk to determine its actual impact on the food safety program. The importance of risk assessment in the food safety program is to evaluate a food safety risk or potential risk based on the requirements of the chosen standard. This information is then monitored by the FST leader and/or the FST either to determine if the hazard or potential hazard can be mitigated or to identify what activity or task must be implemented to protect the safety of the product. In other words, frequently, a situation can be mitigated to either eliminate the risk or reduce it to an acceptable limit (See previous section on mitigation).

RISK MATRIX

A risk matrix is a matrix that is used during the risk assessment to define the various levels of risks related to the likelihood of the occurrence and the severity of the outcome. Many feel that this simple tool, if used effectively, increases the visibility of risks and assists the food safety/HACCP team in making decisions on the significance of a hazard. Specific matrix samples and further explanations as related to the hazard analysis are provided in Chapter 17. It is important that the organization's food safety/HACCP team create its own matrix (decision criteria) based on its operation, justification, and sound judgment based on experience. http://en.wikipedia.org/wiki/Risk_Matrix#cite_note-cox-4#cite_note-cox-4

The use of the risk matrix, as stated throughout this text, can be challenging. If not applied effectively, it can cause more confusion than positive assistance with the decision criteria. In the article *"What's Wrong with Risk Matrices? Risk Analysis,"* Mr. Cox argues that risk matrices have several problematic mathematical features making it harder to assess risks. Risk matrices include the following:

Ambiguous Inputs and Outputs: Categorizations of severity cannot be done with objectively for uncertain consequences. Inputs to risk matrices (e.g., frequency and severity categorizations) and resulting outputs (i.e., risk ratings) require subjective interpretation. Different users may obtain opposite ratings of the same quantitative risks. These limitations suggest that risk matrices should be used with caution, and only with careful explanations of embedded judgments.

However, the risk matrix can be an excellent tool if applied effectively by the organization. Once the risk assessment has been completed, then it must be rated based on the food safety/HACCP team's analysis. From the analysis, the team identifies opportunities for *mitigation* that may be applied to eliminate or reduce the hazard to an acceptable level. If the hazard is rated as significant based on the organization's food safety/HACCP program, but a mitigation activity isn't possible, then the method of control to protect the safety of product must be identified. A method for effective control of the hazard (i.e., CCP) must be defined and implemented.

Figure 15.1 identifies recommended headings for the matrix. It is advisable that the food safety/HACCP team assign the evaluations to subject matter experts as appropriate and then uses the spreadsheet to track the requirements and status of each entry. Mitigation activities can be done strictly based on the organization's inputs and decisions or by software programs that perform various levels of *mitigation*. The application and the extent of detail must be matched with the needs of the organization. Be cautious about becoming overzealous in applying a program that is not value-added for the particular operation.

Once the matrix is created, then the food safety/HACCP team can fine-tune the matrix and start utilizing it in the compliant system.

> * The requirement
> * The hazard or potential hazard
> * The likeliness (frequency) of occurrence
> * The significance (outcome) of occurrence
> * The risk rating
> * The current controls and their effectiveness
> * Any recommended additional controls required
> * Justification related to grading and decision
> * Measure of effectiveness

FIGURE 15.1 Recommended column titles for a risk assessment matrix. (It is up to the organization's food safety/HACCP team to determine what is required to ensure the effectiveness of this process.)

FDA has identified and presented a focused proactive risk assessment that an organization can apply to its processes to aid in the protection against deliberate product contamination. This risk assessment is known as CARVER + Shock.

> CARVER was originally developed by the U.S. military to identify areas that may be vulnerable to an attacker. FDA and the U.S. Department of Agriculture adapted it for the food and agriculture sector. "This approach allows food companies to analyze and identify critical areas that are the most likely targets of an attack," says Donald Kautter, Jr., Acting Supervisor of the Food Defense Oversight Team in FDA's Center for Food Safety and Applied Nutrition. FDA and other agencies have used the method to evaluate potential vulnerabilities in the supply chains of different foods and food processes. (http://www.fda.gov/ForConsumers/ConsumerUpdates/ucm094560.htm)

CARVER is an acronym that stands for six attributes that are appropriate to evaluate *targets of attack*. Figure 15.2 provides the FDA description of these attributes. The referenced FDA website also states that the "CARVER tool also evaluates a seventh attribute—the psychological impacts of an attack or 'shock' attributes of a target. For example, the psychological impact tends to be greater if there is a large number of deaths involved or if the target has historical or cultural significance" (http://www.fda.gov/ForConsumers/ConsumerUpdates/ucm094560.htm).

More information on the process for CARVER + shock can be found at FDA's Food Defense and Terrorism Web (www.FDA.gov). Additional information related to security, food defense, and bioterrorism can be found in "Security, Food Defense, Biovigilance, and Bioterrorism" section (Chapter 14).

Risk assessments and risk matrices are wonderful tools to help guide the organization's decision making, but these are not stand-alone tools. These are tools that provide a guide for performing a consistent and effective risk assessment. It is recommended that the risk assessment, the application of mitigation activities, and the identification of effective control measures are performed by the food safety/HACCP team and applied as the team identifies required PRPs and any aspects of the FSMS-compliant program determined necessary by this team. This process and logic behind it can be applied to the hazard analysis required of the food safety/HACCP program, but the purpose of this section is more to apply this concept and the related tools to not only the hazard analysis (when identified as necessary), but also to other concerns or hazards (potential or existing), related to the FSMS overall. Hazard identification, risk assessment,

---

- Criticality: What impact would an attack have on public health and the economy?
- Accessibility: How easily can a terrorist access a target?
- Recuperability: How well could a system recover from an attack?
- Vulnerability: How easily could an attack be accomplished?
- Effect: What would be the direct loss from an attack, as measured by loss in production?
- Recognizability: How easily could a terrorist identify a target?

FIGURE 15.2 CARVER. (From http://www.fda.gov/ForConsumers/ConsumerUpdates/ucm094560.htm)

risk control, and mitigation are the central components of the structured approach to risk management. In the majority of instances, the risk assessment is merely the identification of what can go wrong and then confirming that the actions have controls that can be put into place to prevent it from going wrong. Records must be maintained (Chapter 7) to confirm compliance, justify decisions, and measure effectiveness. Results of mitigations are recorded through the food safety/ HACCP team meeting minutes (Chapter 16) and reported to top management at management review meetings (Chapter 3).

16 Food Safety Program (HACCP)

Getting Started
(The Five Presteps)

The title of this chapter is not meant to be misleading. In reality, compliance with a food safety management standard is the food safety program; however, many have chosen to link the food safety program to what we used to call our Hazard Analysis and Critical Control Point (HACCP) program or HACCP plan. The food safety management standards have built and expanded their requirements based on the principles of HACCP as defined by Codex Alimentarius; (2003) Hazard Analysis and Critical Control Point [HACCP] guidelines (www.codexalimentarius.org)).

The U.S. version of the HACCP Guidelines is defined by the National Advisory Committee on Microbiological Criteria for Foods (NACMCF) at http://www.fda.gov/. NACMCF is an advisory committee chartered under the U.S. Department of Agriculture (USDA) and comprised of participants from the USDA, Department of Health and Human Services, Center of Disease Control, and several other U.S. government departments or agencies focused on providing guidance and recommendations to the secretary of agriculture and the secretary of health and human services regarding the microbiological safety of foods.

This chapter addresses an overview and best practices related to what are known as the five (5) presteps or the preliminary steps of HACCP, expanding requirements to relate more closely to food safety management standards.

In doing this, the overall HACCP concept does not change:

HACCP is defined as a "logical system designed to identify hazards and/or critical situations and to produce a structured plan to control these situations." (Schmidt 1996)

HACCP is an activity developed to identify and control potential hazards that are critical to consumer safety. (Newslow 2001, p. 364)

HACCP is a management tool directed to control risk and provide safe, quality products while generating profit. (Jon Porter, Newslow, 2001, p. 201)

The focus of HACCP is on product safety ensuring the "control of hazards which are significant for food safety in the segment of the food chain under consideration" (Codex CAC/RCP 1-1969, Rev. 4, 2003).

Dorothy Gittings, an independent consultant and third-party auditor, D.L. Gittings Consulting Services, Inc. explains the value of creating a guidance table:

> A good tool to enhance your understanding and interpretation of ISO 22000 requirement 7—Planning and Realization of Safe Products is to develop a Guidance Table. In one column, list all of the clauses and subclauses of ISO 22000 requirement 7 (or related section of chosen standard or the Planning and Realization of Safe Products), in the next column list tasks/measures to accomplish compliance to that clause/subclause, and finally in the third column, list examples of objective evidence that would confirm that the requirements of the clause/subclause are met. Also, do your *homework* and improve your knowledge regarding recognizing and identifying food safety hazards by researching publications, statutory and regulatory websites, food safety organization websites, and industry group websites. Also visit the Codex website at www.codexalimentarius.org/ to access and review some great information and guidance from Codex Alimentarius.

The term *safe* refers to the processing of food products without contamination from any pathogenic organism or adulteration with harmful chemical or physical material.

Depending on the process or product, similar characteristics (i.e., ingredients, process steps) may be grouped together; however, it is essential that all unique conditions are identified and addressed. Examples of these may relate to a specific ingredient, process, or equipment. Grouping in this manner may also be beneficial when preparing the hazard analysis for the ingredients. Ingredients having comparable characteristics, such as low water activity (A_w), may be grouped together.

Figure 16.1 provides a brief identification of what has become known as the *five presteps* or the *preliminary steps* of the food safety/HACCP program.

THE FOOD SAFETY TEAM LEADER

In discussing the requirements for the food safety team, it is important to first address the food safety team leader (FSTL), also referred to as Safe Quality Food Practitioner in SQF Code, edition 7. ISO 22000:2005 5.5 is clear that "top management [must] appoint a food safety team leader who, irrespective of other responsibilities, [must] have the responsibility and authority to

- Manage a food safety team and organize its work;
- Ensure relevant training and education of the food safety team members;

---

1. Assemble the food safety/HACCP Team
    a. ISO 22000:2005 7.3.2 states that "A food safety team must be appointed";
2. Describe the product
    a. ISO 22000:2005 7.3.3 titles this category as "Product Characteristics";
        i.   ISO 22000:2005 7.3.3.1 Raw materials, ingredients and product-contact materials;
        ii. ISO 22000:2005 7.3.3.2 Characteristics of end products;
3. Identify the intended use;
4. Develop the flow diagram;
5. Verify the flow diagram.

FIGURE 16.1   The preliminary steps.

- Ensure that the food safety management system is established, implemented, maintained, and updated;
- Report to the organization's top management on the effectiveness and suitability of the food safety management system."

It is strongly recommended that both a FSTL and a backup leader be identified. It is critical that these individuals receive extensive training on the requirements of the food safety management system (FSMS) of choice and also a basic HACCP program development workshop approved by a recognized source such as the International HACCP Alliance (IHA). The food safety team members can then be trained in an extended basic HACCP Awareness curriculum, which can be presented by the FSTL or other qualified source (i.e., consultant) approved by the FSTL.

FOOD SAFETY TEAM

Codex requires the assembling of a multidisciplinary HACCP team. ISO 22000:2005 7.3.2 states that:

- "A food safety team must be appointed.
- The food safety team must have a combination of multi-disciplinary knowledge and experience in developing and implementing the food safety management system. This includes, but need not be limited to, the organization's products, processes, equipment, and food safety hazards within the scope of the food safety management system."

The required competencies (i.e., training) both the food safety/HACCP team leaders and team members must be clearly defined with records maintained in compliance with the record control program that provide proof of training and its effectiveness (Chapter 7).

The food safety standards may cause some confusion related to the responsibility for the food safety team versus the HACCP team. In most instances, the food safety team and the HACCP team are the same group of associates. However, in larger companies that may have multiple HACCP programs the team members may vary. The most effective manner to address this is to have two to three members of the food safety team on each HACCP team. Thus, the individual HACCP teams become a subset of the food safety team. It is critical that this is clear in the defined protocol to ensure that responsibilities are addressed and compliant.

The food safety/HACCP team evaluates the hazards and related outputs from the hazard analysis such as the critical control points. Examples of areas of responsibility represented on the team may include receiving, operations, maintenance, warehouse, management, research and development, purchasing, and quality control. It is very important to approach this analysis in a manner that provides insight from associates who are familiar with each aspect of the process and have a combination of experience and knowledge of the organization's products, processes, and equipment. Records related to a team member's qualifications, based on knowledge, education, and training, must also be maintained justifying the reason each person was chosen for the team. Figure 16.2 provides a sample

HACCP/Food Safety Team

The Food Safety Management System is designed and implemented in accordance with the requirements of FSSC 22000 which include ISO 22000:2005 and ISO 22002-1:2009.

The HACCP/Food Safety Team is a multi-disciplinary team. The team is responsible for plan development, verification, validation, and maintenance. Flow diagrams are signed by the HACCP/Food Safety Team leader confirming that these have been verified for accuracy during all stages and hours of operation. The Team Leader has responsibility and authority to manage the team and organize its work to ensure relevant education and training requirements are completed. The Team Leader also ensures that the food safety/HACCP programs are established, implemented, maintained, and updated when necessary and reported to top management.

The HACCP/Food Safety Team Leader and his or her designated back-up person must attend an approved International HACCP Alliance (IHA) HACCP Plan Development workshop. Competency and knowledge of team members is defined by education, training, and experience. Team members must have at least three years of experience in their area of expertise, preferably in the food or beverage industry. Initially and then a minimum of annually, team members are trained in basic food safety/HACCP. Records confirming compliance are maintained as defined for Control of Records.

Name	Title/ Department	Responsibilities	Experience	Employment Duration	Training
Paul Johnson	FS Team Leader	The food safety team leader	30 years in Food and Beverage Operation	2 years	IHA (2011) HACCP, NRFSP, FSSC 22000 Lead Auditor
Jay Ortiz	Maintenance manager	Maintenance activities	25 years in food processing	4 years	HACCP awareness, NRFSP
Steve Elsbury	Operations Manager	All operations activities – reports to plant manager	19 years of operations experience	10 years,	HACCP Awareness

Note: Additional team members would be recorded on subsequent pages.

FIGURE 16.2 An example of a HACCP/food safety team description/identification matrix.

document that can be used to identify and document the experience (justification) for the HACCP/food safety team.

A frequently asked question is "how many members should be on the team?" This is the choice of the organization, but keep in mind that some members could be *ad hoc* and invited because of their areas of expertise, depending on the subject matter of a specific team meeting. If expert advice is not available on site, an expert (i.e., consultant) may be contracted for assistance. Records confirming that all requirements have been met, including qualifications and proposed plans for assistance, must be maintained in compliance with the record control program (Chapter 7). Remember the consultant's company must be identified and managed as a critical food safety and quality consultant's supplier (Chapter 13).

FOOD SAFETY TEAM RESPONSIBILITIES

Depending on the standard of choice, the food safety team (FST) has many responsibilities that are discussed throughout this text. It is recommended that once an organization chooses its FSMS standard, that the team goes through that standard and highlights every reference to the food safety team. This is critical. Records must be available to confirm that FST responsibilities defined by the standard of choice have been completed and are in compliance. At a minimum, the food safety team is responsible to review the food safety/HACCP programs and their application at a defined frequency or whenever a change is made to the process, product, or equipment (see "Change Management" section in Chapter 15) The FST must revise the program if the team deems it necessary to ensure that the food safety hazards remain in control.

NOTIFYING THE FOOD SAFETY TEAM OF PROPOSED CHANGES

The food safety team must be kept informed of proposed changes prior to these changes being implemented. Most organizations choose to manage this through either a process called commercialization or change management. Change management is addressed in more detail in Chapter 15 ("Change Management"). Although the following list has been included in previous chapter, it is important to emphasize the importance of having a structured process for change management that includes, at a minimum, those items identified in this list:

- Products and new products, ingredients, raw materials, and services
- Production systems and equipment, production premises, equipment location, and surrounding environment
- Cleaning and sanitizing programs
- Packaging, storage, and distribution systems
- Personnel qualification levels and/or allocation of responsibilities and authorities
- Legal requirements, knowledge of food safety hazards, and control measures
- Customer and other requirements of the organization
- Relevant inquiries and/or complaints related to food safety hazards
- Any other conditions that may impact food safety

Food safety team meeting minutes must be identified and maintained in compliance with the record control program (Chapter 7).

DESCRIBE THE PRODUCT

Once the team has been identified and trained, the next step must be to describe the product. Early requirements stated that "a full description of the product" must be documented, which included the product, its processing requirements, storage temperature, and characteristics. Characteristics included the information necessary for the evaluation of the hazards, such as compositional factors (i.e., pH, water activity) and processing factors (i.e., heat treatment, chemical agents). The food safety management standards have expanded this into more detail. At a minimum, each food safety/HACCP program must clearly define all of the raw materials, ingredients, processing aids, and product contact materials "in documents to the extent needed to conduct the hazard analysis including the following:

- Biological, chemical, and physical characteristics
- Composition of formulated ingredients, including additives and processing aids
- Origin
- Method of production
- Packaging and delivery methods
- Storage conditions and shelf life
- Preparation and/or handling before use or processing

[This description must include the] food safety–related acceptance criteria and specifications of purchased materials and ingredients related to their intended uses (ISO 22000:2005 7.3.3.1)." This identification expands on the basic requirement for raw ingredients and packaging materials to include processing aids and product contact surfaces. It also includes the identification of any related statutory and regulatory food safety requirements. Any product (i.e., cleaning chemical), contact surface, or item (i.e., stainless steel) must be listed and evaluated for existing or potential hazards. The following examples relate to product contact surfaces:

Example 1: Chemicals used to clean the inside of equipment come into direct contact with the same surface as the product; thus, if a nonfood-grade or unapproved chemical for food contact is used, there becomes a significant risk of cross contamination. Even though every effort may be made to ensure proper rinsing, should something go wrong, then the organization must have controlled the significance of the outcome by using approved chemicals.

Example 2: No one wants to find any amount of lubricant in a product; however, the outcome would be much more serious if the lubricant was nonfood-grade rather than one that was approved for food surfaces. Obviously, none of these chemicals are approved to drink or consume in large quantities. A nonfood-grade chemical could translate as *poison*.

Example 3: There are many types of equipment, for example, copper, that are not generally approved as a food contact surface. The chemical composition of some foods is able to break down copper into a substance that we would not want to consume. Be careful to know the product and its hazards.

Example 4: Stainless steel is thought of as an approved food contact surface; however, not all grades of stainless are approved. The food contact surface (either existing or for a new installation) must be confirmed food safe. In addition, new stainless steel equipment must go through a passivation process. Passivation basically results in a chemical reaction that forms a passive film of chromium oxide that protects against corrosion and rusting. *Passivation* is much more complex than just described, but used here as an example of the hazard analysis of a product contact surface.

Lets review a product contact surface is defined as any surface that may come in direct contact with an exposed product. Examples include pipes, gaskets, agitators, and conveyor belts. A processing aid is defined as any item that may be used in the production of a product but is not actually an ingredient. Examples may be process enzymes, lubricants, cleaning chemicals, steam, and ice. These items may be added to the food for their technical or functional effect in the processing but are not actually ingredients (i.e., identified on the label). In some instances, product contact surfaces and processing aids are combined into one category—product contact surfaces.

DESCRIBE THE END PRODUCTS

The standards require that *characteristics* of end products, to the extent needed to conduct the hazard analysis, be described in documents; the identification of any related statutory and regulatory food safety requirements must be included. At a minimum, this includes the following:

- "Product name or similar identification
- Composition
- Biological, chemical, and physical characteristics relevant for food safety
- Intended shelf life and storage conditions
- Packaging
- Labeling relating to food safety and/or instructions for handling, preparation, and usage
- Method(s) of distribution" (ISO 22000:2005 7.3.3.2)

At a minimum, the standards require that the following are identified for the end product of each specific HACCP program:

- The intended use
- Any unintended but reasonably expected mishandling and misuse of the end product that could result in a food safety event
- Potentially vulnerable groups (i.e., elderly, young children, immunocompromised individuals, etc.)

Where possible, this activity is intended to identify the use of the product, such as consumption by those segments or groups of the population that may be at high risk (i.e., compromised immune system) for getting a food-borne illness. For example, a citrus product such as orange juice is consumed under many different types of situations by children, the elderly, and may be immune-deficient individuals.

FLOW DIAGRAM

Having accurate flow diagrams that provide the necessary information for an effective hazard analysis has always been critical to the food safety/HACCP program overall, but these, too, have been expanded to include more information. It is encouraged that the diagrams remain *simple*, but to the point. The flow diagram must be developed and verified as accurate by the food safety team.

Let's first look at the development of the flow diagram. A flow diagram, created by the food safety/HACCP team, is a systematic representation of the sequence of steps or operations used in the manufacture of a specific product or process. It must reflect the product or process for which the food safety/HACCP program is being developed. It provides a picture of the operation following the steps through the process from first receipt to finished product shipment.

The flow diagram must be clear, accurate, and sufficiently identify the sequence and interaction of all steps in the operation (including location of filters, screens, and magnets). In addition, the following must be identified:

- Sequenced interaction of all steps in the operation
- Any outsourced processes and subcontracted work required by the process
- Ingredients, raw materials, and intermediate products entering the flow
- Reworking and recycling
- End products, intermediate products, by-products, and waste entry and exit points as appropriate to the process

Figure 16.3 is a sample flow diagram, study it closely compared to the standard of choice to ensure that all required information is included.

The food safety team must perform an actual visual verification to confirm the accuracy of the flow diagram to ensure that it is complete and accurate. This verification must be recorded and kept as the objective evidence that the activity has been completed. The food safety team members should initial and date a hard copy that is kept as the record. A notation confirming the activity completed should also be noted in the food safety team meeting minutes. An example of verification would be as follows:

Verified by food safety team on September 1, 2013. Details recorded in food safety team meeting minutes from September 1, 2013. BDL food safety team leader. 9/1/2013 Flow Diagram Issue 1 dated 9/2/2013.

ISO 22000:2005 7.3.5.2 requires that any "existing control measures, process parameters and procedures that may influence food safety are described to the extent needed for conducting the hazard analysis. External requirements, [such as those from legal authorities or customers] that could impact the choice and discipline of the control measures, must also be documented" within the specific food safety/HACCP program. The responsibility for ensuring that current statutory and

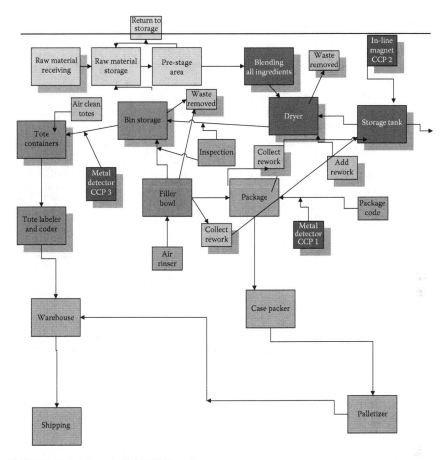

FIGURE 16.3 Sample HACCP/flow diagram.

regulatory requirements are identified and maintained are assigned most frequently to the food safety team leader (FSTL).

COMPARISON BETWEEN THE GFSI-APPROVED FOOD SAFETY SCHEMES

Each of the GFSI-approved schemes requires that a food safety/HACCP program be defined, implemented, and maintained effectively. There are not only similarities between the standards, but also specific differences. Table 17.1, which is located at the conclusion of Chapter 17, provides an excellent comparison between Codex, SQF Issue 6, BRC Edition 7, and FSSC 22000 Version 3. It is recommended that the standard of choice be reviewed using the content of Table 17.1 to gain the best possible understanding of the requirements of the standard of choice.

KEY POINTS

a. The food safety team is responsible for ensuring that the preliminary steps are completed, recorded, and accurate for each food safety/HACCP program.

b. Food safety/HACCP program documents must be identified and maintained in compliance with the requirements for the document control program (Chapter 6).

c. Food safety/HACCP program records must be identified and maintained in compliance with the requirements for the record control program (Chapter 7).

d. The food safety/HACCP team must be identified and must include justification for each member to be on the team, including member's name, experience/training, and responsibilities.

e. Records confirming that the food safety team (individual HACCP teams, if applicable) has been trained and is competent based on the organization's defined requirements must be maintained in compliance with the record control program.

f. Process for identifying, communicating, and being in compliance with federal, state, and local statutory and regulatory requirements must be defined with records maintained confirming compliance with defined requirements.

g. The hazard analysis for each food safety/HACCP program must include all ingredients, raw materials, product contact materials, and processing aids.

h. The FSTL must understand the complete requirements of the standard of choice and communicate this to the food safety/HACCP team and the top management team to ensure continued support, resource allocation, and compliance.

i. Characteristics of end products must be identified and described to the extent needed to conduct a hazard analysis.

j. The intended use, the reasonably expected handling of the end product, and any unintended but reasonably expected mishandling and misuse of the end product must be considered and documented in the food safety/HACCP programs.

k. Records (initialed, dated team meeting minutes) must confirm that the flow diagrams have been completed and verified as accurate by the food safety team.

l. Flow diagrams must include entry and exit steps for rework, recycle, and waste.

m. External requirements (e.g., from regulatory authorities or customers) that may impact the choice and the rigorousness of the control measures must be taken into consideration.

n. The organization must define the process for handling customer specifications and other customer requirements received (and agreed upon) from those customers. Customer specifications and specific requirements as related to the food safety/HACCP programs must be evaluated by the food safety team with related outputs recorded and actions taken as deemed required by the FST.

o. The food safety/HACCP team must be notified/kept informed of proposed changes prior to these changes being implemented (Chapter 15 "Change Management").

COMMON FINDINGS/NONCONFORMANCES

a. Records related to the training of the food safety team leader (FSTL) and the food safety team were not identified and maintained in compliance with the record control program.

b. Records indicated that the flow diagrams were evaluated by the HACCP team, and not the food safety team, as required by the referenced standard. Teams contained different members, and relationships were not defined.

c. It could not be confirmed that requirements for the food safety team vs. the HACCP team to ensure compliance with all defined food safety team responsibilities of the chosen FSMS standard had been implemented.

d. It could not be confirmed that training qualifications for the food safety team had been defined, documented, and maintained in compliance with the referenced standards.

e. Product descriptions had not been completed for all raw materials, ingredients, and product contact materials.

f. Hazard analysis did not include some of the processing aids and product contact surface materials such as air filters, cleaning chemicals, and specific lubricants observed during this evaluation.

g. The description of raw materials, ingredients, and/or product contact materials did not include the food safety–related acceptance criteria and specifications of purchased materials and ingredients related to their intended uses.

h. The HACCP/food safety program did not identify statutory and regulatory food safety requirements related to raw materials, ingredients, and/or product contact materials.

i. Flow diagrams were inaccurate and did not reflect the process flow observed during this evaluation.

j. Unintended, but reasonably expected, mishandling and misuse of the end product that could result in a food safety event had not been identified, considered, and/or described.

k. Records were not available to confirm that the food safety team had verified the accuracy of the flow diagrams by on-site checking.

l. Records were not available to confirm that external requirements from regulatory authorities or customers had been considered and included in the food safety/HACCP program.

m. Records were not available to confirm that the organization had identified and trained a multi-disciplinary food safety team that was compliant with the FSMS standard of choice.

n. Records confirmed that a multidisciplinary food safety team had been identified; however, FST meeting records indicated limited to poor attendance at the meetings (i.e., records of the last eight [8] FST meetings indicated only 25% attendance).

REFERENCE

Newslow, D.L. 2001. HACCP. In: *Food Safety Handbook*, Schmidt, R. and Rodrick, G. (eds.), Wiley-Interscience, Hoboken, NJ.

Schmidt, R.H. 1996. *Hazard Analysis Critical Control Point (HACCP): Overview of Principles and Applications*. Food Science and Human Nutrition Department, University of Florida, Gainesville, FL.

17 Food Safety Program (HACCP)
The Principles

This chapter describes the requirements that have been identified over time as the HACCP principles. There are seven principles (Figure 17.1). As stated in Chapter 16, each of the Global Food Safety Initiative-approved schemes requires that a food safety/HACCP program be defined, implemented, and maintained effectively. Table 17.1 is included at the end of this chapter. This document provides a basic comparison of the food safety/HACCP requirements between Codex, SQF V7, BRC issue 6, and FSSC 22000:2010. For comparisons related to other standards, it is recommended that the specific standard be reviewed by the food safety/HACCP team to ensure compliance with the organization's standard of choice.

Although this text may refer to the HACCP principles as defined by Codex Alimentarius, the food safety management system standards have adopted these steps expanding on their requirements to focus on the planning and production of safe products. The food safety team and the HACCP team may be the same; however, for those organizations with multiple processes, a specific *team* may be required for each process. Also, there are several references throughout this chapter to the food safety team, the HACCP team, and the food safety/HACCP team. The food safety management standards have specific requirements for the *food safety team* such as performing the hazard analysis. It is important that this relationship is clearly defined and that objective evidence is available to demonstrate the food safety team is in compliance with the standard of choice. More detail related to the identification and responsibilities of the food safety/HACCP team(s) is addressed in Chapter 16.

HAZARD ANALYSIS

The first and possibly most challenging principle is the *hazard analysis*.

> "*Principle 1: Conduct a hazard analysis*" is defined as assessing the "hazards associated with growing, harvesting, raw materials and ingredients, processing, manufacturing, distribution, marketing, preparation, and consumption of the food." (Pierson and Corlett 1992)

Once the food safety program foundation (presteps and prerequisite programs (PRPs)) has been established, then the food safety/HACCP team must conduct the hazard analysis. Each food safety management system standard defines the

---

Principle 1: Conduct a hazard analysis;
Principle 2: Identify critical control points;
Principle 3: Establish critical limits at each CCP;
Principle 4: Establish monitoring procedures for critical limits;
Principle 5: Establish corrective actions;
Principle 6: Establish procedures for verification;
Principle 7: Establish documentation and record keeping.

FIGURE 17.1   The seven principles of HACCP. (From [CAC/RCP 1-1969, Rev. 4, 2003])

requirements for performing the hazard analysis in a similar manner. Specific references to a food safety management system standard are incorporated in the following text; however, a complete list of references to each specific standard is listed later in this chapter.

The hazard analysis must be thorough and complete because it provides the scientific basis for the food safety program. It is critical that sufficient resources (i.e., time and energy) are applied aggressively but efficiently. Based on the Codex HACCP principles, performing an effective and efficient hazard analysis is one of the most difficult steps in the development process and generally takes the majority of time.

The food safety/HACCP team is responsible for conducting the *hazard analysis* to identify all possible hazards and to determine which hazards must be controlled, the degree of control required to ensure food safety, and which combination of control measures are required. A *hazard analysis* is defined as the

Process of collecting and evaluating information on hazards and conditions leading to their presence to decide which are significant for food safety and therefore should be addressed in the [Food Safety/HACCP Program] (*CAC/RCP 1-1969, Rev. 4, 2003*).

ISO 22000:2005 7.4.2.1 states that

All food safety hazards that are reasonably expected to occur in relation to the type of product, type of process and actual processing facilities must be identified and recorded. The identification must be based on:

- The preliminary information and data collected for the specific processes, end products, ingredients, processing aid, and packaging as outlined in the flow diagrams;
- Experience;
- External information including, to the extent possible, epidemiological and other historical data;
- Information from the food chain on food safety hazards that may be of relevance for the safety of the intermediate and end products;

The step(s) (from raw materials, processing and distribution to the customer) at which each food safety hazard may be introduced depending on the product must be identified and comments may also be communicated on the product label or other preapproved communication with the customer (i.e., product information sheets).

ISO 22000:2005 7.4.2.2 states that

When identifying the hazards, [the food safety team] must give consideration to:

- The steps preceding and following the specified operation;
- The process equipment, utilities/services, and surroundings;
- The preceding and following links in the food chain.

A hazard is defined by the National Advisory Committee on Microbiological Criteria for Foods as "any biological, chemical, or physical hazard that is reasonably likely to cause illness or injury in the absence of its control."

A biological hazard includes pathogenic microorganisms that either directly or indirectly cause a food safety hazard. Biological hazards include bacteria, fungi, viruses, and parasites. Examples of these *pathogens* include such organisms as *E. coli* O157:H7, *Salmonella*, *Clostridium botulinum*, and *Listeria monocytogenes*. In identifying a pathogenic hazard, the specific organism must be identified and the specific criteria required to control it defined. This category of food hazards poses the greatest risk for illness or injury. Hazards caused from microbiological pathogens affect the largest numbers of individuals per occurrence, thus receiving the most publicity. Thousands of people stretching over great distances can get very sick from one event.

A chemical hazard is defined as a contamination of the ingredient or product that results in either a direct or indirect food safety hazard. There are a wide variety of chemical residues that can occur in foods as a result of chemical usage in food production and processing. Most of these chemicals are regulated substances under federal regulatory agencies. Accidental or environmental chemical residues can also occur in foods under certain conditions. Examples of chemical hazards may be pesticides, antibiotics, food additives, antibiotics, hormones, cleaning and sanitizing agents, industrial chemicals, and pollutants.

A physical hazard is defined as any object or material that is not normally part of the product, such as bones, twigs, seeds, metal, glass, or plastic. A physical hazard such as glass could cause an injury (e.g., broken tooth) or choking if swallowed. Physical hazards occur most frequently, but generally affect only a small number (i.e., one or two) per occurrence.

As discussed throughout this text, the role of the PRP in the food safety/HACCP program is critical. Many existing or potential hazards may be controlled through PRPs. PRPs are activities defined and managed through operational-type programs that effectively eliminate or reduce the likelihood of a food safety hazard. PRPs are the foundation of an effective food safety/HACCP program. Understanding the PRPs and ensuring that each one is effective is critical and must be addressed in the initial development phase of a successful and effective food safety program.

There are many recommendations as to how to conduct a hazard analysis. Most of these involve some version of the following two components:

- *Hazard identification*: Identifying and developing a list of potential hazards
- *Hazard evaluation*: Assessing the relative risks, severity, significance, and likelihood of occurrence of the potential hazard in the specific food processing and handling system

This principle requires the analysis of the significance and likelihood of occurrence of the food hazards included in the hazards list. While this is often not a thorough academic or scientific risk assessment, it does possess elements of risk assessment. During hazard evaluation, the severity of hazards must be evaluated based on the specific conditions and situations anticipated for the specific food products manufactured and the scope of the food safety/HACCP program. At a minimum, the hazard evaluation must include the following:

- Effectiveness of PRPs
- Frequency of association of the hazard with the type of food or ingredients
- Frequency of occurrence within the facility
- Food product composition and other intrinsic factors (e.g., pH, water activity) that may enhance or inhibit growth
- Extrinsic factors (e.g., oxygen level, temperature considerations)
- Processing considerations
- Storage and transportation conditions
- Likely preparation steps by consumers
- Probable consumption information and risk factors (e.g., expected level, target consumer)

HAZARD ANALYSIS AS A RISK ASSESSMENT

The food safety/HACCP team must define the criteria necessary for identifying the hazard. Once the hazard is identified, it must be evaluated for potential risk and the significance of occurrence. *Risk* and *significance* are keys to the analysis. *Risk* is the likelihood that the hazard may occur. *Significance* considers how serious the resulting food hazard would be should the hazard occur. The risk–significance relationship is normally graded by the food safety/HACCP team as high, medium, low, or remote. It is this criteria that is used to determine if a point in the process is significant and must be controlled through a critical control point (CCP) or an operational PRP (OPRP) (FSSC 22000 or ISO 22000 only). More information related to *risk assessment* is provided in Chapter 15 ("Risk Management").

There are many different charts and tables used to compare the various combinations of *risk of occurrence* to *significance of outcome*. For example, a hazard may be a *low* risk to occur, but if it did occur, the outcome may be a *high* significance. Does this translate to a significant hazard that must be controlled via a CCP?

In the grid system, the relative risk of a hazard is assessed by comparing severity (low, medium, high) to the likelihood of occurrence (remote, low, medium, high). It is clear that a hazard with a rating of H/H has a high severity, is highly likely to occur, and should be considered significant; one rated L/L would not be significant. Meanwhile, a rating of H/R would describe a hazard that is very severe, but is only remotely likely to occur in the specific food system; thus, it is not *significant* and need not be addressed in the HACCP program. There can be confusion with regard to the midrange ratings. Figure 17.2 provides a sample of this *grading* or *evaluation* process.

Severity				
High	H/*R*	H/*L*	H/*M*	H/*H*
Medium	H/*M*	M/*L*	M/*M*	M/*H*
Low	H/*L*	L/*L*	L/*M*	L/*H*
	Remote	*Low*	*Medium*	*High*
	Likelihood of occurrence			

FIGURE 17.2   Grid for ranking food hazards. [Codex CAC/RCP 1-1969, Rev. 4, 2003].

Likelihood				
Frequent	5	10	15	20
Likely	4	8	12	16
Occasional	3	6	9	12
Unlikely	2	4	6	8
Improbable	1	2	3	4
	Minor	Medium	High	Very high
Severity of outcome				

FIGURE 17.3   Numbered grid assessment.

Some organizations actually assign numbers to each grid factor to identify the significant hazards. Some subject matter experts recommend using a numerical scale to characterize food hazards to alleviate some of the confusion in using the grid system. For example, in the numbered grid system recommended by Nolan, likeliness (frequency) and severity are each ranked using a 10-point ranking scale. The food safety/HACCP team sets the numerical combined score value that they consider to be significant. This system, though arbitrary, may have merit, especially in training the food safety/HACCP team in the evaluation phase of hazard analysis. Figure 17.3 provides an example of the numbered grid. Using this type of grid, the analysis for a 1–4 rating would be identified as a low risk whereas a 10–20 rating would be recorded as a high risk.

Although these types of grids are available through many sources, it is best that the organization creates its own grid, defining the ranking for a specific number range and also the justification and description for a specific ranking.

The organization must also define the criteria for ranking the likeliness. Frequency may be defined as weekly, whereas improbable may signify that there isn't any evidence of an occurrence. The organization must also define the criteria for ranking severity of the outcome: from minor (limited health issues) to very high (almost certain death) from the hazard. Again, the choice of these tools and how the specifics are defined, are up to the organization. However, it is critical that these defined parameters are realistic for the organization and its processes.

These charts may tend to cause more confusion than clarification. Although they do build consistency for the decisions, it is up to the food safety/HACCP team to evaluate each step in the process and make sound decisions based on the experience, data, and specific processes. These requirements must be defined for the operation to build consistency within its specific decision and evaluation process as related to the organization's food safety/HACCP program(s).

It is the responsibility of the food safety/HACCP team during this analysis to identify and record all potential food safety hazards that could occur in relation to the type of product and specific process for the product. The identification of specific hazards is unique for each operation. What may be a hazard for one operation may not be a hazard for another, even if the company manufactures the same product for the same intended use. This may be due to equipment or other process considerations. It is a good idea to benchmark against similar operations; however, the food safety/HACCP program must be defined specifically for the operation for which it is being developed.

IDENTIFYING THE HAZARDS

All hazards involved with processing the specific product must be identified. This includes not only hazards within the process, but also those related to the raw ingredients, packaging materials, product contact surfaces, and processing aids. Only hazards that can be controlled by the process should be identified. For example, a producer of raw hamburger meat cannot control the cooking temperature used by the end user. This can be identified as a potential hazard of the end product, with communication to the end user (i.e., on the label). It would not be identified as a significant hazard for the raw hamburger processor because it cannot be controlled within the scope of its food safety program.

The ultimate goal is to either eliminate or reduce a significant hazard to an acceptable level. However, with some chemical hazards, there may be an acceptable tolerance level allowed in the end product. This criterion must be identified and records maintained confirming continual compliance. This level must relate to established requirements, legal requirements, customer food safety requirements (as applicable), the intended use by the consumer, and any other relevant data as related to the organization's products and processes. The justification for decisions related to the hazard must be recorded and maintained in compliance with the record control program.

Once the hazard analysis is completed, it is the responsibility of the food safety team to ensure that it remains current. The hazard analysis must be updated, if identified as necessary, by the food safety team prior to any changes in the process. Details related to change management and the requirements for notifying and actions taken by the food safety team are addressed in Chapter 15 ("Change Management").

It is critical to emphasize that the food safety/HACCP program is built upon a thorough hazard analysis. Whichever procedures are followed, it is imperative that this process is based upon current scientifically valid information. If the hazards identified and evaluated in the hazard analysis are incorrect or inappropriate to the organization's process, the food safety/HACCP program is destined for failure. The hazard analysis requires constant review and update based upon evolving science and technology, as well as any modification in the organization's processes.

Based on the hazard assessment, the food safety team must select an appropriate combination of control measures capable of preventing, eliminating, or reducing the identified food safety hazards to an acceptable level. Each control measure must be reviewed with respect to its effectiveness against the specific hazards to be controlled. During hazard analysis, there is often a tendency to jump ahead to identify CCPs. This is inappropriate, as it may jeopardize the scientific validity of the plan.

HAZARD ANALYSIS AND FOOD PACKAGING

As requirements for hazard analysis have applied a more focused emphasis on product contact surfaces, there have been many questions posed related to primary packaging. Before moving into the identification of control measures, it is important to take one more look at the hazard analysis, especially as it applies to other areas than just the process steps. Food safety teams have struggled in understanding exactly how much detail and what records would be required to demonstrate that potential and/or existing hazards related to primary packaging have been evaluated and addressed through the hazard analysis process. *Dr. Tatiana Lorca, manager, Food Safety Education and Training, Ecolab, Food & Beverage Division*, provides the following guideline related to activities evaluating food contact packaging material:

> Clause 5.5 (BRC Global Standard for Food Safety issue 6) states that product packaging shall be appropriate for the intended use. Clause 5.4.1 further elaborates on this by requiring that the processor communicate the characteristics of the food which may affect the suitability of the packaging material (e.g., pH, aw) as well as the intended use of the product (such as microwave in package). The nature of the product or the consumer's use of the product may negatively affect the packaging material and the safety of the product. BRC's requirement is similar to the requirements found in the SQF (Safe Quality Food) Code ed7 and in FSSC 22000 (ISO 22002-1). Food processors need to prove the packaging material that comes into contact with their product (and ingredients) and the conditions of its use do not pose a potential food safety risk. The packaging materials need to be manufactured from materials that meet US regulatory requirements for food contact packaging and its use needs to be suitable for that application. To meet compliance, a food processor should request the following from their primary food contact packaging manufacturer:
>
> 1. Continuing Letter of Guaranty (LOG) that states the packaging materials meet US regulatory requirements (FDA and/or USDA) by stating:
>   a. Raw materials used in the manufacture of the packaging materials meet FDA's requirements for food contact applications and are approved for the food type and condition of use limitations as noted in the LOG.
>   b. Optional: "Condition of Use Statement/Condition of Use Limitations" refers to testing that is conducted by some packaging manufacturers which speaks to its performance (e.g., sealing properties) in common uses of the materials.

2. Property Sheet/Technical Data Sheet for the material [that provides the following recommended] specifications of the packaging material such as moisture, vapor, transmission rate, standard application and other properties as appropriate.

3. If appropriate to the application, a Certificate of Analysis can be obtained which confirms the presence of materials responsible for key attributes (e.g., presence of oxygen).

SELECTION OF THE CONTROL MEASURES

ISO 22000:2005 7.4.4 states that based on the hazard analysis that

An appropriate combination of control measures [must] be selected which is capable of preventing, eliminating or reducing these food safety hazards to defined acceptable levels. The selection and categorization of a control measure [must be performed] using a logical approach that includes assessments with regard to the following:

- Its overall effect on an identified food safety hazard;
- Its feasibility for monitoring in a timely manner to enable immediate corrections;
- Its place within the system relative to other control measures;
- The likelihood of failure in the functioning of a control measure or significant processing variability;
- The severity of the consequence(s) in the case of failure in its functioning;
- Whether the control measure is specifically established and applied to eliminate or significantly reduce the level of hazard(s);
- Synergistic effects (e.g., interaction that occurs between two or more measures resulting in their combined effect being higher than the sum of their individual effects).

CRITICAL CONTROL POINT

Principle 2 states to "identify critical control points." The purpose of principle 2 is to "identify the Critical Control Points required to control significant hazards."

A CCP "is any point in the chain of food production from raw materials to finished product where the loss of control could result in an unacceptable [or potentially unacceptable] food safety risk" (Pierson and Corlett 1992). ISO 22000:2005 3.10 defines a CCP as [a] "food safety step at which control can be applied and is essential to prevent or eliminate a food safety hazard or reduce it to an acceptable level."

A control point (CP) is defined as "any step at which biological, chemical, or physical factors can be controlled" (Stevenson and Bernard 1999). The difference between a CP and a CCP is that if control is lost at the CCP, either a significant food safety hazard or the potential for a significant food safety hazard is created. Loss of control at a CP by itself does not specifically result in a food safety hazard, or there is a subsequent step in the process that will control the hazard. Decisions are based on the food safety team's experience and analysis.

A CCP must be justifiable, validated, measurable, and carefully developed to ensure its effectiveness in preventing, eliminating, or reducing the identified significant food safety hazard to an acceptable level. A CCP must have a defined control measure. If a process step is identified where control is required for a significant hazard, yet no control measure exists, the product or process must be modified at that step, or at any earlier or later stage, to include a control measure. Records related to the control and the effectiveness of this control (or deviations from control) must be maintained in compliance with the record control program (Chapter 7).

ISO 22000:2005 7.6.1 states that the following information must be defined for each CCP:

- "Specific food safety hazard(s) to be controlled:
- Control measure(s);
- Critical limit(s);
- Monitoring procedure(s);
- Corrections and corrective action(s) to be taken if critical limits are exceeded;
- Responsibilities and authorities;
- Monitoring records."

There are numerous decision trees available that aid the food safety team in making decisions related to the identification and control of a CCP. Decision trees basically apply a common sense approach based on the experience of the food safety team. It is recommended that these decision trees are sourced based on the organization's products and processes, reviewed by the food safety team and used appropriately in the food safety program development. In addition, it is also recommended that the food safety team not be shy when researching, asking questions, and including subject matter experts to assist in the evaluation and identification of significant hazards. A word of caution; it is common to be overzealous in the identification of CCPs when first beginning the analysis. The identification of too many CCPs overburdens the process, but too few may not protect the safety of the product. It is important that the food safety/HACCP team give a focused evaluation of the potential outcome of being uncontrolled while considering its function in relationship to other points in the process.

OPERATIONAL PREREQUISITE PROGRAM (OPRP)

ISO 22000:2005 7.4.4, defines that "the control measures selected [must] be categorized as to whether they need to be managed through [either] operational PRP(s) (OPRP) or by a [Critical Control Point (CCP)]." Note that the identification of an OPRP is specific to the ISO 22000:2005 and FSSC 22000:2010 food safety management standards; however, because of its growing popularity as an effective control measure adding an additional layer of effectiveness to a food safety program, further detail related to OPRPs are included as a potentially important tool in the control of existing and potential hazards.

ISO 22000:2005 3.9 defines an OPRP as that "identified by the hazard analysis as essential in order to control the likelihood of introducing food safety hazards to and/ or the contamination or proliferation of food safety hazards in the products or in the processing environment." OPRPs are similar to a CCP except that the standard does not require a defined *limit*, although if there is a *limit* that would enhance control, then this should be included.

PRPs VERSUS OPRPs VERSUS CCPs

Questions on the identification and application of an OPRP are probably the most frequently asked question by both top management and the food safety teams of organizations choosing FSSC 22000 or ISO 22000 as their food safety management system standard. Answers to these questions are not usually the easiest found. Interpretation may also vary depending on the regions where compliance is applied. Basically, it is best to start with the concept that an OPRP, although important to the production of a safe product, is one step below the CCP in its critical status to the program. As mentioned, OPRPs do not require a critical limit (CL), but they must be validated to ensure that the CP is the *right* choice to control the hazards. Understanding what should be a CCP and what should be an OPRP can be a tough choice for the food safety/HACCP team to make. Decisions must be based on experience, data, and the risk. Decision trees related to identifying control risks and making the choice between a PRP, CCP, or OPRP are available. Based on experience, the results obtained from using such tools can be inconsistent causing more confusions than sustainable answers. The application of the food safety/HACCP team's evaluations based on their experience in the industry and the organization's products and processes, is the key to success. Evaluating each individual program making a decision relates to the specific program, food sector, and organization. This has proven to be much more accurate than attempting to search out *generic* type tools that have the answer for everything. The following information provides a basic comparison between a PRP, OPRP, and a CCP. However, it is critical to caution the food safety/HACCP team to make sure that these guidelines are reviewed, and applied with flexibility to the team's specific programs ensuring that decisions are made based on what is best for the specific food safety/HACCP program.

A *prerequisite program* (*PRP*) generally applies across the entire operation such as pest control or good manufacturing practices (GMP). PRPs may contribute to the reduction of a hazard but are not necessarily essential for control. Although there could be critical limits identified for a PRP, having a limit is rare. A failure related to a PRP rarely results in placing product on hold or a recall. Findings must be corrected, but do not necessarily result in a formalized corrective action or correction. PRPs must be documented, implemented and verified as effective. More information related to prerequisite programs is included throughout this text and more specifically in Chapter 10.

Operational prerequisite programs (*OPRPs*) apply to a specific hazard identified for a specific product or process and are essential to reduce the level of the hazard. OPRPs are control measures essential for the control of a hazard, but do not have *absolute* control over the hazard. OPRPs may have defined critical limits but critical limits are not required. A failure in an OPRP may result in product placed on hold or a product recall; however, a recall is not required. Many times there are additional points (CCPs) in the process after the OPRP that may assist with the control of the hazard. The OPRP must be documented, implemented, validated and verified. Deviations in meeting the requirements for an OPRP require formalized corrections and corrective actions. OPRPs may work in combination with other control measures to prevent, eliminate, reduce or maintain a hazard to an acceptable level. Their failure does not automatically imply that a product is hazardous.

A *Critical Control Point* (*CCP*) is usually identified as the last point in the process that is designed to eliminate or reduce a significant food safety hazard to an acceptable limit. A CCP is a process step where a control measure is applied that has *absolute* control over the hazard. A critical limit must be defined with records maintained to confirm that the limit is met. Failure in meeting the critical limit results in a food safety hazard or increases the probability of a significant food safety hazard thus causing the resulting product to be potentially unsafe. The CCP must be documented, implemented, validated and verified. Deviations in meeting the requirements for a CCP require formalized corrections and corrective actions and likely results in a product needed if product has left the content of the organization.

ESTABLISH THE CRITICAL LIMITS

Principle 3 requires that the food safety/HACCP program *establishes* "the critical limits which must be met at each identified Critical Control Point" (Pierson and Corlett 1992). A critical point (CL) is the control parameter that is required to ensure that the product is safe. ISO 22000:2005 7.6.3 defines the following requirements for the identification of a CCP's critical limit:

- "Critical limits must be determined for the monitoring established for each CCP;
- Critical limits must be established to ensure that the identified acceptable level of the food safety hazard in the end product is not exceeded;
- Critical limits must be measurable;
- The rationale for the chosen critical limits must be documented;
- Critical limits based on subjective data (such as visual inspection of product, process, handling, etc.) must be supported by instructions or specifications and/or education and training."

A CL is the control parameter that is required to ensure that the product is safe. An example of a CL would be the required pasteurization time/temperature criteria necessary to deliver a *safe* milk product. A CL should not be confused with an operating limit (OL). For example, a CL set for pasteurization of milk may be 165°F ± 2°F (the legal requirement is 161°F); however, the OL identified to achieve the ultimate-quality product may be 175°F. Setting appropriate OLs may allow room for process adjustment without being out of control from a critical limit or food safety viewpoint.

ESTABLISH THE MONITORING PROCEDURES

Principle 4 requires that the food safety/HACCP program defines monitoring requirements for the critical limits. ISO 22000:2005 7.6.4 defines the following requirements for monitoring CCPs:

- "A monitoring system must be established for each CCP to demonstrate that the CCP is in control. The system must include all scheduled measurements or observations relative to the critical limit(s).
- The monitoring system must consist of relevant procedures, instructions and records that cover the following:
  a. Measurements or observations that provide results within an adequate time frame;
  b. Monitoring devices used;
  c. Applicable calibration methods;
  d. Monitoring frequency;
  e. Responsibility and authority related to monitoring and evaluation of monitoring results;
  f. Record requirements and methods.

The monitoring methods and frequency must be capable of determining when the critical limits have been exceeded in time for the product to be isolated before it is used or consumed."

Monitoring is defined as "the act of conducting a planned sequence of observations of measurements of control parameters to assess whether a CCP is under control." CAC further defines monitoring as a "scheduled measurement or observation of a CCP relative to its critical limits" (CAC/RCP 1-1969, Rev. 4, 2003). Monitoring of a CCP is essential to the overall control process. Information must be available in current time to *control* the hazard. The term *control* has two distinct definitions and one major application as it applies to HACCP (all definitions are taken from CAC/RCP 1-1969, Rev. 4, 2003):

"*Control*" (verb): To take all necessary actions to ensure and maintain compliance with criteria established for the Critical Control Point.

"*Control*" (noun): The state wherein correct procedures are being followed and criteria are being met.

"*Control measure:*" Any action and activity that can be applied and is essential to prevent or eliminate a food safety hazard or reduce it to an acceptable level.

Target limits should be defined as a goal, with any drifts away from these targets (trends in the process) adjusted before the CCP becomes out of control. Ideally, continuous monitoring that provides records (i.e., recording charts) is the means of choice. Records associated with monitoring a CCP must be "signed by the person(s) doing the monitoring and by a responsible reviewing official(s) of the company" (CAC/RCP 1-1969, Rev. 4, 2003). This applies to each identified CCP and must be done on completion of the activity.

ESTABLISH PREPLANNED CORRECTIVE ACTIONS

Principle 5 requires that corrective actions are defined and that the food safety/ HACCP team establishes the "corrective action to be taken when there is a deviation identified by monitoring of a Critical Control Point" (Pierson and Corlett 1992). A deviation is a violation or failure to meet the critical limit (CL). Actions to be performed, should a CCP become out of control, must be defined. The corrective action must be preplanned. *Preplanned* means that the required action, should a deviation occur, is clearly defined to provide immediate action to protect against a food safety hazard. Associates responsible for these activities must have the training and authority to initiate the preplanned corrective action to protect the product and to bring the CCP under control. Actions must also include the proper disposition of the products and records to demonstrate compliance with all defined requirements (Chapter 9).

ISO 22000:2005 7.6.5 states that the following must be met when monitoring results exceed critical limits:

- "Planned corrections and corrective actions to be taken when critical limits are exceeded must be specified for each Critical Control Point. The actions must ensure that the cause of nonconformity is identified, that the parameter(s) controlled at the CCP is (are) brought back under control, and that recurrence is prevented.
- Documented procedures must be established and maintained for the appropriate handling of potentially unsafe products to ensure that they are not released until they have been evaluated."

ESTABLISH REQUIREMENTS FOR VERIFICATION

Principle 6 requires that the food safety/HACCP team establishes "procedures for verification" (CAC/RCP 1-1969, Rev. 3, 1979). This principle requires verification and validation of the CCPs. Verification is defined as "the application of methods, procedures, tests, and other evaluations, in addition to monitoring to determine compliance with the requirements" for each CCP (CAC/RCP 1-1969, Rev. 3, 1979) or those activities, other than monitoring, that establish the validity of each CCP. Verification confirms that all defined requirements of the HACCP plan are being performed. Validation is that element of verification that is defined as the act of "obtaining evidence that the elements of the HACCP plan are effective" (CAC/RCP 1-1969, Rev 4, 2003), or as "that element of verification focused on collecting and evaluating scientific and technical information to determine whether the requirements for [the Critical Control Point], when properly implemented, effectively control the identified food hazards" (NACMCF 1998). Validation confirms that the appropriate activities are being performed to provide a safe product. Procedures must exist to verify that the established control of a CCP is functioning properly. Confirmation that the flow diversion valve is performing as required would be an example of a verification activity. More detail related to verification and validation is discussed in Chapter 18.

DOCUMENTS AND RECORDS

Principle 7 requires that the food safety/HACCP team *establishes* "documentation and record keeping" (CAC/RCP 1-1969 Rev. 4, 2003). Records must be available to provide the objective evidence (proof), not only that the CCPs have been controlled according to the required procedures, but also that the PRPs are being maintained in an effective manner. Examples of records are CCP monitoring activities, deviations and associated corrective actions, and food safety/HACCP team meeting minutes. The food safety/HACCP program and related instructional documents must be identified and maintained in compliance with the document control program (Chapter 6). Objective evidence (records) must be identified and maintained in compliance with the record control program (Chapter 7).

MANAGEMENT COMMITMENT AND THE FOOD SAFETY/HACCP PROGRAM

The CAC HACCP document states that "Management Commitment is necessary for implementation of an effective HACCP system" (CAC/RCP 1-1969, Rev. 4, 2003). It further emphasizes that "the successful application of HACCP requires the full commitment and involvement of management and the work force" (CAC/RCP 1-1969, Rev. 4, 2003). This cannot be overstated. Management must convey a positive message of commitment throughout all levels of the operation. Management responsibility including the role and responsibilities for top management is addressed in Chapter 3.

The food safety/HACCP program must not be considered a static program. As stated previously, the food safety/HACCP program must be reviewed at a defined frequency to ensure its continued effectiveness as related to the operation's ability to ensure the production of a safe product. The required frequency does vary depending on many factors, but it is recommended that the frequency be defined as a minimum of semi-annually or more frequently based on the requirements defined for change management (Chapter 15, "Change Management"). It is through these reviews that the effectiveness of the food safety/HACCP program, changes in technology, scientific data, and any other practical considerations are evaluated. It is essential that the food safety/HACCP team be empowered with the knowledge and authority to establish, maintain, and evaluate the food safety program. Team efforts must be supported by management commitment. Management commitment must be communicated through all levels of the operation and must provide the resources (e.g., money, people, training) to provide the basis of an ongoing effective food safety/HACCP program. Nothing is more important than producing a safe product—a product that each associate feels safe to feed his/her own family. Safety comes first, and although a food safety/HACCP program does not guarantee a safe product, complying with the requirements of a structured food safety management system standard makes it possible to develop and maintain a program that provides confidence while making good business sense.

The *three-legged stool* has been discussed throughout this text in relation to the management system structure. It highlights the importance of the internal audit,

PRPs   Food safety/HACCP
            Management   team
            commitment

FIGURE 17.4   Food safety/HACCP three-legged stool.

management review, and the corrective/preventive action programs to the long-term effectiveness and continuous improvement of the management system (Chapter 1). A version of the *three-legged stool* also applies to the food safety/HACCP program, considering that its overall effectiveness and sustainability are directly linked to the effectiveness of management commitment, the PRPs, and the role of the food safety/HACCP team. Figure 17.4 provides the revised *three-legged stool* as it relates directly to the food safety/HACCP program.

KEY POINTS

a. The food safety team and the HACCP team may be the same team; however, for those organizations with multiple processes, a specific *team* may be required for each process. This relationship of the two teams (if different) and objective evidence confirming that requirements of the standard of choice must be clearly defined (Chapter 16).

b. The food safety team must review and update the food safety/HACCP program prior to any changes requiring an update (related to food safety) being implemented. This also includes changes related to the PRPs. Further requirements and recommendations related to *changes* are addressed in Chapter 15 ("Change Management").

c. As part of the hazard analysis, the food safety/HACCP team must identify and link references to those PRPs or other measures used to control an identified hazard for a material, ingredient, processing aid, product contact surface or at a specific step in the process.

d. The step(s) from raw materials, through processing, and shipping at which a food safety hazard may be introduced must be identified in the flow diagram.

e. Objective evidence confirming that an acceptable level of a potential food safety hazard in the end product is monitored by the food safety team at a defined frequency based on the potential risk must be defined. Records must be maintained in compliance with the record control program (Chapter 7).

f. Food safety team must ensure that where there is more than one hazard identified at a process step, each hazard is assessed based on defined requirements with records maintained in compliance with the record control program (Chapter 7).

g. The hazard analysis for the process steps is based on the flow diagram developed for prestep # 4 (Chapter 16).

h. It is recommended that the food safety/HACCP team use a template for the hazard analysis. There are many different templates available for this activity, which vary depending on the preferences of those creating the template and/or applying to a specific process, ingredient, material, or processing aid. Based on experience, it is recommended that the template or form include the following columns:

- Item (ingredient, material, processing aid, product contact, etc.) or process step.
- Identification of an existing or potential hazard (biological, chemical, physical).
- Likelihood—how likely is it to occur? (remote, low, medium, high).
- Severity or significance—how significant is the outcome? (remote, low, medium, high).
- Justification for decision (i.e., data, research, etc.).
- What measures can be applied to prevent, eliminate, or reduce the identified hazard(s)? Is there a later point in the process that is used to control the hazard that is introduced at this step?
- Is an OPRP (ISO 22000, FSSC 22000 only) or CCP required?

Also, in addition to the template, the use of a *decision tree* may be useful for the food safety/HACCP team to use in making these decisions. The use of a decision tree can, however, be very confusing. It is up to the team to research these to determine if an existing one (i.e., Codex Alimentarius) may be helpful or if it would be more value-added for team to create its own based on its operation and the experience.

i. The food safety/HACCP team must ensure that the severity/significance and likelihood of occurrence of each hazard is assessed, recorded, and justified.

j. The food safety/HACCP programs must clearly identify and where possible link to requirements for ensuring compliance with each required control measure. Justification must be available to explain the basis for decisions made related to the assignment of PRPs, OPRPs, and/or CCPs as the control measure.

k. Records must be identified, maintained, and readily available to confirm that all associates responsible for any food safety responsibility including those for monitoring and compliance with CCPs or OPRPs have the skills (competent) to perform all required activities. This includes those related to monitoring limits, recording results, and planned corrective actions should a deviation occur. Training and competency requirements are defined in Chapter 8.

l. The food safety team must ensure that review of the compliance status (CCPs or no CCPs) is evaluated and recorded in food safety team meeting minutes (Chapter 16).

m. The food safety/HACCP team must ensure that the review of *product contact surfaces* addresses the surface that contacts the product rather than the equipment itself.

n. Food safety/HACCP program documents must be identified and maintained in compliance with the requirements for the document control program (Chapter 6).

COMMON FINDINGS/NONCONFORMANCES

a. It could not be confirmed through review of the hazard analysis that primary packaging materials (direct contact with the product) had been evaluated for potential hazards.

b. Records (i.e., food safety/HACCP team meeting records) were not available to confirm that the food safety team was monitoring and confirming compliance and effectiveness of the food safety/HACCP program.

c. It could not be confirmed through the review of the food safety/HACCP program that control measures for the recorded hazards had been identified.

d. Records did not confirm that the food safety/HACCP team had provided the *justification* (basis for decisions made) related to assigning PRPs, OPRPs (ISO 22000:2005 and FSSC 22000 only), and/or CCPs.

e. Evidence (records) was not available to confirm that the food safety/HACCP team had performed the annual evaluation of the food safety/HACCP (i.e., hazard analysis) program. Note: It was stated that it would be recorded in the food safety/HACCP team meeting minutes; however, review of meeting minutes for this time period did not provide this evidence.

f. The food safety team had not identified any OPRPs or CCPs in its operation. The food safety program stated that this would be monitored by the food safety team during its meetings and recorded in the team meeting minutes; however, records were not available to confirm that the food safety/HACCP team had been performing this review.

g. It could not be demonstrated through the interview process that the operator responsible for CCP #1 (metal detector) had been trained in all related CCP requirements. Note that during the interview process, the operator was unsure of what to do if the CCP (metal detector) failed.

h. It could not be confirmed through review of the food safety/HACCP program and related controlled documents that requirements for what to do if CCP deviations occur had been defined, implemented, and monitored to ensure that the identified hazards were in control.

i. Records maintained to confirm compliance to the CCP operation (metal detector) had unexplained *blank* spaces, thus not providing complete evidence of compliance.

j. Records related to the food safety/HACCP program demonstrating control of identified hazards (potential and existing) and the ultimate production of a safe product were not identified and managed in compliance with the record control program (Chapter 7).

TABLE 17.1

Comparison of HACCP Methods

Codex	Added by These Standards/Guidelines		
	SQF Code (2.4.3)	BRC Food (2.0)	ISO 22000 (7.3)
References: Codex	References: NACMCF/Codex Primary producers can follow HACCP-based reference plan devt. By responsible authority	References: Codex	References: Codex
Step 1—HACCP team	**2.9.4**	**2.1.1**	**7.3.2**
	• HACCP training for all involved in developing, implementing and maintaining food safety plans	• Must include quality, technical, production, maintenance/engineering • Team leader • Team leader to show in-depth knowledge (product, process, hazards) • Does not address scope of the HACCP plan	• Team leader • Records required showing team knowledge and experience

Step 2—Describe product	**No additional requirements**	**2.3**
		• Also consider if applicable:
		  • Composition
		  • Origin of raw materials
		  • Target shelf life
		  • Instructions for use
		• Info be maintained and documented
			7.3.3
			• Describe raw materials, ingredients, product contact materials and include:
			  • Composition of formulated ingredients, processing aides, additives
			  • Origin
			  • Method of production
			  • Storage conditions
			  • Shelf life

	2.4.3	**2.4**	**7.3.4**

Step 3—Intended use	• No additional requirements	• Consider vulnerability of those exposed and to contamination of raw materials, intermediate and finished product
		• Specifically call out allergy-sufferers (**3.5**)
			• Preparation/handling before use/processing
			• Food safety–related acceptance criteria/specs appropriate to their intended use
			  • Identify regulatory requirements
			  • Maintain descriptions
			• For finished products:
			  • Identify regulatory requirements
			  • Maintain descriptions
			• Consider potential misuse

	2.4.3.1	**3.5**	**7.3.5**

Step 4—Process flow diagram	• Rework	• Outsourced processes or subcontracted work
		• Rework and recycling
		• Release/removal of by-products and waste
		• Process parameters
		• Potential for process delay
			• Outsourced processes or subcontracted work
			• Rework and recycling
			• Release/removal of by-products and waste
			• Include as appropriate:
			  • Any outsourced processes and subcontracted work;
			  • Where raw materials, ingredients and intermediate products enter the flow;

(continued)

TABLE 17.1 (continued)
Comparison of HACCP Methods

Codex	SQF Code (2.4.3)	Added by These Standards/Guidelines	
		BRC Food (2.0)	**ISO 22000 (7.3)**
		• Low/high care/high risk area segregation (maps)	• Where reworking and recycling take place; • Where end products, intermediate products, by-products and waste are released or removed. • Control measures, process parameters, (and any customer and regulatory requirements related to these) to be described as needed for the hazard analysis
Step 5—Verify process flow diagram	**2.5.1** • Annual verification	**2.6** • Annual verification	**7.8**
Step 6—Principle 1 hazard analysis	**2.4.3.1** • Assure rework is addressed • All steps and inputs	**2.7** • Determine acceptable levels of hazards documented • State what's controlled by PRPs • Also requires consideration of vulnerability of those allergic to raw materials, intermediate and finished products	**7.4.2** • Hazard identification based on: • Preliminary information and data collected • Experience, • External information (including, to the extent possible, epidemiological and other historical data), and • Information from the food chain on food safety hazards that may be of relevance for the safety of the end products, intermediate products and the food at consumption. • Indicate step(s) (from raw materials, processing and distribution) at which each food safety hazard may be introduced • Consider preceding and following links in the food chain

- Acceptable level of food safety hazard determined (and recorded) where possible
- 7.4.3 Assessment of food safety hazard based on severity, adverse health effects and likelihood of occurrence
- 7.4.4 Classification of control measures in the hazard analysis (PRP, oPRP, CCP) with regard to:
  - Its effect on identified food safety hazards relative to the strictness applied;
  - Its feasibility for monitoring (e.g. ability to be monitored in a timely manner to enable immediate corrections);
  - Its place within the system relative to other control measures;
  - The likelihood of failure in the functioning of a control measure or significant processing variability;
  - The severity of the consequence(s) in the case of failure in its functioning;
  - Whether the control measure is specifically established and applied to eliminate or significantly reduce the level of hazard(s);
  - Synergistic effects (i.e. interaction that occurs between two or more measures resulting in their combined effect being higher than the sum of their individual effects).

7.5

- No CPs recognized
- Recognize oPRPs

2.8

- No CPs recognized

Step 7—Principle 2
Identify CCPs
- No CPs recognized

- Recognizes CPs

(continued)

TABLE 17.1 (continued)
Comparison of HACCP Methods

		Added by These Standards/Guidelines	
Codex	**SQF Code (2.4.3)**	**BRC Food (2.0)**	**ISO 22000 (7.3)**
Step 8—Principle 3 **Establish critical limits**	2.5.2.7 • CLs must be validated	2.9 • CLs must be validated	7.6.3 • CLs must be validated

Step 9—Principle 4 **Establish monitoring**	• No additional requirements	2.10
		• Specific requirements for monitoring records:
			7.6.4
			• Relevant procedures, instructions and records that cover:
			  • Measurements or observations that provide results within an adequate time frame;
			  • Monitoring devices used;
			  • Applicable calibration methods (see 8.3);
			  • Monitoring frequency;
			  • Responsibility and authority related to monitoring and evaluation of monitoring results;
			  • Record requirements and methods.
			  • The monitoring methods and frequency shall be capable of determining when the critical limits have been
			  • Exceeded in time for the product to be isolated before it is used or consumed.

Step 10—Establish **corrective action**	• No additional requirements	2.11	7.10.2 • Evaluation of data from PRP and CCP monitoring • Corrective action on oPRPs and non-conforming product • Non-conforming product to be handled as potentially unsafe

		2.12	**8.4.2**
Step 11—Establish verification	• No additional requirements	• Independent verification by 3rd party is not required • Review – IA, complaints, Incidents • Communicate reviews to HACCP Team	• Verification to be planned and address HACCP + PRPs + relevant docs
	2.2.2	**2.10.2, 2.13**	**4.2.3, 7.8**
Step 12—Establish record keeping	• Records controlled through SQF system requirements (completion, retention, accessibility) • Allowance for electronic records	• Monitoring records must be signed by person doing activity, dated, result included and be verified by authorized person • Records controlled through BRC system requirements (completion, retention, accessibility) • If electronic records, evidence records have been checked and verified	• Records controlled through ISO 22000 system (completion, retention, accessibility) • Maintain all info and records needed to do hazard analysis
Additional requirements	• 2.1.2.5 SQF Practitioner HACCP trained • 2.4.4 Requires use of HACCP methodology to develop Food Quality Plan and identify CQPs and QPs	• 2.14.1 Review HACCP Plan incl. PRPs • Reviews to be conducted include following events: 　• Change in raw material or its supplier 　• Change in ingredient/recipe 　• Change in processing conditions or equipment	• 7.3.5 Description of external requirements • 7.5 oPRPs documented and include: 　• Food safety hazard(s) to be 　• Control measure(s) 　• Monitoring procedures showing oPRPs implemented; 　• Corrections and corrective actions for oPRPs

(continued)

TABLE 17.1 (continued)
Comparison of HACCP Methods

Codex	SQF Code (2.4.3)	Added by These Standards/Guidelines	
		BRC Food (2.0)	**ISO 22000 (7.3)**
		• Change in packaging, storage or distribution • Change in consumer use • Emergence of a new risk, for example adulteration of an ingredient • Developments in scientific information associated with ingredients, process or product. • 2.14.1 Changes resulting from review incorporated into the HACCP plan and/or PRPs, fully documented and validation recorded.	• Responsibilities and authorities; • Monitoring records

Sources: NACMCF. Hazard Analysis and Critical Control Point Principles and Application Guidelines. 1997. http://www.fda.gov/food/foodsafety/HazardAnalysisCriticalControlPointsHACCP/ucm114868.htm; Codex Alimentarius FOOD HYGIENE (BASIC TEXTS), 4th edition. 2009; British Retail Consortium. 2011. BRC Global Standard for Food Safety. Issue 6; SQFI. 2012. SQF Code. Edition 7; ISO 22000:2005.ISO. 2005; Table 17.1 printed by permission from Ecolab USA, Inc.

Note: All approaches follow the same methodology and have the same basic requirements derived from NACMCF or Codex. The differences in the approaches compared to the NACMCF is highlighted in the following table. Blanks indicate no significant difference to NACMCF/Codex.

REFERENCES

Pierson, M. and Corlett, D.A. 1992. *HACCP Principles and Applications*, AVI, Van Nostrand Reinhold, New York, NY.

Stevenson, K. and Bernard, D. 1999. *HACCP A Systematic Approach to Food Safety*, Food Processors Institute, Washington DC.

18 Verification and Validation

An effective program for the verification and validation of the food safety program is, of course, a critical requirement of every food safety program whether developing a simplified HACCP program based on Codex guidelines or developing a structured and disciplined food safety management system. Verification is a term that has been addressed specifically through Codex HACCP principle 6, which requires that procedures are established for the verification to confirm that the HACCP system is working effectively. Note that older versions of HACCP texts and articles may have principles 6 and 7 reversed. Requirements did not significantly change just the order of the principles.

VERIFICATION PLANNING

ISO 22000:2005 7.8 addresses verification planning stating that it must "define the purpose, methods, frequencies, and responsibilities for the verification activities and that these activities must confirm that:

a. The prerequisite programs (PRPs) are implemented
b. Input to the hazard analysis is continually updated
c. The operational PRP(s) (FSSC 22000, ISO 22000 only) and critical control points (CCPs) are implemented and effective
d. Hazard levels are within identified acceptable levels
e. The organization's related procedures are implemented and effective"

ISO 22000:2005 is very clear in stating that the "output of this planning must be in a form suitable for the organization's method of operations. Verification results must be recorded and must be communicated to the food safety team. Verification results must be provided to enable the analysis of the results of the verification activities."

ISO 22000:2005 further addresses verification requirements in Section 8.1 stating that "The food safety team must plan and implement the processes needed to validate control measures and/or control measure combinations, and to verify and improve the food safety management system."

Verification is also an action that overlaps into many other food safety–related activities. In addition to this discussion, information on verification is discussed as part of specific topics throughout this text (see Chapter 5 (Internal Audit), Chapter 4 (Corrective Action/Preventive Action [CAPA] and Continuous Improvement), Chapter 15 (Change Management) and Chapter 17 (Food Safety Program (HACCP)—The Principles).

Although the term *validation* is a popular discussion point, it is considered a subactivity of verification.

VERIFICATION VERSUS VALIDATION

Although these terms are addressed and discussed throughout various sections of this text, it is important to provide a specifically focused discussion on their meaning and application to an effective food safety management system. Many find it interesting how the food safety management standards have chosen to address verification and validation. These requirements are located throughout the standards in a manner that emphasizes their importance. Before looking further into how the food safety management system standards discuss these two very important terms, it is important to review what these terms truly mean.

1. *Verification*: *Verify* that activities are being completed as defined. Verification is designed to check the validation process.
2. *Validation*: The HACCP program, as written, effectively controls the hazards as established by the team. Validation is part of verification.

Verification is an ongoing activity to determine that, control measures have been implemented as intended and includes reviewing (*observing*) actual monitoring activities and the records that confirm the control measures (activities) are being performed as defined. In other words, verification confirms activities are being done as defined; however, validation confirms that the correct activities are being used to control the hazard.

EXAMPLE OF VALIDATION

One popular example is applying the control of the hazard *E. coli* 0157 in pasteurized concentrated orange juice. Scientific data, as a result of structured testing, is available stating that this juice must be heated to 161°F for 3 seconds. The hazard analysis for this process would state the specific hazard (*E. coli* 0157) parameters required to control (kill) the hazard and ensure that the pathogen has been eliminated. Having this information as part of the organization's food safety program is validation. Proof that the correct activity is being performed to control or eliminate the hazard. Verification would include records such as a recording chart confirming that the equipment is operating at the defined temperature. Most operations would actually process at a higher temperature to provide verification (proof) that the operation is being done as defined. A variance would be if the organization identifies the same hazard, but chooses to process at 150°F for 3 seconds. The same records can be created and maintained showing evidence of processing at this temperature (verification). However, validation, which is proof that the right activity is being done to control the hazard, would not be available simply because the organization is not processing at the temperature identified as required to ensure control (kill) of the pathogen. The process must provide not only evidence that activities are being performed as defined, but also validation information (doing the correct activity for controlling the hazard).

VERIFICATION ACTIVITIES

Examples of some verification activities may include:

- Confirming the accuracy of product descriptions and the flow chart
- Confirming that CCPs are being monitored in compliance with the defined HACCP program
- Confirming that processes are operating within the established critical limits (CLs)
- Monitoring consumer complaints to determine if they may be related to food safety
- Confirming that requirements for the CCP and the HACCP program overall are being followed, compliant, and the records maintained as defined
- Random sample collection and analysis
- Calibration activities.

REQUIRED CCP RECORD REVIEW

Verification and record reviews related to a CCP must confirm that required records are completed accurately, at defined time intervals, and reviewed and maintained per defined requirements. Record review must confirm that:

- Monitoring activities have been performed at the locations as defined
- Monitoring activities have been performed at the frequencies defined
- Corrective actions have been performed whenever monitoring indicated a deviation from CLs
- Product manufactured when CLs were out of control has been managed as defined to prevent product from being consumed
- Equipment has been calibrated at defined frequencies and that records confirm the accuracy of the equipment per defined requirements (i.e., calibration procedure)

VALIDATION IN MORE DETAIL

As previously stated, validation is considered part of the verification process. Validation according to Codex Alimentarius is defined as "obtaining evidence that a control measure or combination of control measures, if properly implemented, is capable of controlling the hazard to a specified outcome" (*Codex CAC/ GL 69-2008 Guidelines for the Validation of Food Safety Control Measures*) (www.codexalimentarius.org/input/download/standards/11022/cxg_069e.pdf).

The validation process demonstrates that selected control measures are actually capable, on a consistent basis, of achieving the intended level of hazard control. Validation focuses on the collection and evaluation of scientific, technical, and observable information to determine whether control measures are capable of achieving their specified purpose in terms of hazard control. It involves measuring performance against a desired food safety outcome or target to a required level of hazard control. Validation must be performed as the food safety program is being designed or whenever changes indicate the need for revalidation. The change management program (Chapter 15) provides examples of change activities that may trigger a review and possibly a revision to the food safety program and the means to track and record these changes to ensure the continued production of a safe product.

VALIDATING CONTROL MEASURE EXAMPLES

Codex CAC/GL 69-2008 *Guidelines for the Validation of Food Safety Control Measures* (www.codexalimentarius.org/input/download/standards/11022/cxg_069e.pdf) provides some excellent examples of some effective means to ensure the effectiveness of the validation process.

The following information from the referenced Codex document provides additional insight into an effective validation process:

1. *Reference to scientific or technical literature, previous validation studies, or historical knowledge of the performance of the control measure.* These sources may include:
   a. Scientific journals or technical literature
   b. Previous validation studies
   c. Historical knowledge of the performance of the control measures
   d. Government documents, regulations, guidelines directives, performance standards, tolerances, action levels
   e. Trade association guidelines
   f. University extension publications and university extension agents
   g. International standards or guidance (Codex)
   h. In-plant studies or research
   i. Processing authorities
   j. Industry experts and consultants
   k. Equipment manufacturers

2. *Scientifically valid experimental data that demonstrates the adequacy of the control measure, which* requires appropriately experimental trails such as laboratory challenge testing designed to mimic process conditions and industrial or pilot plant trails of particular aspect of a food processing system. Examples may include
   a. Mapping temperature distribution
   b. Thermal death time studies
   c. Inoculated pack or microbial challenge studies

3. *Mathematical modeling,* which is defined as mathematically integrating scientific data on how factors affecting the performance of a control measure or combination of control measures affect their ability to achieve the intended food safety outcome. An example would include pathogen growth

models that assess the impact of changes in pH and water activity on the control of pathogen growth. Mathematical modeling requires that a model be appropriately validated for a specific food application.

Results of the validation process demonstrate that a control measure or a combination of control measures is either:

- Capable of controlling the hazard to the specific outcome if properly implemented, and thus should be implemented, or
- Not capable of controlling the hazard to the specified outcome and should not be implemented

INITIAL VALIDATION

As stated previously, validation is performed initially when the food safety program is being developed. The purpose, at this point in the process, is to ensure that the food safety program and, more specifically, the identified CCPs are valid for controlling food safety hazards and functioning as intended.

This initial evaluation typically includes, at a minimum, the following:

- Review of the hazard analysis by the food safety team to confirm that all significant hazards are identified
- Confirmation that control measures specified are appropriate to control the specific hazards
- Review confirming that CCPs, CLs, monitoring activities, record keeping, and corrective actions are effectively controlling the hazards

REVALIDATION

Revalidation occurs at a defined frequency (i.e., annually) or when any changes occur (Chapter 15 "Change Management"). The frequency for reevaluation may depend on the standard of choice or if the organization falls under one of the USDA or FDA-regulated food segments (i.e., meat, poultry, seafood, or 100% juice). Reevaluation must also be performed when a change occurs that could affect the hazard analysis, CCPs, or the food safety program overall. Examples of process changes that could trigger revalidation may include new equipment, product formulation, raw materials, packaging, suppliers, employee practices, storage/distribution, and/or labeling. Other changes may include

- System failure such as multiple deviations from control limit, inadequate record keeping, recalls, withdrawals, consumer complaints
- New scientific or regulatory information such as new information concerning the safety of the product or an ingredient, product, or product category linked to a food-borne disease outbreak, and/or regulatory alerts related to the process or products

Change management (Chapter 15) provides more detailed information related to process changes.

The extent of required validation activities may relate directly to whether the food safety program has identified any CCPs or OPRPs (FSSC 22000, ISO 22000 only). Technically, validation is required for CCPs and Operational PRPs (OPRP). However, the food safety team must understand validation and its requirements and should apply the same technique and logic throughout the process to ensure that the *correct* activities are being performed to control existing or potential hazards.

ISO 22000:2005 8.2 defines that

> Prior to implementation of control measures for OPRPs or CCPs and after any change, the organization must validate that:
>
> a. the selected control measures are capable of achieving the intended control of the food safety hazard(s) for which they are designated;
> b. the control measures are effective and capable of, in combination, ensuring control of the identified food safety hazard(s) to obtain end products that meet the defined acceptable levels.

> If the result of the validation shows that one or both of the above elements cannot be confirmed, the control measure and/or combinations thereof must be modified and re-assessed. Modifications may include changes in control measures (i.e. process parameters, rigorousness and/or their combination) and/or change(s) in the raw materials, manufacturing technologies, and end product characteristics, methods of distribution and/or intended use of the end product.

The internal audit process is an excellent tool that is considered as part of the verification process. The internal audit process is addressed in more detail in Chapter 5. ISO 22000:2005 8.4.2 specifically addresses the requirements for the evaluation of individual verification results stating that

> The food safety team must systematically evaluate the individual results of planned verification. If verification does not demonstrate conformity with the planned arrangements, the organization must take action to achieve the conformity. Such action must include, but is not limited to, review of
>
> a. Existing procedures and communication channels;
> b. The conclusions of the hazard analysis, the established operational PRP(s) and CCP(s);
> c. The PRP(s);
> d. The effectiveness of human resource management;
> e. The effectiveness of human resource management and of training activities.

ISO 22000:2005 8.4.3 states that "the food safety team must analyze the results of verification activities, including the results of the internal audits and external audits. The analysis must be carried out in order to:

a. Confirm that the overall performance of the system meets the planned arrangements and the food safety management system requirements established by the organization;
b. Identify the need for updating or improving the food safety management system;

c. Identify trends which indicate a higher incidence of potentially unsafe products;
d. Establish information for planning of the internal audit program concerning the status and importance of areas to be audited, and;
e. Provide evidence that any corrections and corrective actions that have been taken are effective.

The results of the analysis and the resulting activities must be recorded and must be reported to top management as input to the management review. It must also be used as an input for updating the food safety management system."

The food safety team/top management must be able to show evidence (records, interviews, etc.) that improvements are monitored and that changes to the system are evaluated. Examples of some of the processes and records that may provide this evidence include, but are not limited to:

- Management commitment and management responsibility
- Management review meetings
- Food safety team meetings
- Continuous improvement program
- Internal audits
- Hazard analysis
- Change management

The most important aspect related to verification and validation is that the organization's management and team associates understand the requirements for food safety and ensure that a continued focus provides the necessary proactive tools that ensure that a safe product is produced every day.

KEY POINTS

a. Site must determine, implement, and document verification activities. Verification must be evaluated to confirm effectiveness.
b. Objective evidence must be available to confirm that if verification does not demonstrate conformity with the planned arrangements that action is taken according to the organization's defined requirements.
c. There must be objective evidence of verification planning for input to the hazard analysis.
d. Verification results must be recorded and communicated to the food safety team.
e. Records must be maintained to demonstrate that verification planning is being performed.
f. Effectiveness of verification activities must be confirmed.
g. Evaluation and analysis to confirm verification and analysis of PRPs may be performed through GMP inspections, internal audits, food safety program review, and management reviews.
h. Analysis of the results of verification activities must be recorded with actions applied to ensure effectiveness.

i. Records must be maintained in compliance with the procedure for record control (Chapter 7).

j. Validation must be performed for CCPs and OPRPs (ISO 22000, FSSC 22000 only) and records maintained to confirm compliance (Chapter 7).

k. Management must ensure that associates are trained to understand the difference between verification and validation. Records for associates performing validation activities must clearly define the parameters and justifications for the related decisions.

l. The food safety program must include either actual information or link to the location of the scientific, historical, or in-house data used to validate control points.

m. The food safety team must ensure that the selected control measures are capable of achieving the intended control of the food safety hazard(s) for which they are designated.

n. Verification is actually verifying that activities are being completed as defined. Verification is designed to check the validation process.

o. Validation is confirming that the food safety program, as written, effectively controls the hazards as established by the team. Validation is part of verification.

COMMON FINDINGS/NONCONFORMANCES

a. The organization did not have a formalized process to validate control measures and/or control measure combinations.

b. The organization did not have a formalized program for verifying and improving the food safety management system. Site must determine, implement, and document verification activities. Verification must be evaluated to confirm effectiveness.

c. Objective evidence must be available to confirm that if verification did not demonstrate conformity with the planned arrangements that action was taken according to the organization's defined requirements.

d. Evaluation and analysis to confirm verification of PRPs through GMP inspections, internal audits, food safety program review, and management reviews were not available.

e. It could not be confirmed, through review of records and interviews, that an effective program for change management had been defined, implemented, and maintained as related to food safety.

f. The food safety team leader presented *validation* records for the CCP metal detector; however, data only provided represented verification that the equipment was operating.

g. Records were not available to confirm that associates responsible for performing validation of the thermal units had been trained in defined competency requirements.

h. It was stated that the metal detector was validated semi-annually per corporate requirements; however, records were not available to confirm this.

i. Validation studies related to the justification and control of related pathogens controlled through pasteurization were not available.

j. Documented procedures for PRP verification stated that results of verification activities would be reported and reviewed at the management review meetings; however, minutes from the last two meetings did not confirm review of these items.

k. The hazard analysis stated that the potential physical hazard for metal fragments was controlled by the PRP programs for managing the magnets, metal detectors, and screens; however, records were not available to confirm that this PRP had been verified for effectiveness.

l. The organization had a robust change management program; however, records were not available to confirm that this program included the evaluation of change related to validation of the control limits for the CCPs.

m. Records were not available to confirm that the organization had implemented a formalized program for verification.

n. Records were not available to confirm that the food safety team had formally and systematically evaluated the individual results of planned verification for PRPs.

o. Records were not available to confirm that verification activities, including those for PRPs, were in compliance.

p. Evidence was not available confirming that management had implemented a PRP verification planning process that defined the purpose, methods, frequencies, and responsibilities of the verification activities related to the ISO 22000:2005 7.2.3. (a) → (k) and ISO 22002-1:2009.

q. A process for verification planning in compliance with the requirements of the FSMS had not been defined and implemented.

19 A Case for Integrating Your Management Systems*

There are basically only two reasons to establish and maintain a formal management system: (1) earn an ISO certificate to hang in your lobby to (hopefully) keep customers off your back, or (2) you really do want to improve your business operations. If you choose reason number one, then the words of that revered quality guru, Mr. T, comes to mind—"pity the fool!"

In corporate America today, *quality* is an unfortunate word. Unfortunate in the sense that in many industrial and commercial sectors, including the food and beverage sector, the term *quality* still translates narrowly to *quality control* run by a group of specialized (yes, and somewhat geeky) folks flitting around the production lines with thick eyeglasses, micrometers, and statistical tables in hand. These represent a group of employees that top management secretly wishes it could do without. "Dang overhead!" Again, Mr. T's hallowed words come to mind.

Not that quality professionals aren't sometimes our own worst enemies. We absolutely are. Many among us were no doubt drawn to quality as a profession as a direct outgrowth of our love and competency in science, technology, engineering, and math. And micrometers and statistical tables absolutely have a front-and-center role in establishing and maintaining a viable quality system. The profession's problem is that far too many of us are enamored with, and are the face of, the tools of the trade instead of the raison d'être—a successful organization. If that organization is a for-profit business, then at least a part of its reason for being is to pay all the bills and have funds to distribute (profit). Even an outwardly socially conscious firm like Ben and Jerry's can't give to its favorite cause what it doesn't have in the bank.

Many corporate managers wish their staff of quality professionals had more of an understanding of what drives business success—and the ability and focus to elevate their individual and collective contributions toward that success. To help *lead* the organization in that sense.

There are no magic bullets. But quality professionals can absolutely find ways to impact their organizations beyond their current (and perhaps too traditional) roles. I relish the following excerpt from a classified ad for a quality assurance manager at

---

* Contributed by Dennis Sasseville, an independent consultant, and third party auditor, president of Cobalt Advisors, Bedford, NH.

a prominent and progressive food processing firm that appeared within the last year. The ad read in part:

> Participate in a World Class Manufacturing program and work toward continuous process improvement in all aspects of Plant Quality ... provide coaching constantly seeking improvement through data gathering, problem solving, and initiation of appropriate actions. Partner with the manufacturing team by providing relevant data and analysis to identify key issues and the identification of solutions. Communicate with departments, shifts, maintenance, production, and engineering on conditions which require attention or decisions. Work with these departments to determine root cause for specific quality concerns and lead implementation of solutions.

The company that placed this ad seeks an individual to help promote a desired approach (or better, *framework*) that goes beyond the traditional role for the quality professional. Without using the term explicitly, the company seeks an *integrator*—someone to help knit the disparate business components of the organization together. Essentially, someone to help foster success.

I don't know if the company's top management found the candidate they were looking for, but I applaud them for knowing where they need to head and recognizing the type of talent that would aid in reaching that destination, an integrator with a melding of both technical and people skills—representing the face of quality in today's rapidly evolving organization.

WORKING TOWARD AN INTEGRATED MANAGEMENT SYSTEM

Over the past four decades, we've taken an ISO system-type focus on the various facets of our organizations: production and products (ISO 9001, ISO 22000, HACCP), environmental and safety matters (ISO 14001 and OHSAS 18001), and even more recently, energy management (ISO 50001). These standards, and more, are typically lumped under a general quality-based *plan–do–check–act* approach, and the organization is usually quite cognizant of its commonalities. But of course, recognition by a certification body comes separately, standard-by-standard, and typically sequentially because, of course, there is no integrated ISO standard! Even though ISO and its committees look to promote and help facilitate system integration, the actual integration effort is, of course, entirely up to you, the individual companies.

The reality is that the individual stand-alone standards *can* be integrated together if an organization chooses to pursue that approach. Also, keep in mind, that the goal of most forward-thinking companies is not just a mechanized style of integration for the sake of declaring "we've arrived." The desired goal for progressive firms is an integrated *business* system—a framework that solidly supports *all* organizational goals. The age-old adage that the whole is much more than just the sum of the parts applies here. You wouldn't see some 12,263 new book titles (2011 www.bowker.com) on business, published annually in the United States alone if business wasn't still obsessed with improvement and optimization.

So the challenge of integrating systems is more than a bit daunting, even for those who can see where they want to head (i.e., *have the vision*). All in all, it is a little like driving the narrow streets of downtown Boston—called "the Hub" for good reason.

It may be better termed a spaghetti bowl when you examine its throughways first-hand. Between the tall buildings, one can see the desired destination ("Ah yes, the John Hancock Building, right there!") but just try to get from where you are at the moment to where you desire to be. One-way streets cascade into more one-way streets, triple parked cars imitate D-Day coastal barriers, and drivers seem to be purposely aiming their vehicles at you as in some perverse game of chicken! Likewise, management system integration is not for the faint of heart.

Can a company paint-by-the-numbers in instituting an integration of its various management system components and standards? Certainly, and the results will produce *some* value in terms of organizational efficiency, common language for monitoring and measurement, and a framework for setting integrated goals.

Many confuse systems with plans, programs, or tools. The latter support a system, but are not systems in and of themselves. Plans and programs are usually created for a specific goal or objective and have an associated finite time span. Systems have continuity and longevity. Tools are just that, tools. Different tools serve different and useful purposes—they help build a system, support a system to be sure, but are not the sum total of the system.

At this juncture, we could slip into a discussion of the mechanisms that can be used when starting a by-the-book system integration initiative: the tools, the comparative tables, and the cross-reference matrix of common elements for each ISO standard. But those tools are already out there and readily available to the reader. Rather than rehash and replicate those tables here, let's look at just two aspects of management system integration that will determine success or failure: the person and the plan.

PERSON

"In our early years we operated more intuitively—'What seems to need to be worked on?' We relied mostly on 'tribal knowledge.'" These are the words of William, a mid-level manager in a food and beverage processing company. Several years ago, William was challenged by top management to help improve the firm's business position and operational efficiencies. The focus was both on certification to individual ISO standards and on initiatives to integrated systems internally. At that point in time, William was not an experienced *ISO guy*. In fact, he had relatively little formal exposure to ISO or any other standards up to that point. But he had worked with diverse global suppliers in representing his company's initiatives at developing a socially responsible supply chain. He was used to melding aspirational stretch goals, project implementation details, and working with a wide range of people and personalities. He was a great choice as an *integrator-in-chief.*

Xanterra Parks and Resorts is a privately held premier owner and manager of hospitality properties, especially those located in National and State Parks. The firm with a twenty-first-century name really has a 130-year-old hospitality legacy as it is the successor to the venerable and legendary Fred Harvey Company.

Xanterra's current business encompasses some 7400 employees managing 21 properties, mostly in the western United States, with some Eastern holdings as well. Xanterra's operations include lodging (at Yellowstone National Park and the Grand

Canyon, for instance), restaurants, retail stores, marinas, golf courses, camping and RV sites, and even a functional steam train. In 2010, the firm handled over 1.9 million guests overnights.

Many of the company's operations are not only situated on public lands, but in some of the jewel national parks. It must compete against other firms, large and small for the ability to be a National Park Service concessioner. Business efficiency, delivering stellar customer service, and treading with a light environmental footprint wherever it operates are crucial to Xanterra's business success.

> Xanterra's mission is to be the industry leader in park and resort hospitality. We are committed to practicing integrity and quality, maintaining positive relationships with our employees and clients, leading in environmental stewardship and creating unforgettable memories for our guests.

How does Xanterra fulfill its commitments and foster business success? The company developed a robust management system approach to addressing all of its environmental regulatory and sustainability commitments, and it strives to integrate those systems into its large business management systems. The environmental and sustainability components are branded as *Ecologix* and certified to the ISO 14001 standard. For most of the past 10 years, the system and its business-like approach were in the hands of a dedicated manager, Chris Lane, who was enthusiastically supported by top management. Chris was the vigilant *integrator* and made Xanterra's management systems approach highly visible to the National Park Service, vendors and suppliers, its employees, and customers. In fact, Chris always placed a great emphasis on employees as system integrators and facilitators of Xanterra's business philosophy and approach.

I can personally attest to Xanterra's system integration success. Just recently while on vacation in the West, I was at the front desk of a Xanterra lodge in one of our large national parks waiting for the next available desk agent. It was there that I overheard one of the lodge's desk attendants enthusiastically explaining to one of the other lodge guests about their *Ecologix* management system and what they were trying to accomplish, an example of business system integration that touches all the way to the end user—the customer.

You may be the product of a technical background like engineering or quality control or, like William or Chris Lane, come from an entirely different field. Either way, through your outlook and chosen approach, you can step into the role of championing an integrated approach to your organization's management systems.

PLAN

No two organizations are exactly alike, of course. In spite of the myriad of business books published each year and the how-to guides, concepts as basic to business as strategic planning, goal setting, implementation of action plans will look different among two competitors within the same industry making a similar product. So there is no single formula of what is the best approach to a management system integration initiative. I simply offer the observation that without an actionable plan that is blessed by top management, partial success is likely to be the best that can be hoped for, and that's being kind.

I am a proponent of organizations about to embark on a new or particularly challenging venture, developing a formal project charter. This seems an especially worthwhile approach where the buy-in from multiple parties (senior management, middle management, production supervisors, support staffs, and rank-and-file employees) will make or break the initiative. When is cooperation and buy-in not an important factor in the success equation?

A prominent and rapidly growing company in the food and beverage sector recently used a project charter approach when it desired to take its disparate management system components—ISO 22000 certification, organic product certification, and a somewhat effective home-grown Site Continuous Improvement system—and build it into a workable integrated management framework. For good measure, the company implemented an ISO 14001-type environmental management system to be eventually supplemented with an OHSAS 18001 health and safety management system.

Are they there yet? No, but they are constantly fighting fires, distractions common to virtually all fast-growth companies. They have the quality tools, the champion with team support, and the project charter in place. They are well on their way.

EXAMPLE OF A PROJECT CHARTER

Effective date: October 22, 2012.

Project: Certify to ISO 14001, while integrating with existing food safety and quality systems with our Site Continuous Improvement procedures to provide an effective business systems approach that enables the achievement of current and future corporate goals.

Business need: ISO 14001 certification is now required by two of our major customers and by a third potential large customer. Our two current customers have indicated an expectation that we will achieve such certification by the end of calendar year 2013. Questions relating to our certification status and/or plans for certification have been received as part of the request for quotation from our potential customer. Certification to ISO 14001 is thus seen as a business requirement.

In addition, current energy costs have risen significantly over the last 2 years and are forecast to increase by another 25% by the end of CY 2013. Waste generation for both 2010 and 2011 increased by over 15% each year. As such, development of an integrated management system that focuses on controlling and reducing the costs associated with these expenses is seen as an operational necessity impacting on our ability to execute our strategy of being the low-cost producer within our industry.

Finally, environmental stewardship is predicted to become a greater decision point in the general market as concerns with global warming and environmental degradation continue. ISO 14001 certification is thus seen as a useful marketing avenue to environmentally conscious consumers.

Expected project outcomes: Deployment and certification of an ISO 14001 environmental management system that contains and/or reduces our environmental costs while integrating fully with our food safety and quality systems and our Site Continuous Improvement program.

Cost and schedule: The project must be completed by December 31, 2013. The budget authorized for this project is $60,000, not including internal personnel time (detailed analysis presented in Appendix A).

Personnel and support: Chuck Bennefitt has been assigned as project manager and has been authorized to dedicate up to 50% of his time to this project through December 2013. Each department will provide support as needed and will assign a department lead to work with Chuck on this project. The primary project sponsor is John Stamford, president and CEO.

John Q. Stamford, President and CEO

Of course, a project charter could be much more lengthy and complex than this example. But the presented example is only a slightly modified version (with changed particulars) of an actual charter that guides a fairly involved initiative. Documentation complexity isn't always warranted, or desired. The actual integration efforts are complex and involved enough—expend your efforts there by following this project charter with a project action plan including a schedule. This is identical to how a company normally approaches the development of a new production line, product initiation, or even a boiler replacement. Improving your business management systems is, no less, an effort.

ACKNOWLEDGMENTS

Mr. Sasseville, president of Cobalt Advisors, Bedford, NH and also a senior associate with Normandeau Associates, Inc., with 35 years of experience as an environmental management consultant. His consulting and training assignments have taken him from "Millinocket to the Mohave." He has extensive experience with Environmental Management Systems (EMS), Environmental Health & Safety (EH&S) performance improvements, Quality Management Systems, and sustainability services. Previously, Dennis was a senior manager in the ISO Management Systems practice and the Environmental Advisory Practice of the global consulting firm, KPMG, LLP. While at KPMG, he served as a member of the firm's 26-country International Sustainability Network (ISN) and as an advisor to the firm's U.S. Integrity Management Practice. As a lead auditor in the ISO Registration practice of KPMG, he registered over 50 facilities to the ISO 14001 standard. Dennis is an RABQSA EMS Lead Auditor, a certified hazardous materials manager, a member of the Auditing Roundtable serving on the association's Conflict Minerals interest group, and an author of several John Wiley and Sons' books on environmental management. Dennis speaks frequently at conferences and workshops. (dsasseville@ normandeau.com).

20 Example of Integrating ISO 22000 Requirements into a Compliant Quality Management System*

"This Standard is too generic; I need a more prescriptive list of requirements." We hear complaints like this often, and most of the time, the sources of these are totally understandable. If a document is too generic, then it can lead to misinterpretation and/or misunderstandings. It is important to take a step back and look at this situation from a different perspective. A management system standard should be generic enough to prevent sector-specific variances. If we try to forget (only for a while) the requirements that we need to implement and try to focus on the expected results, things become clearer and the requirements make more sense. The scope of ISO 22000:2005 clearly specifies: "… the requirements for a Food Safety Management System where an organization in the food chain needs to demonstrate its ability to control food safety hazards in order to ensure that food is safe at the time of human consumption." It is likely that the organization has several activities and procedures in place that are already compliant with many of the ISO 22000:2005 requirements. Having a compliant HACCP program may be the first step. That said, it is critical to keep in mind that implementing HACCP is not done just by having a 4 inch binder full of information.

It was mid-2006, during a HACCP seminar, when I first met this quality assurance manager. She was the QA manager of a well-known glass bottle manufacturer. We exchanged business cards but shortly lost contact. A few years later, I was visiting a different client in the same city and decided to contact her and visit the facility.

The facility is a fairly large site with some very important clients, mainly breweries and soft drink bottling companies. The large display case in the reception area showcased several bottles from its very important recognized international brands. The QA manager told me that they had frequent customer audits.

That facility had been registered as ISO 9001 for several years. The audit's results were good, but two or three important customers were demanding that they implement a compliant HACCP program. "I'm feeling totally lost and disoriented," the QA manager said. She went on to state that "I took a HACCP training session that was mainly a microbiology course and we recently hired a consultant that wanted to start teaching us how to clean and disinfect the lines with water and cleaning

* Contributed by Erasmo Salazar, Food Safety Lead Assessor, and Food Industry Technical Advisor for Lloyds Register Quality Assurance (LRQA) North America, Houston, TX.

agents. We cannot use water in a glass furnace or the glass forming equipment!"
(Sometimes, some consultants have only one tool in their toolbox [a hammer], and
for them, every problem or opportunity looks like a nail).

I started working as a consultant with this organization a few months after that
meeting, beginning our implementation project with a 2 day training session integrating their current system with ISO 22000:2005. This initial training included a
training session with 15 associates (the future food safety team [FST], several mid-
dle-management personnel, and some already qualified ISO 9001 internal auditors)
who had no previous experience on food safety. Frequently, a consultant can estimate
the duration of the course depending on the audience. If the attendees are not familiar
with the subject, there is a high possibility of having low participation with limited
questions and, most of the times, the session is awkwardly short. I was expecting a
situation like that, but it turned out to be the opposite! During the training session, the
attendees started identifying the elements in their current quality management system
that complied or partially complied with the ISO 22000:2005 standard. This interaction also included suggestions and/or changes to integrate documentation.

During the explanation of *hazard identification*, it was remarkable how fast the
food safety hazards were identified. It was simple for them since they already had
a list of what they called *critical defects*. Those defects that could represent a food
safety hazard (internal loose glass, false bottom, bird swing, etc.) had been previously considered during quality planning. Also, the activities relating to control measures were already included in the quality plan. This experience was great! I was in

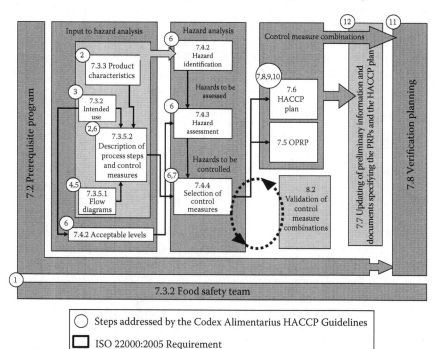

FIGURE 20.1 Integration of food safety requirements.

front of an unbiased group with no previous knowledge of food safety, but a significant knowledge of the management system structure and format.

The concept of *integration* instead of *implementation* was understood in the first couple of hours. The quality manual that is mandatory for ISO 9001:2008 4.2.2 was used as the main framework for their quality management system. Specification PAS 99:2006 (Specification of Common Management System Requirements as a Framework for Integration) was also used as a relevant reference for the integration process.

To explain with more detail how the basic 12 steps of the HACCP methodology were integrated into the *planning and realization* element of their quality management system, let's use Figure 20.1 adapted from the first ISO 22000:2005 draft (ISO/DIS 22000:2004 "Food Safety Management Systems—Requirements for any organization throughout the food chain" [Draft International Standard] [2004-06-03]).

This figure represents the 12-step HACCP methodology along with the basic HACCP prerequisite programs (PRPs). It also includes references to ISO 22000:2005 elements.

PRESTEP 1: FOOD SAFETY TEAM AND THE PREREQUISITE PROGRAMS

Many of the basic HACCP PRPs (ISO 22000:2005 7.2) were already defined and implemented. A document identifying the required basic work environment conditions had also been defined and implemented. Other already defined PRP programs included those related to infrastructure such as good manufacturing practices, warehouse storage practices, employee behavior, sanitation, and pest control. In addition, the PRP verification methods were already defined and actually met the requirements of ISO 22000:2005 7.8.

The FST (ISO 22000:2005 7.3.2) not previously in existence was appointed the first week of the project and included experienced associates from various departments. All FST members received specific training on the requirements of ISO 22000:2005 and their role in compliance.

PRESTEPS 2–5: THE PRELIMINARY STEPS TO ENABLE THE HAZARD ANALYSIS

Raw material specifications previously identified were used to document the description of ingredients, raw materials, processing aids, and food contact surfaces. The FST completed a gap analysis comparing their specifications with the requirements of ISO 22000:2005 7.3.3.1 (raw materials, ingredients, and product contact materials). The FST identified only a few necessary changes to complete the integration with ISO 22000:2005 requirements.

The same process was used to define the description of the end products. This resulted in very detailed end-product specifications that included drawings, pictures, and defined requirements.

All required details for *intended use* were included in the end-product specifications.

The process description that includes the process steps and control measures (ISO 22000:2005 7.3.5.2) were integrated with the existing flow diagrams created as part of the quality management system. References to work instructions and procedures were included in the flow diagrams to better illustrate and describe process steps and control measures. Additional details as appropriate for the operation such as rework, waste removal, and outsourced activities were also included in the revised flow diagrams.

STEP 6: THE HAZARD ANALYSIS

One important element that was not considered in their quality management system was the hazard assessment (evaluating the identified hazards according to the possible severity of adverse health effect and the likelihood of occurrence). ISO 9001:2008 does not have general requirements to identify and assess hazards, and for that reason, that part was not implemented. The methodology was defined. The second FST meeting included a multidisciplinary FST providing an excellent foundation for identifying and evaluating the hazards and potential hazards. It was incredible to watch this team come together with *no* major discussions or disagreements being generated.

Being a nonfood organization, it was a unique opportunity to use the decision tree proposed in ISO/TS 22004:2005 (Food Safety Management Systems—Guidance in the Application of ISO 22000). The use of the Codex Alimentarius decision tree to select the critical control points (CCPs) is most common in food manufacturing, but with the ISO/TS 22004:2005 decision tree, instead of analyzing *process step by process step*, the analysis team decided to first define the protocols for the validation for the control measures (either CCPs or operational PRPs). The method was defined and documented. A generic form was designed and included in the management system. ISO 9001:2008 7.5.2 (Validation of Processes for Production and Service Provision) was no longer an ISO 9001:2008 exclusion.

STEPS 7–10: CCPS, CRITICAL LIMITS, MONITORING SYSTEM, CORRECTIONS, CORRECTIVE ACTIONS

A second decision tree was utilized for the classification of the control measures. Instead of having the identified hazards as inputs for the first decision tree, the inputs for the second decision tree were the control measures. As a result of this, the FST identified one control measure as a CCP, which was the electronic bottle inspecting machine that detects and rejects defects. The required monitoring activities for that equipment had already been defined and documented in the quality plan. The quality plan defined the responsibilities, frequency, responses, associated documentation, and records. This was basically in compliance with ISO 22000:2005 7.6 for the HACCP plan. A *quality plan* as defined in the ISO 9000:2008 standard (3.7.5) "is a document that is used to specify the procedures and resources that will be needed to carry out a project, perform a process, realize a product, or manage a contract. Quality plans also specify who will do what and when." The differences are then minimal!

Another remarkable surprise was the fact that the food safety team (FST) had no problem with differentiating between *corrections* and *corrective actions*, which is a very common problem in the food sector. The current procedure for *corrective action and preventive action* had minimal changes. The procedure for nonconforming material and product was modified to integrate the requirements of ISO 22000:2005 7.10.1 for *corrections*, which is a mandatory procedure for ISO 22000:2005.

The FST asked: "Do we need to have a different document for the HACCP plan or can we integrate the requirements of clause 7.6 into our Quality Plan?" Actually, the FST answered themselves within 20 seconds, quickly identifying the fact that there is no requirement to have two separate documents.

STEPS 11 AND 12: VERIFICATION, PLANNING AND DOCUMENTATION, UPDATING OF PRELIMINARY INFORMATION

The FST identified that there would be a new document required for the verification plan (ISO 22000:2005 7.8). At first, there was also confusion between the terms *monitoring* and *verification* (remarkably, the term *validation* was never an issue). The FST created a table that defined the related methods, frequency, responsibilities, and records. In many instances, references to existing work instructions or procedures were all that was necessary.

Updating of preliminary information in compliance with ISO 22000:2005 7.7 was direct and simple. Only a few changes were necessary for the process descriptions and flow diagrams. A control measure that was reclassified as a PRP resulted in a change in the verification plan.

The document control and record control processes were already implemented in compliance with ISO 9001:2008 4.2.3 and 4.2.4 (control of documents/control of records). These documents required only minimal changes to ensure compliance with ISO 22000:2005 4.2.2 and 4.2.3.

CORE ELEMENTS AND INTEGRATION OF OTHER REQUIREMENTS

Integration of the HACCP methodology is not enough to be totally compliant with ISO 22000:2005. The quality policy was modified to include the organization's commitment to food safety, and the *measurable* quality objectives were modified to include food safety–related measurable objectives. External communication had been poorly addressed in the beginning of the process, but was upgraded in a more robust and appropriate manner. Minor revisions were necessary with internal communications. The FST along with top management also had to define, integrate where possible, and implement requirements related to elements for the "analysis of results of verification activities" (ISO 22000:2005 8.4.3), "updating of the FSMS" (ISO 22000:2005 8.5.2), and "analysis of data" (ISO 9001:2008 8.4).

The organization already had procedures for recalls, withdrawals, traceability, and the performance of mock recalls that were all required by some of their customers. The required activities required only some modification, which was completed, implemented, and confirmed effective by the FST.

The organization had an extensive procedure for crisis management, which had to be revised to ensure compliance with ISO 22000:2005 5.7: "Emergency preparedness and response." This involved assurance that defined requirements addressed actions necessary after a crisis to confirm and protect the safety of the products.

The first internal audit was a totally integrated QMS/FSMS audit with the use of already qualified internal auditors. The organization ensured independency by having the unit planning element (the hazard analysis) audited by a person not on the FST.

The mandatory inputs to the management review were identified by the QA manager, having been assigned as the FST leader. A revised management review agenda was designed and implemented to include the additional food safety–focused inputs and outputs for the meeting function. Although many were already being addressed, the revised agenda highlighted these to ensure effective focus by the top management team during the meetings.

At this point, the organization was ready to begin the stage one audit and very close, if not already totally compliant, to the requirements of ISO 22000:2005. Remember, integrating the initial step, is critical, no matter which standard is chosen.

ACKNOWLEDGMENTS

Erasmo has 16 years of experience with ISO standards and more specifically the world of food safety as both a consultant and an auditor. Erasmo is currently a food safety lead assessor and food industry technical adviser for a leading ISO registrar.

During our many discussions, the topic of effective integration as a result of companies performing an in-depth GAP analysis prior to starting an implementation process has led to many stories of companies reinventing their system to become ISO certified. The old saying that "hindsight is 20/20" holds true. So many folks tell us after the fact that they spent time doing things that they really didn't need to do or made changes that were not necessary. In Erasmo's example outlined in this chapter, he shares a true story that occurred while consulting for a large glass bottle manufacturing company. Take heed from this and remember—Do not reengineer but integrate.

The importance of performing an effective gap analysis can never be understated. I use the example of setting a broken leg. The doctor cannot set the leg until an x-ray is taken. The same is true of your management system. The gap analysis is the x-ray to know exactly what exists, partially exists, or doesn't exist. From this information, the system is built integrating the requirements into the organization's current processes, adding and changing where necessary, and taking advantage of current processes that are already compliant or close to it.

Mr. Salazar provided an excellent example of the importance of doing a *GAP analysis* and *integrating* the requirements of the new standard wherever possible. Many companies do not take this approach, and many times end up doing much more work than necessary and spending a lot of money either through the redesign or having to redo activities. One of the most critical and effective aspects of developing a system is to know the standard you are implementing, first evaluate what you already have and wherever possible integrate. The management system standards are purposely generic in nature so that the requirements can be applied and integrated into various food sectors. Remember "from farm to fork."

21 Summary, Conclusions, and Next Steps

As the information contained in this text comes together in this conclusion, it is exciting but also difficult to look into the future. In the recent past, a large segment of the food industry did not see the value of a structured management system. In today's world customers and consumers are demanding a safe food supply compared to possibly in the past *assuming* we had a safe food supply.

Whether an organization moves forward to achieve certification, becomes compliant but does not *take the final exam*, or chooses to operate as always, without the management system, production of food that is safe (free of hazards) is a must.

As an organization and its team review the different management standards, it must be stressed one more time that the emphasis must focus on the organization and what is best for the specific operation. What works for one company may not be the best choice for another. It is equally as important that management from the top learns the standard and takes an active role in the journey through system development and implementation. Before drawing specific conclusions on the next steps and best approaches to decision-making choices, let's review more information from our industry expert surveys. Remember that this includes a sampling of individuals representing many companies and industry sectors; however, some were forced to remain anonymous due to company policies.

Yolanda Nader, the CFO and CEO of Dosal Tobacco Company, demonstrates support and commitment through her organization's strong focus on compliance and continuous improvement. Ms. Nader shares the following when asked about her organization's choice to develop and certify a compliant management system:

> The organization needed to develop a standardized program of processes in order to ensure a consistent quality product to deliver to the consumer as well as to be prepared to comply with a new regulatory regime being imposed on the industry. The process helped the organization self-evaluate its procedures; brought input from all levels of the organization forging more ownership and investment from personnel; and ultimately yielded standardized procedures which have resulted in improved efficiencies.

WHAT WERE SOME OF THE *DIFFICULT* CHALLENGES?

When asked about challenges, there are seldom any surprises: resources, document control, communication, understanding the requirements, sustainability, calibration, training, internal audit availability, timely completion of corrective actions, and more have been highlighted as *challenges*. Let's first review some of the difficult challenges identified during our information-gathering survey. Industry professionals were asked what the most difficult challenges were during the implementation stage. Samples of their responses follow:

> The main challenges were changing the status quo, organizing and prioritizing the implementation, and maintaining focus and commitment. In essences, it boils down to a commitment and support from Senior Management, which we have received and continue to receive. Our success and the progress made has been directly proportional to the commitment of our associates led by top management's example. (*Guido Abreu, Quality System Analyst for Dosal Tobacco Corporation*)

> Understanding the certification process with SQF was very confusing to us, with a first audit of only paperwork followed by corrective actions, and then another audit of paperwork and the facility implementation. It was also a challenge to understand the auditor and the relationship of the certifying body. The entire process of going through the audit, addressing findings, and then being awarded our certificate was very confusing. Implementation efforts are tough enough without having to figure out the registration process. (*Anonymous*)

> At first it was very difficult to get employees to buy-in; however, this was resolved through extensive training in the relevant standard, highlighting its benefits. A top down approach was necessary beginning with the executive management team. Having a senior VP directly involved in the training and implementation process was a major advantage to communicating the importance and value of the certification process. (*Andy Fowler, former VP Administration and Environmental Compliance, Bacardi & Company Limited*)

> One challenge was the allocation of time and resources; however, we quickly realized that they were time & resources well-spent. We had complete buy-in for the process; the facilities that initially went through certification first were motivated to expedite training & implementation for the remaining facilities. Through continuous improvement and enhanced problem-solving skills, our strategic approach to issue resolution is more structured, consistent, and business relevant. (*Mike Burness, Vice President, Global Quality and Food Safety, Chiquita Brands International, Inc.*)

> The most challenging opportunity was developing a comprehensive HACCP program that complied with both our state food safety regulations and those of the food safety management standard. Due to the size of our factory, which is one of the largest and most complex in our business unit, and putting together all of the prerequisite programs to support a solid food safety plan to include OPRPs and CCPs as well as training, documentation control, record control and ensuring continuous improvement were very challenging. There was a significant amount of time invested in assuring our factory success; however, it has been well worth it. (*Anonymous*)

> One of the challenges was to delineate each of the procedures and complete the instructions, middle management did not embrace the process and it required significant involvement from top management to carry the message that this would be accomplished. Another was to help the employees comprehend why we were doing this and the need to follow the procedures/instructions, many have been with the company over 10 or 15 years and were accustomed to doing things in their own way, and lastly the time and resources required for putting together the procedures and the training of the personnel while still meeting production demands and everyday operational requirements. (*Yolanda Nader, the CFO and CEO of Dosal Tobacco Company*)

Our toughest challenge was listening to many different ideas on what directions that we should go, and then we were told that we were ready for an audit. We had a certification audit that turned into a pre-assessment because we were not ready. The decision to hire an outside company to help direct us should have been our 1st decision, would have saved us money in long run by moving in right direction instead of going round and round like we were doing for almost a year. We hired an excellent consulting team who guided us onto the right track. It was a hard process, but by working together and doing our own programs, we really learned the standards and how they work with what we do every day in our jobs. Having the structured system provided us an excellent foundation for continuous improvement overall; once we had the system established, we saw our progress every day. (*Russ Patty, an independent consultant and former quality manager of a beverage operation*)

Preparing the documentation was very difficult because we just did not have the required team resources. (*Anonymous*)

One of our biggest challenges was that we had to validate processes and activities which had been in existence for many years. Identifying applicable regulations and proving adherence to same. Establishing communication links along the supply chain. The food safety/HACCP team struggled with the new concept of an Operational PRP and whether this would apply to our processes. (*Anonymous*)

One of our toughest challenges was document control. At first, we gave it a great attempt to manage manually but it was just overwhelming. We purchased document control software, Master Control, and once we were trained and able to use this program effectively, it allowed us to organize and track documents so that we know we are using the most current version/revision. We also received different interpretations on what exactly was required by the standards which we resolved by ensuring that we based our programs and procedures on what we actually did. Also, resource allocation was a huge challenge with the implementation task taking a lot more time than we had originally planned. (*Joe Hembd, Management Systems Manager, Chesterman Company*)

Our first challenge was the *fear of the unknown*. One of the drawbacks of using a less prescriptive standard like FSSC 22000 is that we were open to wonder if the actions or mitigations taken were enough to satisfy the standard, or even more importantly, keep the consumer safe. The second significant challenge was the required *culture change*. Many of our employees were of the firm opinion that "we had always done it that way so why should we change". This was a mindset that was, obviously, deemed unacceptable and was (and to some small extent still is) very hard to change. One way that we conquered this was by breaking another cultural *mantra*, which was that of managers and supervisors who had the opinion, "the employees will never change". Once top management saw the value of system implementation, then these thought patterns were identified as unacceptable because without support from all areas and levels of the organization, the change must be embraced in order for the systems to work. The third challenge was getting total buy-in from the top down. We found that once this was accomplished, that system implementation and sustainability became much more realistic and attainable. (*Eric Putnam, Food Safety, Quality & Training Systems Manager, Wixon, Inc.*)

Time and resources. (Times 3!) As Operations Manager, I would not necessarily be called upon to write the program, but as it turned out, with over 25 years of previous Quality Practitioner experience, including the implementation of ISO at multiple facilities, I was most qualified for that task. Already working 50+ hours a week, this just added to the time spent *working*, both at home, as well as in the office. The timetable for achieving certification was aggressive and this required significant project management, over and above what is normally required. This was handled by delegating additional duties to my immediate reports throughout the duration of the project. (*Alan Lane, former packaging manager, now an independent consultant*)

Our toughest challenge was learning and understanding the standard. Our implementation team spent several months reading the standard and performing a gap analysis in an attempt to learn by doing. However, once we engaged the assistance of an experienced consultant who was able to explain and provide examples of what the standard expected the sites to have, everything started to make sense. (*Anonymous*)

UNDERSTANDING THE STANDARDS: GAP ANALYSIS

Almost every aspect of management system development, implementation, and maintenance relates to effective use of resources in today's world of lean operations. Operations, both large and small, are just trying to survive in an unfriendly economy that has a direct impact on resource allocation.

Several quoted industry experts did mention that their initial understanding of the requirements of the standards of choice and the team's basic misinterpretation of the standards were a challenge. Actually, those thoughts are what prompted this book. Although one never has all of the answers, the key is to know where to find them. Sharing experiences and best practices based on experience is very helpful and makes the journey more effective. It is very important and cost effective in the long run to engage the assistance of a knowledgeable industry expert to assist the implementation team in training, identifying, and closing any gaps an organization may have in meeting the food safety management standard of choice. Once the implementation team is established, the most effective next step must be the performance of a gap analysis. It is recommended that this be done by an independent external consultant, or in a larger company that may have corporate resources, possibly by a trained (in the standard of choice) company associate who is external to the operation. In many instances, the organization does not make decisions on standards or time tables for completion until after the gap analysis. A favorite example is that a gap analysis is like taking an x-ray. It is absolutely foolish and ineffective to set a broken leg until an x-ray is taken. This is the same with a gap analysis. Evaluate the existing organization and its activities, identify the gaps, and get assistance in closing these gaps to achieve compliance. A gap analysis can be performed by the registrar of choice, but the auditor representing the registrar is very limited on how much assistance can be provided in closing the gaps. This is considered a conflict of interest since an auditor must not audit his/her own work. Good auditors may be able to adjust their answers so that it is more of a sharing of best practices seen in similar operations; however, this is a very tenuous line to walk, and auditors must be very careful never to communicate or do anything that could be interpreted as a conflict of interest. When choosing a consultant or auditor, be certain to check his or her experience. It is critical to have the assistance of someone with expert credentials and experience in the industry sector and standard of choice.

Let's discuss the resource issue. Based on experience, the absolute best and most effective process in addressing the availability resources issue is to empower the entire associate team. Many times, management has stated that they just did not have the resources to do what was needed, but some of these organizations had 200 plus associates. If every one of those associates spent 1 hour a week assisting the implementation teams in writing, reviewing, and implementing the process, positive progress would be achieved while the associates become part of the team, learning and enhancing individual knowledge of the system and its requirements. Documentation and other activities move forward more efficiently because more individuals who have firsthand knowledge of the processes are assisting the process and much more.

It is also very important that the top management team, along with the food safety team leader, SQF practitioner, and/or other identified system leaders, ensure that resources are appropriate for the certification project plan. Defining implementation as a *journey* is just that. If we plan a vacation or business trip, we plan our journey. Although I did miss a turn years ago on my journey from Boston to Gainesville, Florida, and ended up in Cape May, most of the time we would never embark on a trip without some sort of plan. Driving might require a map or a trip tik. Flying requires us to know where we are and where we are flying to. Developing the *journey* toward compliance is not any different. Those who started earlier and then wished they had known the standard better are like us arriving in a strange city with an address but no map or GPS. An industry professional with experience in your standard and your industry is like a good GPS. Enter the destination, plan the journey, empower the entire associate team, and start the journey 1 day at a time.

SURVIVING THE AUDIT?

We asked our experienced industry professionals for insight into the actual certification audit. Did they have concerns during the audit and were there any surprises that they had wished they knew prior to the audit?

> Time management was certainly something we needed to effectively manage. The certification audit requires several days, so adequate and appropriate resources must be available to effectively complete the process. (*Mike Burness, Vice President, Global Quality and Food Safety, Chiquita Brands International, Inc.*)

> The toughest challenge was coordinating all the different managers to interface with the auditors. Also, the fact we had 2 auditors that audited on divergent paths in our factory. This was difficult to manage. (*Anonymous*)

Tom Marchisello, Industry Development Director, Grocery Retail, GS1 US, has assisted numerous organizations with multiple sights in the development of food safety and management systems. In his role in the development and implementation of these programs, Tom shares the following:

> The most challenging and critical part of certification of an organization with multiple facilities is auditor consistency and interpretation of the application of the audit standard. This is most important for capital planning, cost projections and standardization of deployment of policies and practices through an organization's supply chain network.

> It was tough because initially we did not totally understand the requirements; we ended up going right up to the deadline for approval. If we had not made it, then our customer

would have found other sources of product. We did not want to, nor could we afford, to lose case production. (*Russ Patty, an independent consultant and former quality manager of a beverage operation*)

We had a language barrier with the auditor; because of his background, we also had what we felt were differences in interpretation of the standards. In hindsight, if we had had more time before our deadline, we would have delayed the audit until someone more familiar with our industry was available. (*Joe Hembd, Management Systems Manager, Chesterman Company*)

We found that we had an issue with metal detection validation which related to on-going compliance with having a HACCP program that met all of the required aspects according to the standard (hazard/risk assessments). We had had a HACCP program for a long time, but it was just not at the level required by the standard. This definitely was an advantage to our program overall, just had to focus on the understanding of the food safety/HACCP team, and also, the employees responsible for the PRP programs. (*Eric Putnam, Food Safety, Quality & Training Systems Manager, Wixon, Inc.*)

As we chose to utilize the *Preassessment Audit* approach, we really didn't have any tough challenges at audit time. The issues identified during the *Preassessment Audit* were handled effectively and efficiently, and when audit time came, it was more of a formality. I highly recommend the additional cost of this approach, as the value proposition more than justifies the outcome. (*Alan Lane, former packaging manager now an independent consultant*)

SO NOW WE'RE CERTIFIED: WHAT DO WE DO NOW? WHAT ARE OUR NEXT STEPS?

Next, a sampling of industry experts with management system experience were asked what the toughest challenges were to maintain certification and continuously improve their system. The range for certification duration was between systems recently certified and those which had been approved for five (5) plus years.

Achieving the certification is an all out, (hopefully) management supported attack; maintaining it requires a programmed approach to auditing, identifying compliance, and "opportunities for improvement" issues, following up with appropriate action items, including the assignment of ownership for those actions, and then follow up. Having a strong, well facilitated audit team, with someone specifically assigned to own that activity, is critical to continued compliance and the continued improvement of the system. (*Alan Lane, former packaging manager now an independent consultant*)

It has been challenging to keep up with the changing/updating of the GFSI code and then the subsequent changes of the schemes to meet the code. (*Anonymous*)

Keeping the interest and excitement level up, over time, the required semi-annual surveillance became very taxing on the Internal Auditing team, especially if they had other core responsibilities. (*Andy Fowler, former VP Administration and Environmental Compliance, Bacardi & Company Limited*)

We aligned our facility with PDCA (Plan, Do, Check, Act) and the FSSC 22000 standard which provided us the logic and the tools to help maintain certification and continuous improvement. Based on our efforts, I recommend using the PAS 99 standard to understand the relationship between Plan, Do, Check, Act (PDCA), Total Productive

Maintenance (TPM), and the ISO standards. Being able to be certified but using these additional documents to aid us in communicating and linking these concepts together into our system was very helpful to both management and the associates. We were able to educate and understand the true advantage to our operation overall without certification becoming just another thing that quality will bug us about. (*Anonymous*)

Our system was built as a growing, improving system and just because we have a certification, the system tasks are not complete. It is never actually done which was something we had a hard time understanding, but the advantage is that it is a living and growing system that continues to change as the system continues to mature. The best part of a true system is that it starts to manage your duties instead of you trying to manage the system. This is a hard thing to truly understand until you actually live and experience it. (*Russ Patty, an independent consultant former quality manager of a beverage operation*)

We find that we still struggle with staying on top of or ahead of the system. We have not been in it long enough to develop lasting habits. We are still distracted by being in the fire-fighting mode. (*Joe Hembd, Management Systems Manager, Chesterman Company*)

We learned that we must continue with effective internal audits, effective training, and continued executive buy in and support from top management and middle management. Once the audits are completed, then we have found that we must keep reminding management that just because we have the certificate, the journey is not over. Top management have learned through our maturing process that it must continue to reinforce their support and continually strive toward compliance as a way of life. (*Eric Putnam, Food Safety, Quality & Training Systems Manager, Wixon, Inc.*)

Currently, the toughest challenge is fighting the human element of letting things slip. In order to live in a continuous improvement environment and uphold the FSSC 22000 Certification, we have found it a necessary to continue training as well as holding associates and management accountable for maintaining compliance in their areas of responsibility. (*Anonymous*)

QUALITY MANAGEMENT AND FOOD SAFETY TOGETHER?

The next question to our industry professionals was related to combining (maintaining or creating) a quality management system with the food safety management system. Based on experience, can a system have an effective quality management system without including food safety or a food safety program without a quality program? Members of this group who had led their organizations through the development, implementation, and certification of a compliant food safety management system were asked: Did your management decide to maintain (or create) a quality management system in combination with your food safety management program? "Did your organization already have a quality management system, and how did this affect the development of the food safety management system and decision whether or not to include this in the certification?"

We chose to pursue only Level II SQF certification, as we had a mature QMS that had been developed and tested to be adequate over many years. Additionally, as we were only packaging, and the standards at the time were only food oriented, we felt that this was the best bang for the buck. We had a robust and mature QMS previous to pursuing

SQF certification. I believe this was a distinct advantage as much of the current QMS was transferable to the requirements of SQF. Without having to develop the entire management system, achieving SQF was easier and achieved in a shorter time frame. (*Alan Lane, former packaging manager now an independent consultant*)

We have recently chosen to include quality (SQF level III). Main reason for implementation of a scheme with quality was adaption of a standard for all facilities across our enterprise. We had quality management systems prior to our implementation of SQF. The advantage this gave us was only minor adjustments were needed to meet the scheme requirements and it allowed us for standardization. The disadvantage was only related to changing of the system name or including different requirements of the code, thus causing of some confusion with our associates. (*Anonymous*)

We initially focused on application from a food safety perspective, yet easily integrated quality as the system evolved. I believe that is one of the unique aspects of the process, the standards are strategic enough that they can be applied to multiple aspects of the business. We did have an existing food safety management system, which was both comprehensive and established. An advantage of this was that it was relatively easy to assimilate it into a food safety management system framework. A disadvantage was that our existing system was so established: any minor tweaks we made to it required us to really focus on communicating the changes and incorporating them into work instructions and training materials. (*Mike Burness, Vice President, Global Quality and Food Safety, Chiquita Brands International, Inc.*)

Yes, we included both food safety and quality because it aligns with our parent company standards. This makes sense because food safety and quality complement each other and with both systems in place provides a stronger overall quality system. We had a quality management system in place at our facility prior to embarking upon the FSSC 22000 certification process. The quality system was enhanced and improved by implementing FSSC 22000. (*Anonymous*)

Although we had portions of an effective quality management system, we choose to focus on the food safety management standard because of our customer's requirements and short deadline; however, I feel that every plant that is in production should have an effective quality management system in place that is not just word of mouth and not documented. Our quality management system should have been stronger; if we had had a certified ISO 9001 based quality management system first it would have made it easier going to FSSC 22000. (*Russ Patty, an independent consultant former quality manager of a beverage operation*)

We had an internal quality system that was close to being compliant to ISO 9001; however, it was not as strong as it should have been. We did have work teams established and work instructions already written for most of our procedures which overlapped into the food safety requirements; however, management decided to only officially seek certification to FSSC 22000. (*Joe Hembd, Management Systems Manager, Chesterman Company*)

We are currently reviewing our options as to which direction to take—only food safety (with design and development still following 9001) or an integrated system. We have been certified to and remain compliant to the requirements of ISO 9001:2008 for several years. It was a great advantage to us because our documentation and mindset was already in alignment with what was expected by FSSC 22000. (*Eric Putnam, Food Safety, Quality & Training Systems Manager, Wixon, Inc.*)

"WHAT WOULD YOU DO DIFFERENTLY IF YOU COULD DO IT ALL OVER AGAIN AND WHY?"

SQF is a good system; the only thing we would do differently is go for level III in the beginning versus starting with level II first. (*Anonymous*)

During our journey of becoming FSSC 22000 certified, we endured many challenges ranging from resources to understanding the audit process. The only thing I personally would have changed would have been to get a proven consultant on board in the beginning rather than wait until after the 1st audit. Although there were many struggles and frustrating moments, I feel that the journey taught everybody on the team something. I feel it helps strengthen our understanding as well as our commitment to upholding this system. (*Anonymous*)

[We would] have a dedicated internal quality team whose primary focus would be to maintain the quality management system. (*Andy Fowler, former VP Administration and Environmental Compliance, Bacardi & Company Limited*)

FSSC 22000 certification has been a great tool to fully integrate our food safety management system into the business and get everyone to "speak the same language". We wouldn't do things differently! (*Mike Burness, Vice President, Global Quality and Food Safety, Chiquita Brands International, Inc.*)

I would not change anything. I do recommend getting a "great" consultant to help prior to certification. (*Anonymous*)

I would start out by hiring an external consultant to assist us with development and implementation. Most production plants know their business but don't know how to tie it all together and look at compliance with the ISO standards as just more paper work instead of an effective tool to enhance our foundation for producing safe food products for everyone. Being compliant to these requirements forced us to design additional structure to our entire process from receiving raw materials and ingredients through shipping the finished product. (*Russ Patty, an independent consultant former quality manager of a beverage operation*)

As the person responsible for guiding the implementation, I would definitely receive formal training to fully appreciate the nuances of the standard prior to embarking on implementation. I would also request that top management allow more time for research and documentation which would add additional foundation to the complete management system. (*Anonymous*)

What would have really helped would have been if we had a larger implementation team and more consistent active participation from all members. However, that said, we are back to resources again which I believe have always been and will always be every organization's biggest challenge. (*Joe Hembd, Management Systems Manager, Chesterman Company*)

At the onset of the project, I would hire a consultant that is not only good at finding the gaps but also able to suggest obtainable solutions and mitigations to those gaps. Our consultant kept telling us that he/she was the coach and that we needed to be the active players. (*Eric Putnam, Food Safety, Quality & Training Systems Manager, Wixon, Inc.*)

If I were to undertake the process again, I truly would not do anything differently. As a quality practitioner, it has always been my practice to "plan the work. and work the plan", and that's exactly what we did. Additionally, we had complete upper management support, making the project a priority with everyone. This alone, is critical to any project of this scope, and I would be very hesitant to begin any project if this were not in place. (*Alan Lane, former packaging manager, now an independent consultant*)

The common thread from these comments is that organizations must make the commitment that a compliant food safety (and hopefully quality) management system must be developed, defined, implemented, and maintained. As stated earlier, the key is to become familiar with the standard of choice, then assess the organization's current status (i.e., gap), and then begin the process. It is critical that top management understands the commitment required to successfully implement a compliant food safety management system. In understanding this commitment, it also means that this is not a one-person or one-department project, but it has to be a team effort with the entire team engaged and active in the process.

Time is also a limited resource in today's world. There never seems to be enough time no matter how hard we try to plan efficiently; management must be realistic in defining the time frame. Don't procrastinate. Start the process today. Develop the plan and track the progress, adjusting resources and focus as required, but be careful not to be overzealous. It is critical that the system is developed in a manner that will grow and sustain into a great management system that truly adds value and efficiency to the entire operation. Keep in mind that taking short cuts, or rushing the process or just trying to get something done to get through the audit, increases the likelihood of not passing the audit (which will require more time and expense). Passing the audit with a system likely to not sustain through the first approval cycle jeopardizes the certificate and requires additional resources to patch and maintain compliance. Time is a precious commodity, but taking the extra time even if just a month to build a more solid foundation for compliance pays dividends in the long run. A good example of this is an individual rushed, short of time, and deciding he just didn't have time to read the driver test manual prior to taking the written test. Well, guess what? He didn't pass, had to study again, pay again, and actually could not travel until he passed the test and received his new driver's license. He did admit that if he had taken only 30 minutes to read the booklet, it would have saved him not only in cost but also in that precious commodity of time that he didn't have enough of to prepare for the test; he ended up having to find even more time to study and retake the test.

At the conclusion of the survey, the participants were asked if they had any other thoughts or comments that they would like to share:

Having now a thorough understanding of both systems, after employing both ISO and SQF successfully, it's clear that they are not only compatible, but parallel in requirements and intent. For food facilities, I can't think of any reason not to employ the GFSI, (in particular, SQF), as the primary QMS for a developing business. (*Alan Lane, former packaging manager now an independent consultant*)

The elements of a food safety management system are easily transferable to other aspects of the business, incorporating all aspects of company's goals, actions, measurements & continuous improvement. (*Mike Burness, Vice President, Global Quality and Food Safety, Chiquita Brands International, Inc.*)

An effective food safety management system does not just include the quality department but must include all departments to ensure a robust quality and food safety system in place. (*Anonymous*)

Important wisdom that we learned after the fact is that an organization must approach the implementation process with an open focus on what having an effective and compliant management system if done correctly will do for your production. The system does and will add valve if it is implemented effectively with the correct attitude. (*Russ Patty, an independent consultant former quality manager of a beverage operation*)

All hands must be on board, from top management to line personnel. FSSC 22000 elaborates a comprehensive FSMS which if implemented will assist companies to achieve food safety objectives. Implementation is strenuous and costly but entirely worthwhile. (*Anonymous*)

Implementation, maintaining, and continuously improving a structured and disciplined management system may require a culture change; that the entire operation must be ready to commit to and stick to using it. (*Joe Hembd, Management Systems Manager, Chesterman Company*)

Never settle for the status quo. If the system is not moving forward and improving food safety and the business, it is not working the way it is intended. Remember to "reserve the right to get smarter" and leverage the system for all it can provide to positively impact the business and add value. (*Mike Burness, Vice President, Global Quality and Food Safety, Chiquita Brands International, Inc*)

TECHNOLOGY AND ITS ROLE IN THE FUTURE OF FOOD SAFETY

Dan Bernkopf, Vice President, FSQA Applications SafetyChain Software, shares his thoughts on technology and its role in compliance and continuous improvement of our effective food safety management systems: "Today, the food & beverage industry is widely adopting technology to effectively and efficiently execute—via organization, simplification and authorized accessibility—on prerequisite programs and document control initiatives. Technologies have been proven to not only save time and costs but to enhance the management systems overall. Examples of what an effective program can accomplish would include the following:

- Transform current hard-copy paper, binder and file systems with a user-friendly, highly secure (think online banking!), web-based repository of documents—automating document storage as well as access to prerequisite programs, inspection tasks and records.
- Administer revisions while creating secure access to current program documents and forms to other company stakeholders—with 21 CFR Part 11 compliance for security, control and electronic signatures.
- Complete—on mobile devices—inspections and other verification tasks, or schedule tasks with automatic email/text alerts.
- Automate creation of email/text alerts to stakeholders when non-compliant results are detected to promote real time corrective actions.
- Save time and labor while facilitating trend analysis and creating compliance efficiencies."

FUTURE OF FOOD SAFETY

The group of subject matter experts and industry professionals were asked the following question on the future of food safety: "What do you feel is the future of food safety and what do you feel will be required to continually improve the overall safety of our food supply?" The response was very interesting:

> Clearly, food safety is becoming more and more critical throughout the world. We get fresh fruit and vegetables throughout the year that are brought in from all over the world. Organizations developing systems compliant with a GFSI approved food safety scheme will fill the role that ISO has filled for manufacturing sites in that businesses are doing business in a trusted manner and can provide quality product regardless where they are, or who they are, if they have a compliant system that can be trusted. *(Alan Lane, former packaging manager now an independent consultant)*

> *Tom Marchisello, Industry Development Director, Grocery Retail, GS1 US*, related to the future of food safety "believes that food safety will become more and more regulated and prescriptive across all food commodity categories. Government agencies charge with enforcement and verifying regulatory compliance will drive certification to generally recognized standards such as FSSC 22000 as well as partner with independent third party audit bodies."

> One of the most advantageous parts of any of the food safety management system standards meeting the GFSI food safety scheme is that it does require a structured and recorded approach to continuous improvement, so my hope is that this scheme will continually meet the code and thus the organizations of the future will continue to be compliant and grow internally as their FSMS mature and continually improve. *(Anonymous)*

> The future of food safety will be more regulated and scrutinized by the public. People want safe food and manufacturers of food and consumable goods must ensure the safety of their products to avoid potential adverse business outcomes. *(Anonymous)*

> The consumers know more in today's time, so are asking for more information. Consumers are learning more and more about ingredients and sanitation chemicals. I see the nutrition facts changing on labeling to include not only ingredients in that product but what cleaning chemicals, oils, lubricants that were used to help produce that product. I see manufacturing sites continue to enhance their systems, documenting, checking, verifying, and having 3rd party audits to confirm compliance with certification standards. The more information the consumer has the more that they will ask for. I feel in the future as the consumers and the customers learn more, the food companies will be forced to share more, but then education has never been a negative thing not for the companies with integrity. *(Russ Patty, an independent consultant former quality manager of a beverage operation)*

> There is no doubt that food safety will gain momentum as consumer concerns heighten. Legislation coupled with education to ensure that all players in the food chain embrace food safety. *(Anonymous)*

> Food safety standards will continue to evolve. In order to continually improve, businesses will have to include food safety in their resource allocations (budget, personnel, etc.). *(Joe Hembd, Management Systems Manager, Chesterman Company)*

Food safety is going to be the next big litigious hot-spot very similar to the state of medical malpractice is now. There is going to need to be some grand changes in the way food companies of all sizes think and operate. If this shift in philosophy does not happen even the food giants will be Goliath to the consumer watchdogs' David. To avoid this I feel that the training of food handlers all the way up to top management needs to be in depth and meaningful so that all affected truly understand the advantages and the consequences of becoming complacent. (*Eric Putnam, Food Safety, Quality & Training Systems Manager, Wixon, Inc.*)

In my opinion, the future of food safety is bright - Why? "Threats to the food supply are constantly being recognized and consumers are genuinely concerned and speaking out about the safety of foods they eat therefore. Food safety awareness is at an all-time high. Consumers are pushing back and putting the responsibility on food establishments, retailers and producers to ensure proper food safety practices are followed. To ensure food safety practices are being followed, more and more companies recognize the importance of implementing *food safety programs*. Though food safety programs alone do not guarantee *safe foods*, it does promote commitment to food safety and more importantly, a *cultural* shift towards a food safe environment at all levels within a business". (*Bob Wihl, independent consultant, D.L. Newslow & Associates, Inc.*)

Frank Yiannas, Vice President of Food Safety for Walmart in his book Food Safety Culture Creating a Behavior-Based Food Safety Management System (Springer, c2010, pp. 84–85), provides us with the four critical success factors that he feels is significant for today's industry to "make significant leaps in food safety:"

1. *"To make significant leaps in food safety, we need creativity and innovation:"....* The bottom line is that creativity and innovation lead to change and change can lead to even great reductions in the risk of food borne disease. Simply put, "it is impossible to advance food safety without change." There is no better way to say this than as Mr. Yiannas has done in this statement. Change is tough for many, but without it, our food safety efforts are going to continue to tread water.

2. *"To make significant leaps in food safety, we need leadership:"* Mr. Yiannas's concept of leadership vs. management was discussed in the earlier chapter on management responsibility (Chapter 3). Mr. Yiannas quotes Stephen Covey, Merrill and Merrill (1994), stating that "Management works within the system; leadership works on the system. ... [Food] safety managers deal with planning, directing, and overseeing specific details for the system. Food safety leaders, in contrast, see the need for improvement, create a compelling vision for change, and inspire innovation, all of which lead to even greater reductions in food borne disease. To advance food safety, some of us need to be courageous pioneers and help lead the way." These words provide the emphasis to management and the entire associate team that success depends on leadership. This book has addressed many compliant-type *best practices* and proven approaches for defining, implementing, and maintaining a compliant food safety management system; however, no matter how great our *tools* may be, the foundation for success must begin and continue through

top management's commitment and leadership. Remember the three-legged stool as it related to the management system (Chapter 1).

3. *"To make significant leaps in food safety, we need more research:"* Mr. Yiannas states that "we need to be continual learners and more research is needed to answer some of the food safety questions of our day." Mr. Yiannas also shares his innovative thinking, which many of us are already experiencing by learning "from other disciplines such as the medical, information technology, and biotechnology fields to name just a few." Mr. Yiannas states that he believes "that some of our greatest future food safety solutions may not even come from within the field of food safety."

4. *"To make significant leaps in food safety, we need better collaboration:"* Mr. Yiannas explains that "Today more than ever, food safety is truly a shared responsibility. Regulators, academicians, consumers, and industry professionals must recognize that we can do more to advance food safety by working together than by working alone."

There is no doubt based on the experience of this industry professional that the future of food safety is much more than just developing and maintaining a compliant food safety management system. This is of course critical because we must have systems that are structured, disciplined, and consistent in both its application and the outputs of the process (end products). But in addition, we must communicate, educate, and be ready to think outside of the box to move forward. The journey is more than just a cookbook; tell us what to do, and we will do it and then we will be done. We can't lose sight that the management system and all of its components are proactive tools that when applied and managed effectively provide the organization with a means to identify existing and potential hazards and control these through its defined processes to build confidence in the production of a safe food product. Compliance does not guarantee food safety and remember a tool is truly only as effective as the person or team using it. Our food safety processes are only as effective as their weakest link. We must strive to continually produce a safe product that meets defined requirements within a process that is consistent, compliant, and continues to improve.

John Kotter in Leading Change (Harvest Press, c1996, p. 25) devotes 200 pages to the topic of *culture change* applying his eight-point process. Building on Mr. Yiannas's emphasis on *leadership* vs. *management*, Mr. Kotter provides the following description:

Management is a set of processes that can keep a complicated system of people and technology running smoothly. The most important aspects of management include planning, budgeting, organizing, staffing, controlling, and problem solving. Leadership is a set of processes that creates organizations in the first place or adapts them to significantly changing circumstances. Leadership defines what the future should look like, aligns, people with that vision and inspires them to make it happen despite the obstacles.

It is very interesting that Mr. Kotter wrote his description in 1996, and 14 years later, Mr. Yiannas emphasizes to everyone in the food industry that we must change and that we must be ready to make these changes and apply some of the logic and theory that others have applied for years. It is no longer effective to just keep doing it the same way because we always have. It is this concept of leadership and commitment that the new generation of management must accept, embrace, and basically develop as a way of life for today's organizations.

Dennis Sasseville, an independent consultant, third-party auditor, and president of Cobalt Advisors specializing in the ISO standards shares the following thoughts on the future of food safety:

> One the best things about working in the food safety and quality profession is that it never stands still. In my opinion, the future of food safety will be focused in a few key areas (in no particular order):
>
> - Research—we continue to learn more and more about food safety, whether it be pathogen resistance, foreign material contamination, or stealth chemical contamination. This knowledge is, and will always be required, as the pathogens, foreign objects, and chemicals themselves evolve and change. We can never rest on what we know today; we need to continue to be out in front, working to prevent the next potential issue.
> - Technology—technology is changing more rapidly now than ever. To that end, we need to embrace the development of technological solutions to food safety challenges. Whether new ways to eliminate pathogens, nanotechnology that helps deliver micro-nutrients, social media techniques to educate society on food safety, we need to challenge ourselves to continually raise the bar on the safety of the global food supply.
> - Globalization and food security—with the continuing globalization of the food supply, we need to ensure we work to level the food safety playing field. The food supply chain gets more complex every day, and with that comes the challenges of ensuring that regardless of origin, destination, or target population, food safety is a fundamental requirement. Additionally, as the population grows to an estimated 9 billion people by 2050, we need to figure out ways to ensure we are able to grow, produce, and ship enough food to provide them with a healthy and nutritious diet.

> Understanding food safety fills a critical need in the must-have resources available to food safety and management professionals today. Professionals in the field of food safety and quality are challenged by the myriad of demands to develop and maintain integrated systems that will robustly serve production and process operations, as well as the upstream and downstream supply chain. The second decade of this century is experiencing an accelerated pace of customer requirements and the pressures on industry professionals have never been greater. Superior approaches and tools are required to keep pace with this acceleration.

Rick Biros, founder and former president of Carpe Diem publisher of Food Quality magazine (1994–2006) and now the founder and president of Innovative Publishing Company, LLC, the provider of the on line magazine Food Safety Tech (http://www.foodsafetytech.com/FoodSafetyTech/Blogs/392.aspx), was asked recently about the role of technology in food safety. Rick's response was

Yes. I believe food safety technology has advanced significantly empowering the industry with the ability to detect contaminants, faster and more accurately. Food safety awareness in the food industry as well as education has increased. Food safety professionals have earned the respect from the C-level suite and are seeing increased support. That's a big step forward from 1994, however, food safety is a marathon with no finish line in sight and there is still much more work to be done! The food industry certainly has its challenges ahead and there is a growing need for information, education and knowledge, on a global scale.

As mentioned in a previous chapter, *Jon Porter*, 10 plus years ago, stated that "without the internet HACCP would not be possible." If only Jon was still with us, he would be amazed at the progress made in the past 6 years since he left us. However, through this evolution what has not changed is the consistent need for, as Mr. Biros states, "information, education and knowledge." The future of food safety depends on creating and sustaining an effective foundation to be able to consistently share information and knowledge while consistently providing the basis for *education* through the industry. Jon Porter was firm on his beliefs on *education* compared to *training*. Education is sharing knowledge and reasons why an activity must be done in the manner presented. Training is communicating the required actions and stating why these are important. Chapter 8 Training, Awareness and Competency discusses this concept in more detail.

At some point, the information presented in this text must be concluded and the manuscript submitted to the publisher. If this author were to be asked what was the most challenging part of writing this book, it would definitely be identifying the point where it was completed. In discussing food safety, reviewing years of experience with management systems, and working with so many knowledgeable professionals in this field, it is very hard to not just keep researching and adding more content. There are numerous reference materials identified, many of which include a link to a website or web page. It is important that the astute reader understands that learning and experiences are not going to stop. On behalf of this auditor, our team, and the numerous professionals who shared their experiences many of who cannot be recognized because of company policies but still took the time to share, we encourage industry professionals to stay up-to-date with current happenings and be patient through the learning curve. A good rule of thumb is that the organization supported by a committed top management team will get as much out (return) as they put into the development and continued compliance of the management system. This is true of whichever standard is chosen.

Keep in mind that there is an advantage to being familiar with the specific requirements of the chosen standards, but it is not necessary to memorize every word. The key is to know how to use them. Remember, one does not have to memorize a dictionary to know how to use it. There are excellent courses, excellent written materials, and articles. Do not be shy, ask questions. Learn the system, justify the decisions based on facts, and have fun with the process. A compliant and effective system that adds value as the system matures replaces everyday panics associated with *fire fighting*.

The strength of this text has been the experience and best practices that have been shared through research, experience, and contributions of subject matter experts. It is up to the organization to make its choice on the standard that is best for its operation. Each learn, embrace the requirements, and then move forward in defining, implementing, and maintaining a food safety (and hopefully) quality management system. Being proactive toward the production of a safe food product while meeting the organization's, customers', and legal requirements, adding value through compliance and continuous improvement to the organization is an attainable and worthwhile goal. It is appropriate to conclude with the following quotation:

Quality means doing it right when no one is looking. (*Henry Ford*)

Appendix A: Definitions

This addendum is being included upon the request of our reviewers. It includes *definitions* or a description of words, and phrases used throughout this text. It focuses on clarification of terminology that is related specifically to the subject matter of this text. Note: where appropriate for the term, the chapter that may provide more detail or background is referenced. A separate chapter follows that identifies commonly used industry and/or management standard acronyms.

Allergen: A food protein substance that an individual may have a reaction to causing the body's immune system to react thus causing an allergic reaction.

Association of American Feed Control Officials (AAFCO): AAFCO's goal is to provide a mechanism for developing and implementing uniform and equitable laws, regulations, standards, and enforcement policies for regulating the manufacture, distribution, and sale of animal feeds resulting in safe, effective, and useful feeds (www.FDA.gov).

Audit: An evaluation to determine if activities and results comply with defined requirements. A systematic and independent examination of a management system or portion of the management system (i.e., process or element audits) to determine whether activities are being performed in compliance with defined requirements (standard of choice, organization's, regulatory, etc.) (Chapter 6).

Audit Guide: Representative from the area being audited who accompanies the auditor. It is very important that a representative of the area being audited (supervisor, manager, team leader, etc.) accompanies the auditor in order to experience and see firsthand what the auditor sees. An internal audit must never be conducted without a representative of the area accompanying the auditor (Chapter 6).

Audit Team: When an audit is conducted by more than one auditor, this group of auditors is referred to as the audit team. It may also be referred to as the *internal audit team*, which is a group of the organization's associates trained in auditing techniques and the standard of choice. Members of this team are assigned process and element audits to be performed according to defined requirements for the internal audit program (Chapter 6).

Auditee: The individual or group being audited.

Benchmarking: Benchmarking is a methodology that may be used to search for best practices. It can be applied to strategies, policies, populations, processes, products, and organizational structures. The Global Food Safety Initiative (GFSI) board of directors applies the benchmarking process for its approval of food safety management system schemes comparing these to the most current version of the GFSI Guidance Document (http://www.praxiom. com/iso-definition.htm).

Biological Hazard: A biological pathogenic hazard relating to microorganisms that either directly or indirectly cause a food safety hazard. Examples of these types of hazards include such organisms as *E. coli* O157:H7, *Salmonella*, *Clostridium botulinum*, and *Listeria monocytogenes*.

Bioterrorism: This is terrorism involving the intentional release or dissemination of biological agents. These agents are bacteria, viruses, or toxins and may be in a naturally occurring or a human-modified form (http://en.wikipedia. org/wiki/Bioterrorism). Bioterrorism threats related to the food supply can be a concern; however, being proactive in protecting our food supply is an effective tool in prevention (Chapter 14).

Bioterrorism Act (BT Act): As a result of the events of September 11, 2001, Congress responded by passing the Public Health Security and Bioterrorism Preparedness and Response Act of 2002 (the BT Act), which President Bush signed into law on June 12, 2002. The Food and Drug Administration (FDA) is responsible for carrying out certain provisions of the BT Act, particularly Title III, Subtitle A (Protection of Food Supply) [www.FDA.gov] (Chapter 14).

BRC Global Standard for Food Safety: The British Retail Consortium (BRC) Global Standards are widely used by suppliers and global retailers. They facilitate standardization of quality, safety, operational criteria, and manufacturers' fulfillment of legal obligations. The BRC Global Standard for Food Safety is a GFSI approved scheme widely applied throughout over 90 countries by suppliers and global retailers. This standard is designed to provide protection to the consumer (Chapter 1) (http://www.brcglobalstandards.com/ GlobalStandards/Home.aspx).

Brittle Plastic: Plastic that is made from acrylic resins that break into pieces when subjected to blows beyond its impact resistance (Chapter 10).

Calibration: Term used to reflect equipment that is used to control, measure, test, and confirm the accuracy of results (Chapter 11).

Canada Good Agricultural Practices (Canada GAP): Canada Good Agricultural Practices (GAP) is the name of the Canadian Horticultural Council's (CHC) On-Farm Food Safety (OFFS) Program. This program consists of national food safety standards and a certification system for the safe production, storage, and packing of fresh fruits and vegetables (Chapter 1) (http://www. canadagap.ca/).

Canadian Food Inspection Agency (CFIA): Canada's governmental agency responsible for food safety (www.inspection.gc.ca).

CARVER: An acronym denoting criticality, accessibility, recuperability, vulnerability, effect, recognizability: A risk assessment tool to conduct vulnerability assessments by determining the *critical nodes* that are the vulnerable targets for terrorist attacks and lead to the identification of steps or countermeasures that may reduce the risk to the production of that product. The intent of the assessments is to comprehensively assess FDA-regulated products and thus improve the safety and security of the food supply. Upon completion, the results are shared with industry and individual companies to apply as appropriate within their production system. A joint effort is underway entitled Strategic Partnership Program Agroterrorism (SPPA) that employs CARVER (www.FDA.gov).

Centers for Disease Control and Prevention (CDC): The nation's premiere health promotion, prevention, and preparedness agency and a global leader in public health. CDC's mission is to promote health and quality of life by preventing and controlling disease, injury, and disability (www.FDA.gov).

Center for Food Safety and Applied Nutrition (CFSAN): One of six product-oriented centers in FDA that is responsible for ensuring that the nation's food supply is safe, sanitary, wholesome, and honestly labeled. It also includes assurance that cosmetic products are safe and properly labeled (www.FDA.gov).

Certificate of Analysis (COA): A document provided to confirm that the referenced ingredient, material, or product meets product specifications, which may be defined by the supplier or the customer. The COA generally includes the specification target and the test results performed on a specific batch or lot.

Certificate of Conformance (COC): A COC, which may also be referred to as a *certificate of compliance*, is a document that a supplier provides its customer confirming that the referenced lot number or batch meets the required specifications.

Clean in Place (CIP): The cleaning and sanitizing system where detergents, sanitizers, and water are circulated through equipment and lines by pumping or spraying without the need to dismantle and hand clean (Chapter 10).

Cleaning: The removal of soil, food residue, dirt, grease, or other objectionable matter (Chapter 10).

Codex Alimentarius Commission: The internationally recognized entity whose purpose is to guide and promote the elaboration and establishment of definitions, standards, and requirements for foods and to assist in its harmonization and, in doing so, to facilitate international trade [SQF Code, Edition 7, Appendix 2: Glossary B].

Codex HACCP: Codex Alimentarius: Hazard Analysis Critical Control Point as defined by Codex Alimentarius (2003). (*Hazard analysis and critical control point [HACCP] system and guidelines for its application*) (CAC/RCP 1-1969, Rev. 4, 2003).

Color Code: Color coding system may be established to identify cleaning utensils for designated areas such as raw or cooked product and allergens or nonallergens. A color coding process may also be used for the identification of recycle, rework, and trash containers. An effective color coding process aids in the prevention of cross contamination issues.

Commercialization: A term used by some organizations to reflect new product, equipment, or other changes introduced into a process or system.

Consumer: Intended person or animal actually consuming the product or using the service.

Contaminant: Any biological or chemical agent, foreign matter, or other substances not intentionally added to food, which may compromise food safety or suitability (CAC/RCP 1-1969, Rev. 4, 2003).

Contamination: The introduction or occurrence of a contaminant in food or food environment (CAC/RCP 1-1969, Rev. 4, 2003).

Correction: "Action to eliminate a detected nonconformity" (ISO 22000:2005 3.13). Typically, a correction does not require a root cause and is a single or limited event such as a lost cap or a case of leaking containers. The situation is corrected without a full analysis; however, corrections must be monitored for trends. Corrections that result in trends should be recorded and tracked through the formalized corrective and preventive action (CAPA) program (Chapter 4).

Corrective Action: Related to the *system*, corrective action (CA) is "an action taken to eliminate the causes of an existing nonconformance. This is the action taken to eliminate the cause of a detected nonconformity or other undesirable situation." Corrective action includes root cause analysis resulting in actions taken to prevent recurrence (ISO 22000:2005 Section 3.14). This is a formalized process that requires root cause analysis, identified actions to be taken to correct the situation in a timely manner based on the risk of the situation, actions required to minimize or prevent its recurrence, and the evaluation of effectiveness to ensure that the action or actions applied are effective (Chapter 4).

Crisis: A critical event or potential event that if not addressed or protected against could cause a catastrophic or otherwise critical outcome. A crisis can be anything from an adverse weather situation (i.e., tornado, hurricane, earthquake, etc.) to a fire, bomb scare, flood, or product recall. Associates must be trained to know what to do immediately if a crisis occurs (Chapter 14).

Critical Control Points (CCPs): A step at which control can be applied and is essential to prevent or eliminate a significant food safety hazard or reduce it to an acceptable level (CAC/RCP 1-1969, Rev. 4, 2003).

Critical Equipment: Equipment that is used to verify food safety limits and also equipment used to confirm compliance for quality, regulatory, and customer requirements.

Critical Limit (CL): Identified limits that separate acceptability from unacceptability as related to a food safety hazard. A criterion that separates acceptability from unacceptability (CAC/RCP 1-1969, Rev. 4, 2003).

Cross-Contamination: Contamination by a substance or article that is not part of the original product.

Customer: May be the same as the consumer; however, may also be a broker or food chain sector that purchases directly from the organization and then sells to the consumer.

Decision Tree: A sequence of questions that can be applied to a process step with an identified hazard to determine if the hazard is significant and that control at that step must be managed as a critical control point (CCP) (Chapter 17) (CAC/RCP 1-1969, Rev. 4, 2003).

Deviation: Failure in meeting a critical limit of a critical control point (CCP) (Chapter 17) (CAC/RCP 1-1969, Rev. 4, 2003).

Disposition: The decision related to how and what happens to items that have been placed on hold due to food safety or quality concerns. The act of disposing or disposition of items is done according to defined requirements (Chapter 9).

Document Control: The process used to ensure that the most current version of instructions, requirements, external documents and procedures are readily available to those who must have the information to perform required activities and maintain effective levels of responsibility (Chapter 6).

Education: Sharing knowledge and reasons why an activity must be done in the manner presented.

End Product: Product that will not undergo any further processing or change prior to delivery to the customer or consumer. The term *end product* may also be referred to as a *finished product* and varies depending on the organization's products. An end product for an ingredient supplier is the ingredient whereas the end product of a fluid milk plant is the milk sold to the supermarket to sell to the consumer.

External Communication: Communication external to the organization including communication throughout the food chain, which includes the suppliers, customers, and consumers (Chapter 3).

External Documents: A document required by the system but not created and controlled by the system such as a standard method testing procedure, operating manual, or food safety management system standard. External documents must be identified and managed in compliance with the organization's document control program (Chapter 6).

Failure Mode Effect Analysis (FMEA): A risk assessment tool that is used to systematically analyze failures and the effect the failure has on the operation and/or system. There are many ways to use such a tool, which is usually incorporated in an organization's process for troubleshooting or project management (Chapter 15).

Federal Food, Drug and Cosmetic Act (FDCA): The U.S. Federal Food, Drug, and Cosmetic Act (abbreviated as FFDCA, FDCA, or FD&C) is a set of laws passed by Congress in 1938 giving authority to the U.S. FDA to oversee the safety of food, drugs, and cosmetics. FFDCA is the primary law, among others, that FDA regulates and enforces (http://en.wikipedia.org/wiki/Federal_Food,_Drug,_and_Cosmetic_Act).

Federal Meat Inspection Act (FMIA): The FMIA of 1906 is a U.S. Congress Act that works to prevent adulterated or misbranded meat and meat products from being sold as food and to ensure that meat and meat products are slaughtered and processed under sanitary conditions. These requirements also apply to imported meat products, which must be inspected under equivalent foreign standards (http://en.wikipedia.org/wiki/Federal_Meat_Inspection_Act).

Finding: Also may be referred to as a nonconformance. A nonconformance or finding is defined as a situation that violates a requirement of the standard or a defined requirement of the system (Chapter 5).

First In First Expired (FIFE): Ingredients, materials, and finished products are used, stored, and shipped in the order of the expiration date (Chapter 10).

First In First Out (FIFO): Ingredients, materials, and finished products are used, stored, and shipped in the order that they were received (Chapter 10).

Flow Diagram: Created by the Food Safety/HACCP team, this is a systematic representation of the sequence of steps or processes used in the production or manufacture of a particular food item (CAC/RCP 1-1969, Rev. 4, 2003).

Food Allergen Labeling and Consumer Protection Act (FALCPA): The U.S. FDA is responsible for most food allergen labeling as defined by the FALCPA of 2004. This law requires that the top eight food allergens be identified (labeled) using their common names. The top eight allergens are milk, eggs, fish, crustacean shellfish, tree nuts, peanuts, wheat, and soybeans (Chapter 10).

Food and Drug Administration (FDA): FDA is the federal agency responsible for ensuring that foods are safe, wholesome, and sanitary; human and veterinary drugs, biological products, and medical devices are safe and effective; cosmetics are safe; and electronic products that emit radiation are safe. FDA also ensures that these products are honestly, accurately, and informatively represented to the public (Chapter 1). The FDA is also responsible for advancing the public health by helping to speed innovations that make medicines and foods more effective, safer, and more affordable, and helping the public get the accurate, science-based information they need to use medicines and foods to improve their health (http://www.fda.gov/RegulatoryInformation/RulesRegulations/default.htm).

Food Defense: Food defense is the collective term used by the FDA, USDA, DHS, etc., to encompass activities associated with protecting the nation's food supply from deliberate acts of contamination or tampering. This term encompasses other similar verbiage (i.e., bioterrorism (BT), counterterrorism (CT), etc.) (Chapter 14) (http://www.fda.gov/Food/FoodDefense/EducationOutreach/).

Food Defense Plan: A Food Defense Plan identifies steps to minimize the risk that food products in your establishment will be tampered with or intentionally contaminated (www.fda.gov/fooddefense) (Chapter 10).

Food Hygiene: All conditions and measures necessary to ensure the safety and the suitability of food at all stages of the food chain (CAC/RCP 1-1969, Rev. 4, 2003).

Food Processors Association: A trade association for the food and beverage industry in the United States and worldwide and provides technical and regulatory assistance to member companies and represents the food industry on scientific and public policy issues involving food safety, food security, nutrition, consumer affairs, and international trade.

Food Safety: Terminology used that relates to finished product assuming that it does not contain a food hazard when used according to its intended use. Also, as defined in Codex Alimentarius, "Assurance that food will not cause harm to the consumer when it is prepared and/or eaten according to its intended use" (CAC/RCP 1-1969, Rev. 4, 2003).

Food Safety Critical Supplier: A food safety critical supplier provides ingredients, packaging material, product contact/processing aids, and/or services that are required for the production of safe products and also for meeting regulatory requirements related to food safety (Chapter 13).

Food Safety Hazard: Biological, chemical, or physical agent in food, or condition of food, with the potential to cause a significant health issue.

Food Safety and Inspection Service: The public health agency in the USDA responsible for ensuring that the nation's commercial supply of meat, poultry, and egg products is safe, wholesome, and correctly labeled and packaged.

Food Safety Management System (FSMS): Management system developed in compliance with the requirements of a food safety standard such as FSSC 22000, ISO 22000, SQF, BRC, IFS, or other GFSI-approved scheme (Chapter 1).

Food Safety Modernization Act (FSMA): "The U.S. Food and Drug Administration (FDA) Food Safety Modernization Act (FSMA) was signed into law by President Obama on January 4, 2011. However, as of the release of this text, not all of the enabling regulations have been implemented. [This law focuses on ensuring that] the U.S. food supply is safe by shifting the focus of federal regulators from responding to contamination to prevention" (Chapter 1) (http://en.wikipedia.org/wiki/FDA_Food_Safety_Modernization_Act).

Food Safety System Certification (FSSC): The foundation for Food Safety Certification, which was founded in 2004, developed FSSC 22000 (ISO 22000, ISO 22002-1, FSSC 22000 additional requirements). This standard is a recognized GFSI food safety approved scheme. (http://www.fssc22000.com).

Food Safety Team: A chosen team with the responsibility for developing, implementing, and monitoring the food safety management system under the guidance of the food safety team leader, SQF practitioner, HACCP team leader, or other referenced leader. Note: This position may be referred to by a different name, depending on the standard of choice.

Food Safety Team Leader (FSTL): The food safety team leader is a position required by ISO 22000:2005 Section 5.5. This position is defined to have ultimate responsibility for managing the food safety team, organizing its activities, and ensuring that the food safety management system has been established, implemented, and is being maintained per defined requirements (Chapters 3 and 16).

Food Suitability: As defined by Codex Alimentarius, "assurance that food is acceptable for human consumption according to its intended use" (CAC/RCP 1-1969, Rev. 4, 2003).

Gas Chromatography (GC): May be referred to as Gas Liquid Chromatography (GLC) is laboratory test equipment used in analytical chemistry for identifying and analyzing specific compounds. This type of equipment may be used in analyzing a product or ingredient to identify its composition or as a monitoring method to ensure that products or ingredients have not been altered (i.e., an artificial ingredient substituted for an all-natural ingredient).

Global Aquaculture Alliance (GAA): The GAA is an international nonprofit trade association dedicated to advancing environmentally and socially responsible aquaculture. The Global Aquaculture Alliance (GAA) Seafood Processing Standard is a GFSI-recognized food safety scheme that focuses on aquaculture ensuring the continued sustainability and growth of the supply of seafood (Chapter 1) (http://www.gaalliance.org/).

Global Food Safety Initiative (GFSI): The GFSI is a nonprofit foundation managed by the Consumer Goods Forum (Chapter 1) (http://mygfsi.com/).

GLOBAL G.A.P.: This is a private sector body that sets voluntary standards for the certification of production processes of agricultural (including aquaculture) products around the globe (Chapter 1) (http://www.globalgap.org/).

Global Red Meat Standard (GRMS): This is a recognized GFSI food safety scheme developed by the Danish Agriculture & Food Council. This standard focuses on the transport, stunning, slaughtering, deboning, cutting, and handling of meat and meat products (http://www.grms.org).

Good Agricultural Practices (GAP): Practices on farms that define the essential elements for the development of best practice for production, incorporating integrated crop management, integrated pest management, and integrated agricultural hygienic practices (CAC/RCP 1-1969, Rev. 4, 2003).

Good Manufacturing Practices (GMP): The combination of management and manufacturing practices designed to ensure that food products are consistently produced to meet relevant legislative and customer specifications (SQF Code, Edition 7, Appendix 2: Glossary).

Grandfathered Suppliers: Related to critical quality suppliers, existing suppliers with a good performance history can be *grandfathered* into the approved supplier program. The criteria or qualifications for *grandfathering* must be defined with records maintained to demonstrate compliance. Critical food safety suppliers cannot be *grandfathered*.

HACCP Plan: Some organizations use the terminology of *HACCP plan* and *HACCP program* interchangeably. Some literature, such as the ISO standard, refers to the HACCP plan as a direct link to a CCP. Some feel that an organization cannot have a HACCP plan without having a CCP because to be a HACCP plan, the organization must address all seven principles. Without a CCP, principles 2–5 would not be addressed. Review Chapters 16 and 17 for more detail and an explanation on how to address this situation. Definition for a *HACCP Program* may also apply.

HACCP Program: A document created in accordance with the principles of HACCP (Codex Alimentarius) or in compliance with a specific food safety management system standard (i.e., FSSC 22000:2005 7.0), which identifies, evaluates, and ensures the control of hazards that are significant to the safety (food) of the products and processes being produced. The definition for an *HACCP plan* may also apply (Chapters 16 and 17).

Hazard: A biological, pathogen, chemical, and/or physical agent that is reasonably likely to cause illness or injury in the absence of its control. Codex Alimentarius defines a *hazard* as "a biological, chemical, or physical agent in, or condition of, food with the potential to cause an adverse health effect" (Chapters 16 and 17) (CAC/RCP 1-1969, Rev. 4, 2003). The additional hazards identified under FSMA (Food Safety Modernization Act, 2011) are allergens and radiological.

Hazard Analysis: The *hazard analysis* is the process for collecting and evaluating information on hazards and conditions to determine which are significant and must be addressed as either a CCP or OPRP for a specific food safety/HACCP program. Codex Alimentarius defines the *hazard analysis* as "the process of collecting and evaluating information on hazards and conditions leading to their presence to decide which are significant for food safety and therefore should be addressed in the HACCP plan" (CAC/RCP 1-1969, Rev. 4, 2003) (Chapters 16 and 17).

Hazard Analysis Critical Control Point (HACCP): The process that is used to identify, evaluate, and control or prevent hazards that are significant for food safety as defined in Codex Alimentarius. Considered a process tool to identify and assess existing and potential food safety hazards to ensure the production of a safe product.

Hazard Identification: The process of listing all existing or potential hazards that may occur for the specific process step or item. This includes ingredients, raw materials, processing aids, product contact materials, packaging, and process steps during the manufacture of the product and its distribution (Chapters 16 and 17).

Hazardous Chemicals and Toxic Substances: Solids, liquids, or gases that are radioactive, flammable, explosive, corrosive, oxidizing, asphyxiating, pathogenic, or allergenic, including but not restricted to detergents, sanitizers, pest control chemicals, lubricants, paints, processing aids, and biochemical additives, which if used or handled incorrectly or in increased dosage may cause harm to the handler and/or consumer. Hazardous or toxic chemicals may be described by regulation as *dangerous goods* and may carry a *poison*, *Hazmat*, or *Hazchem* label depending on the jurisdiction (SQF Code, Edition 7, Appendix 2: Glossary).

High-Risk Area: A segregated room or area where high-risk food processes are performed and may require a higher level of sanitation to prevent or eliminate contamination of a high-risk food by pathogens.

High-Risk Food: A food that may contain pathogenic microorganisms and may support formation of toxins or growth of pathogenic microorganisms and may have a significant likelihood of growth causing illness or injury to a consumer if not properly produced, processed, distributed, and/or prepared for consumption. It may also apply to a food that is deemed high risk by a customer or declared high risk by the relevant food regulation.

High-Risk Food Process(es): The production, handling, storage, processing, manufacturing, and/or preparation of high-risk food (SQF Code, Edition 7, Appendix 2: Glossary).

Internal Audit: Activity designed to apply a systematic, disciplined approach assessing compliance and performance of an organization's policies and procedures to a specific quality and/or food safety standard, procedure, and/or guideline(s). *Internal audits* are performed within an organization to measure its own performance, strengths, and weaknesses against its own established procedures, processes, and in management systems of choice.

Internal Communication: Communication to associates within the scope of a program. An example would be a plant-wide meeting by top management communicating about the management system, status in achieving measurable goals, and any proposed or process changes (Chapters 3 and 15).

International Features Standard (IFS): (Previously known as the International Food Standard.) IFS is a GFSI approved scheme for auditing food safety and quality of processes and products of food manufacturers (Chapter 1) (http://www.ifs-certification.com/).

International Organization for Standardization (ISO): ISO means *equal in Greek*. International Organization for Standardization (ISO) is the world's largest developer of voluntary International Standards. International Standards give state-of-the-art specifications for products, services, and good practice, helping to make industry more efficient and effective. Developed through global consensus, they help to break down barriers to international trade. It was founded in 1947 and, since then, has published more than 19,500 International Standards covering almost all aspects of technology and business, from food safety to computers, and agriculture to healthcare (http://www.iso.org/iso/home/about.htm).

ISO 9001:2008 Quality Management Systems: ISO 9001:2008 defines the requirements for a quality management system and is the only certification standard in the ISO 9000 family of standards. It is purposely generic in nature and can be used by any organization, large or small, regardless of its field of activity. In fact, ISO 9001:2008 is implemented by over one million companies and organizations in over 170 countries. ISO 9001 is part of the ISO 9000 family of standards, which address various aspects of quality management. The standards provide guidance and tools for companies and organizations who want to ensure that their products and services consistently meet customer's requirements and that quality is consistently improved (http://www.iso.org/iso/home/standards/management-standards/iso_9000.htm).

ISO 14001:2004 Environmental Management: An environmental management system standard. The ISO 14000 family of standards addresses various aspects of environmental management providing practical tools for organizations looking to identify and control their environmental impact and constantly improve their environmental performance. ISO 14001:2004 and ISO 14004:2004 focus on environmental management systems. The other standards in the family focus on specific environmental aspects such as life cycle analysis, communication, and auditing (http://www.iso.org/iso/home/standards/management-standards/iso14000.htm).

ISO 19001:2011 Guidelines for Auditing Management Systems: An ISO guideline developed to provide guidance when auditing management systems (Chapter 5) (http://www.iso.org/iso/catalogue_detail?csnumber=50675).

ISO 22000:2005 Food Safety Management Systems: A certifiable food safety management system standard. ISO 22000:2005 in combination with

ISO 22002-1:2009 becomes FSSC 22000: Version 3, GFSI approved food safety scheme.

ISO 22002-1:2009 Prerequisite Programmes on Food Safety: This document is equivalent to the former PAS 220, which has been withdrawn and replaced by the ISO document. ISO/TS 22002-1 is applicable to all organizations, regardless of size or complexity, which are involved in the manufacturing step of the food chain and wish to implement prerequisite programs in such a way as to address the requirements specified in ISO 22000:2005, Clause 7.

ISO 22004:2005 Food Safety Management Systems Guidance on the Application of ISO 22000:2005: A generic guideline on the application of ISO 22000:2005.

Kosher: The word means fit or proper in Hebrew. Kosher is used to describe food that has been prepared under and meets Jewish dietary laws. Kosher product labeling requires the product be kosher certified by a recognized rabbinical body. Kosher certification is provided based upon the rabbinical body after a rabbi reviews and approves the product ingredients, all manufacturing processes used, and direct contact equipment. Examples of non-Kosher foods include animals (pork, rabbit, and horse meat), fowl (owl and stork), fish (catfish, shellfish), and insects.

Management Review: The defined process for top management to manage, track, and ensure compliance, suitability, effectiveness, and improvement of the management system. Management review meetings are performed at defined intervals in compliance with a specific quality and/or food safety standard (Chapter 3).

Material Safety and Data Sheet (MSDS): An MSDS is a document that provides health and safety information about products, substances, or chemicals (http://www.worksafe.vic.gov.au/safety-and-prevention/health-and-safety-topics/material-safety-data-sheets).

Mitigation: The activity for evaluating a situation in order to eliminate the severity of the impact or likelihood of a risk or an event (Chapter 15).

Mock Recall: A practice recall exercise (Chapter 14).

Monitoring: Related to a HACCP/food safety program it is conducting a planned sequence of observations and measurements to assess whether control measures are operating as intended (Chapters 16 and 17).

National Advisory Committee on Microbiological Criteria for Foods (NACMCF): NACMCF is an advisory committee chartered under the USDA and comprised of participants from the USDA, Department of Health and Human Services, CDC, and several other U.S. government departments or agencies focused on providing guidance and recommendation to the secretary of agriculture and the secretary of health and human services regarding the microbiological safety of foods (http://www.fda.gov).

National Institute of Standard Technology (NIST): NIST is the federal technology agency that works with industry to develop and apply technology, measurements, and standards (www.nist.gov/).

Nonconformance or Findings: A situation that violates a requirement of the standard or a defined requirement of the system (Chapters 4 and 9).

Nonconforming Material and Product: A material or product that does not meet specifications (Chapter 9).

Nonconforming Situation: A failure or potential failure to meet a specified requirement (Chapters 4 and 9).

Observation: A situation that if not addressed would likely result in a nonconformance (Chapter 5).

OHSAS 18000: An international occupational health and safety management system specification. It comprises two parts, 18001 and 18002. OSHA was created by the U.S. Occupational Safety and Health Act and is the Occupational Safety & Health Administration (Chapter 1).

Operational Prerequisite Program (OPRP): An OPRP is a point in the process identified in the hazard analysis as essential in order to control the likelihood of introducing food safety hazards to and/or the contamination or proliferation of food safety hazards in the products or in the processing environment. OPRPs are frequently steps that provide a measure of food safety control while also being an integral step to alter the physical or chemical composition at that stage of the process. OPRPs apply to FSSC 22000 or ISO 22000 FSMS only.

Organic Foods: Organic foods are made in a way that comply with organic standards set by national governments and international organizations. In the United States, organic production is a system that is managed in accordance with the Organic Foods Production Act (OFPA) of 1990 and regulations in Title 7, Part 205, of the Code of Federal Regulations to respond to site-specific conditions by integrating cultural, biological, and mechanical practices that foster cycling of resources, promote ecological balance, and conserve biodiversity (http://encyclopedia.thefreedictionary.com/Organic+foods).

Outsourcing: This is an external operation that provides an activity (i.e., external warehousing, special packaging) for the operation defined within a management system. Top management and designated department managers or designees define, ensure control and compliance based on the relationship of the activity to the scope of a management system. Services such as external sanitation services and storage facilities may be considered outsourcing. These suppliers are managed in compliance with the critical food safety and quality approved supplier programs (Chapter 13).

PAS 96:2010 Defending Food and Drink: A guidance document with information for food companies related to avoiding, identifying, and addressing a deliberate attack on the food and beverage industries (Chapter 14).

PAS 220:2008 Prerequisite Program on Food Safety for Food Manufacturing: This Publicly Available Specification (PAS) document has been withdrawn and replaced by ISO/TS 22002-1. This change is the first of many related to the family of prerequisite programs for food safety. PAS documents have been expanded to include design and manufacture of food packaging, manufacture of animal food, as well as related logistical and distribution programs.

PAS 222: 2011 Prerequisite Programmes on Food Safety for Manufacturing Food and Feed for Animals: PAS 222 was developed to be applied along with ISO 22000 specific to the manufacture of commercial animal food, feed, and ingredients.

PAS 223:2011 Prerequisite Programmes and Design Requirements for Food Safety in the Manufacture and Provision of Food Packaging: 223 PAS was developed to be used along with ISO 22000 for the food packaging industry.

Pathogen: An infectious agent that can cause disease in the host. Disease-causing microorganisms, such as bacteria, fungi, and viruses are found commonly in sewage, hospital waste, run-off water from farms, and in water used for swimming. Some pathogens are parasites (live off the host), and the diseases they cause are an indirect result of their obtaining food from, or shelter in, the host. Larger parasites (such as worms) are not considered pathogens (http://www.businessdictionary.com/definition/pathogens.html#ixzz2L2FiSq4S).

Prerequisite Program (PRP): A program put into place that, when maintained effectively, controls, eliminates, or reduces a hazard or potential hazard to an acceptable limit. PRPs provide the foundation for the HACCP program and address the basic conditions and activities required for food safety that are necessary to either control or eliminate a potential hazard, address a quality parameter, and/or maintain a hygienic environment throughout the operation [SQF Code, Edition 7, Appendix 2: Glossary].

Preventive Action: Preventive actions are steps that are taken to remove the causes of potential nonconformities or potential situations that are undesirable. The preventive action process is designed to prevent the occurrence of nonconformities or situations that do not yet exist. It tries to prevent occurrence by eliminating causes.

Preventive Maintenance (PM): The care and servicing by personnel for the purpose of maintaining equipment and facilities in satisfactory operating condition by providing for systematic inspection, detection, and correction of incipient failures either before they occur or before they develop into major defects (http://www.thefreedictionary.com/preventive+maintenance).

Process: "A process is a set of activities that are interrelated or that interact with one another. Processes use resources to transform inputs into outputs. Processes are interconnected because the output from one process becomes the input for another process. In effect, processes are 'glued' together by means of such input output relationships" (http://www.praxiom.com/iso-definition.htm).

Process Approach: The process approach is a management strategy. When managers use a process approach, it means that they manage the processes that make up their organization, the interaction between these processes, and the inputs and outputs that tie these processes together (http://www.praxiom.com/iso-definition.htm).

Processing Aids: Items that may be used in the production of the product, but are not actual ingredients. Substances that are added to a food for their technical or functional effect in the processing, but are present in the finished food at insignificant levels and do not have any technical or functional effect in that food. Examples may include cleaning chemicals, enzyme treatments, water, lubricants, etc. (www.fda.gov).

Product Contact Surface: A product contact surface is any surface that may come in direct contact with exposed product. Examples would include storage tanks, conveyor belts, table tops, saw blades, augers, and stuffers (Chapters 16 and 17) (www.usda.gov/).

Product Recall: A product recall related to food safety is defined as an action performed by the manufacturer and/or the distributor to remove product from the market that may be unsafe, having the potential to cause health problems, or more seriously death (Chapter 14).

Publicly Available Specification (PAS): PAS is a sponsored fast-track standard driven by the needs of the client organizations and developed according to guidelines set out by BSI. After 2 years, the PAS is reviewed and a decision is made as to whether it should be taken forward to become a formal British Standard (http://shop.bsigroup.com/Navigate-by/PAS/).

Quality Critical Supplier: A quality critical supplier is a supplier who provides ingredients, packaging material, product contact/processing aids, and/or services that are required for the production of a product that may influence whether the product (in process, finished product, etc.) meets quality, customer, and any nonfood safety and regulatory requirements.

Quality Management System (QMS): Overall, defined organizational structure, procedures, processes, and resources put into place to define, implement, and maintain compliance to a specific quality management system standard.

Quality Manual and/or Food Safety Manual: A document that defines the scope of and compliance with the quality and/or food safety management system in compliance with a specific quality and/or food safety management system standard.

Record: Objective evidence that provides proof that an activity is maintained to demonstrate conformance to defined requirements.

Risk Assessment: The process of determining the likelihood and the severity of a specific negative event. The hazard analysis conducted for the food safety/ HACCP program may be considered a type of risk assessment (Chapters 4, 15, and 17).

Root Cause: A root cause is a *cause* (harmful factor) that is *rooted* (deep, basic, fundamental, underlying, or the like) of why something goes wrong. (http://en.wikipedia.org/wiki/root_cause) (Chapter 4).

Root Cause Analysis (RCA): An RCA is a method of problem solving that tries to identify the root causes of faults or problems that cause operating events. RCA practice tries to solve problems by attempting to identify and correct the root causes of events, as opposed to simply addressing their symptoms. By focusing correction on root causes, problem recurrence can be prevented. Root Cause Failure Analysis (RCFA) recognizes that complete prevention of recurrence by one corrective action is not always possible (http://en.wikipedia.org/wiki/root_cause_analysis) (Chapter 4).

Safe: Related to the food safety management system, a term used to identify a food product that is free of biological, physical, and/or chemical hazards.

Safe Quality Food (SQF): SQF is an approved GFSI scheme. More information related to the SQF code can be found at http://www.sqfi.com/ (Chapter 1).

Sanitation: Treatment of a cleaned surface to eliminate microbial activity.

Sanitation Standard Operating Procedure (SSOP): Procedure that includes detailed steps required to clean and sanitize.

Sanitizing: The treatment of clean surfaces and equipment by a process that reduces microorganisms to levels considered safe from a public health viewpoint.

Second-Party Audit: Also known as an external audit, which is usually performed by a company representative from an outside source, such as a customer that has an interest in the organization being audited.

Significant Food Safety Hazard: Food safety hazard that has been identified with such significant importance that its elimination or reduction to acceptable levels is essential to the production of a safe food.

SQF Practitioner: An individual designated by a producer/supplier to develop, validate, verify, implement, and maintain that producer's/supplier's own SQF system [SQF Code, Edition 7, Appendix 2: Glossary].

SQFI: The SQFI Institute, a division of the Food Marketing Institute (FMI) [SQF Code, Edition 7, Appendix 2: Glossary].

Standard Operating Procedure (SOP): An SOP is a term used to relate to procedure-level documents. An SOP normally provides the detail on how a requirement is addressed including the who, what, where, when, and how the task is to be performed. Some systems may refer to the second level of documentation as a procedure rather than an SOP. The term *SOP* has been used in the food industry for many years (Chapter 6).

Step: A point, procedure, operation, or stage in the food processing chain many times noted as including, from primary production to final consumption.

Surveillance Audit: An audit (frequency varies depending on the scope and contract with the certification body) of part of a supplier's management system where that system has previously been certified or recertified and whose certification is current [SQF Code, Edition 7, Appendix 2: Glossary].

Third-Party Auditor: Auditor who is totally independent of the system being audited, generally representing the third party certification body auditing against an independent standard.

Third-Party Registration Audit: This is performed by an auditor representing an accredited registrar, independent of any involvement with the company business. Note: the registrar has been accredited to perform the registration audit of the specific standard of choice.

Top Management: Management position or team, identified as ultimately responsible for the management system.

Top Management Team: Led by the highest ranking manager within the scope of the standard usually includes those areas of responsibility who report directly to the top management position.

Total Productive Maintenance (TPM): "Total productive maintenance (TPM) originated in Japan in 1971 as a method for improved machine availability through better utilization of maintenance and production resources. One way to think of TPM is 'deterioration prevention': deterioration is what happens naturally to anything that is not 'taken care of'. For this reason many people refer to TPM as 'total productive manufacturing' or 'total process management'. TPM is a proactive approach that essentially aims to identify issues as soon as possible and plan to prevent any issues before occurrence. One motto is 'zero error, zero work-related accident, and zero loss'". http://en.wikipedia.org/wiki/Total_productive_maintenance

Traceability: Related to food safety and quality management systems, the ability to trace the history and delivery of a product or service from the raw product and ingredients through to the finished product at its distribution location (Chapter 14).

Training: Communicating the required actions to responsible associates.

Turtle Diagram: A tool that aids in performing process audits. The *turtle diagram* creates a *picture* of the process including any related system elements along with related inputs to the process being audited (Chapter 5).

Validation: Obtaining evidence that a control measure or combination of control measures if properly implemented, is capable of controlling the hazard to a specified outcome (CAC/GL 69-2008) (Chapter 17).

Verification: "Ongoing activity to determine that control measures have been implemented as intended. Includes observation of monitoring activities, review of records to confirm, that activities are being performed as defined" (CAC/GL 69-2008) (Chapter 17).

Verification Schedule: A schedule outlining the frequency and the responsibility for carrying out the methods, procedures, or tests required to confirm effectiveness of required activities.

Water Treatment: The microbiological, chemical, and/or physical treatment of water for use in processing or cleaning required to ensure its suitability for use (SQF Code, Edition 7, Appendix 2: Glossary]).

Appendix B: Acronyms

AIB	American Institute of Banking Auditing and Standards Organization
ATP	Adenosine triphosphate
AWWA	American Water Works Association
BRC	British Retail Consortium
Canada GAP	Canada Good Agricultural Practices
CAPA	Corrective action/preventive action
CAR	Corrective action request
CCP	Critical control points
CFIA	Canadian Food Inspection Agency
cGMP	Current good manufacturing practices
CIP	Clean in place
CL	Critical limits
COA	Certificate of analysis
COC	Certificate of conformance
COP	Clean out of place
FALCPA	Food Allergen Labeling and Consumer Protection Act
FDA	Food and Drug Administration
FDCA	Federal Food, Drug, and Cosmetic Act
FIFE	First in first expired
FIFO	First in first out
FMEA	Failure mode and effect analysis
FMIA	Federal Meat Inspection Act
FSMA	Food Safety Modernization Act
FSMS	Food safety management system
FSSC	Food safety system certification
FST	Food safety team
FSTL	Food safety team leader
GAA	Global Aquaculture Alliance
GAP	Good agricultural practices
GFSI	Global Food Safety Initiative
GHPs	Good housekeeping practices
Global GAP	Global good agricultural practices
GMP	Good manufacturing practices
GRMS	Global Red Meat Standard
HACCP	Hazard analysis and critical control point
IHA	International HACCP Alliance
IFS	International Featured Standards

ISO	Means *equal* in Greek; International Organization for Standardization
MCS	Master cleaning schedule
MSDS	Material safety and data sheet
NACMCF	National Advisory Committee on Microbiological Criteria for Foods
NIST	National Institute of Standard and Technology
OL	Operating limit
OPRP	Operational Prerequisite Program
PAR	Preventive Action Request
PAS	Publicly available specification
PCA	Peanut Corporation of America
PCO	Pest control operator
PDCA	Plan, Do, Check, Act
PM	Preventive maintenance
PRP	Prerequisite program
QA	Quality assurance
QC	Quality control
QMS	Quality management system
R&D	Research and development
RCA	Root cause analysis
RCFA	Root cause failure analysis
ROI	Return on investment
SOP	Standard operating procedure
SQF	Safe quality food
SSOP	Sanitation standard operating procedure
TPM	Total productive maintenance
USDA	United States Department of Agriculture

Appendix C: References

Biros, R. (President). *Food Safety Tech*, Innovative Publishing Company, LLC, Pennsylvania, PA. http://www.foodsafetytech.com/FoodSafetyTech/Blogs/392.aspx.

BRC Global Standard for Food Safety, 6th edn. http://www.brcglobalstandards.com/GlobalStandards/Home.aspx.

CanadaGAP (Canadian Horticultural Council On-Farm Food Safety Program) (Version 3.0 2012). http://www.canadagap.ca/.

Canadian Food Inspection Agency CFIA. http://www.inspection.gc.ca/english/fssa/recarapp/rap/mg1e.shtml. Also www.inspection.gc.ca.

CARVER + shock. http://www.fda.gov/ForConsumers/ConsumerUpdates/ucm094560.htm.

Codex Alimentarius. 2003. Hazard analysis and critical control point (HACCP) system and guidelines for its application. ANNEX to Recommended International Code of Practice/General Principles of Food Hygiene. CAC/RCP 1-1969, Rev. 4. FAO/WHO Codex Alimentarius Commission.

Codex Alimentarius Commission Recommended International Code of Practice—*General Principles of Food Hygiene*, CAC/RCP1-1969, Rev. 4-2003. www.codexalimentarius.org.

Codex CAC/GL 69-2008. Guidelines for the validation of food safety control measures. www.codexalimentarius.org/input/download/standards/11022/cxg_069e.pdf.

Cox, L.A. Jr. 2008. What's wrong with risk matrices? *Risk Analysis*, 28(2), 497–511. doi:10.1111/j.1539-6924.2008.01030.x.

Current Good Manufacturing Practices, *Code of Federal Regulations*, title 21, section 110. www.cgmp.com.

Drexler, M. October 2011. Why your food isn't safe, *Good Housekeeping*.

FDA Food Allergen Labeling and Consumer Protection Act (FALCPA). www.cfsan.fda.gov/~dms/alrguid3.html.

FDA Recall information. http://www.fda.gov/safety/recalls/ucm165546.htm.

FSSC 22000:2010 Food Products. http://www.fssc22000.com/en/.

GFSI. http://www.mygfsi.com/gfsifiles/Guidance_Document_Sixth_Edition_Version_6.1.pdf.

Global Aquaculture Alliance Seafood Processing Standard (BAP), Issue 2, May 2009. http://www.gaalliance.org/.

GLOBAL G.A.P., V 4.0 Edition 4.0 February 2012. http://www.globalgap.org.

Global Red Meat Standard (GRMS), 4th edn., V4.1, April 2012. http://www.grms.org/.

Groenveld, C. and Pillay, V. 2009. Food safety goes global, *Food Quality*, February–March, 2009. www.foodquality.com/mag/02012009.03012009/fq_03012009_FE3.htm.

http://www.fda.gov/alert

http://www.fda.gov/Food/FoodDefense/ToolsEducationalMaterials/ucm295898.htm

http://www.cfsan.fda.gov/fooddefense.

http://dictionary.reference.com.

http://en.wikipedia.org/wiki/Risk_assessment.

http://en.wikipedia.org/wiki/Risk_Matrix#cite_note-cox-4#cite_note-cox-4.

http://en.wikipedia.org/wiki/Total_productive_maintenance.

http://www.businessdictionary.com/definition/mitigation.html#ixzz2LPmhXYBk.

http://www.iso.org/iso/home/standards/management-standards.htm.

http://www.qualitydigest.com/inside/quality-insider-article/auditors-turtle-diagrams-and-waste.html.

IFS Food Version 6. http://www.ifs-certification.com/.

ISO 9000, 9001 and 9004: Quality Management Definitions. http://www.praxiom.com/iso-definition.htm.

ISO 9000:2005: Quality Management Systems—Fundamentals and Vocabulary. http://www. iso.org.

ISO 9001:2008: Quality Management Systems—Requirements. http://www.iso.org.

ISO 19001:2011: Guidelines for Auditing Management Systems. http://www.iso.org.

ISO 22000:2005: Food Safety Management Systems—Requirements for Any Organization in the Food Chain. http://www.iso.org.

Kotter, J. 1996. *Leading Change*, Harvard Business School Press, Boston, MA, p. 25.

Link, E. 2008a. *An Audit of the System, Not of the People: An ISO 9001:2008 Pocket Guide for Every Employee*, Quality Pursuit, Inc., 47 Woodcliff Terrace Fairport, NY 14450. www.qualitypursuit.com.

Link, E. 2008b. *An Audit of the System, Not of the People: An ISO 22000:2005 Pocket Guide for Every Employee*, Quality Pursuit, Inc., Fairport, NY. www.qualitypursuit.com.

Mead, P.S., Slutsker, L., Dietz, V. et al. 1999. Food-related illness and death in the United States. Retrieved June 29, 2007. http://www.cdc.gov/ncidod/eid/vol5no5/mead.htm.

Meat and Poultry HACCP. http://fsrio.nal.usda.gov/haccp/meat-and-poultry-haccp.

Micklewright, M. 2007. Auditors, turtle diagrams and waste, *Quality Digest*, 1/8/2007.

Muliyil, V. October 2012. (Technical Manager for North America Food Safety Services, SGS) and Supreeya Sansawat (Global Food Business Manager SGS) in their white paper, *Comparing Global Food Safety Initiative (GFSI) Recognized Standards*.

National Advisory Committee on Microbiological Criteria for Foods (NACMCF). 1992. Hazard analysis critical control point systems, *International Journal of Food Microbiology*, 16, 1–23.

National Advisory Committee on Microbiological Criteria for Foods. August 1997. *Hazard Analysis and Critical Control Point Principles and Application Guidelines*.

Newslow, D.L. 2001a. Hazard Analysis Critical Control Point (HACCP). In: *Food Safety Handbook*, Schmidt, R. and Rodrick, G. (eds.), Wiley-Interscience, Hoboken, NJ, pp. 363–378.

Newslow, D.L. 2001b. *The ISO 9000 Quality System Applications in Food and Technology*, Wiley-Interscience, A John Wiley & Sons, Inc. Publication, New York. www.wiley.com.

Newslow, D.L. 2003. Sanitation and its place in your quality management system, *Food Quality Magazine*, November/December, 2003 issue.

Nolan, M. 2007. Personal communication, S.A.F.E. Food Consulting Services.

Okes, D. 2009. *Root Cause Analysis: The Core of Problem Solving and Corrective Action*, ASQ, Milwaukee, WI.

PAS 99:2012. October 2012. *Specification for Common Management System Requirements as a Framework for Integration*.

Peach, R. 1997. *The ISO 9000 Handbook*, 3rd edn., Irwin, Homewood, IL.

Pierson, M.D. and Corlett, D.A. Jr. 1992. *HACCP Principles and Applications*, AVI, Van Nostrand Reinhold, New York.

PrimusGFS V 1.6. http://www.primusgfs.com/.

"Risk matrix." http://en.wikipedia.org/wiki/Risk_Matrix#cite_note-cox-4#cite_note-cox-4.

Russell, J.P. 2003. *The Process Auditing Technique Guide*, ASQ Quality Press, Milwaukee, WI.

Safe Quality Food (SQF) Code edition 7. http://www.sqfi.com/. (Note that information quoted from the SQF code is credited to and under copyright by the Food Marketing Institute.)

Schmidt, R.H., Goodrich, R.M., Archer, D.L., and Schneider, K.R. 2003. *General Overview of the Causative Agents of Foodborne Illness (FSHN033)*, Department of Family, Youth, and Community Sciences, Florida Cooperative Extension Service, Institute of Food and Agricultural Sciences, University of Florida, Gainesville, FL. Retrieved June 29, 2007. http://edis.ifas.ufl.edu/FS099.

Schmidt, R.H. 1996. *Hazard Analysis Critical Control Point (HACCP): Overview of Principles and Applications*, Food Science and Human Nutrition Department, University of Florida, Gainesville, FL.

Schmidt, R.H. and Newslow, D.L. 2007a. *Hazard Analysis Critical Control Points (HACCP)— Getting Started, Preliminary Step (FSHN0701)*, Department of Family, Youth, and Community Sciences, Florida Cooperative Extension Service, Institute of Food and Agricultural Sciences, University of Florida, Gainesville, FL. Retrieved July 20, 2007. http://edis.ifas.ufl.edu/FS137.

Schmidt, R.H. and Newslow, D.L. 2007b. *Hazard Analysis Critical Control Points (HACCP)— Prerequisite Programs (FSHN0702)*, Department of Family, Youth, and Community Sciences, Florida Cooperative Extension Service, Institute of Food and Agricultural Sciences, University of Florida, Gainesville, FL. Retrieved July 20, 2007. http://edis.ifas.ufl.edu/FS138.

Seafood HACCP (CFR 123 Fish and Fishery Products). http://www.fda.gov/Food/FoodSafety/HazardAnalysisCriticalControlPointsHACCP/SeafoodHACCP/default.htm.

SQF Code Edition 7, *Appendix 2: Glossary*. http://www.sqfi.com/wp-content/uploads/SQF-Code-Ed-7-Final-8-13-12.pdf.

SQFI Guidance. 2012. RE: 2.8.3 Allergen Cleaning and Sanitation Practices, October 25, 2012.

Stevenson, K.E. and Bernard, D.T. (eds.). 2006a. *HACCP: A Systematic Approach to Food Safety*, The Food Processors Institute, Washington, DC. www.fpi-food.org.

Stevenson, K.E. and Bernard, D.T. (eds.). 2006b. *HACCP: Verification and Validation*, The Food Processors Institute, Washington, DC. www.fpi-food.org.

Surak, J.G. 2001. International Organization for Standardization ISO 9000 and related standards. In: *Food Safety Handbook*, Schmidt, R. and Rodrick, G., (eds.), Wiley-Interscience, Hoboken, NJ.

Three (3)—A standards. http://www.3-a.org/.

Sertkaya, A. et al. 2006. Top ten food safety problems in the United States food processing industry, *Food Protection Trends*, 26(5), 310–315.

U.S. Food and Drug Association. http://www.fda.gov/Food/FoodSafety/default.htm, also, www.fda.gov.

U.S. Food and Drug Administration. 2009. Food Code, 5-202.12. www.fda.gov/Food/FoodSafety/RetailFoodProtection/FoodCode/.

U.S. Food and Drug Administration, Center for Food Safety and Applied Nutrition. 2001. *Fish and Fisheries Products Hazards and Controls Guidance*, 3rd edn. Retrieved June 29, 2007. http://www.cfsan.fda.gov/~comm/haccp4.html.

U.S. Food and Drug Administration, Center for Food Safety and Applied Nutrition. 2004. *Juice HACCP Hazards and Controls Guidance*, 1st edn. Retrieved June 29, 2007. http://www.cfsan.fda.gov/~dms/juicgu10.html.

U.S. Food and Drug Administration, Center for Food Safety and Applied Nutrition. 2006. *Foodborne Pathogenic Microorganisms and Natural Toxins Handbook: The Bad Bug Book*. Retrieved June 29, 2007. http://www.cfsan.fda.gov/~mow/intro.html.

Weber, C. President, Vinca, LLC. http://www.22000-tools.com.

Yiannas, F.C. 2010. *Food Safety Culture Creating a Behavior-Based Food Safety Management System*, Springer Science & Business Media, New York.

Index